AF358482

# Power System Analysis Control and Operation

# Power System Analysis Control and Operation


Guest Editors

**Bing Yan**
**Marialaura Di Somma**
**Jianxiao Wang**

Basel • Beijing • Wuhan • Barcelona • Belgrade • Novi Sad • Cluj • Manchester

*Guest Editors*

Bing Yan
Department of Electrical and
Microelectronic Engineering
Rochester Institute of
Technology
Rochester
USA

Marialaura Di Somma
Department of Energy
Technologies and Renewable
Sources
ENEA
Rome
Italy

Jianxiao Wang
National Engineering
Laboratory for Big Data
Analysis and Applications
Peking University
Beijing
China

*Editorial Office*
MDPI AG
Grosspeteranlage 5
4052 Basel, Switzerland

This is a reprint of the Special Issue, published open access by the journal *Energies* (ISSN 1996-1073), freely accessible at: https://www.mdpi.com/journal/energies/special_issues/813CRUR697.

For citation purposes, cite each article independently as indicated on the article page online and as indicated below:

Lastname, A.A.; Lastname, B.B. Article Title. *Journal Name* **Year**, *Volume Number*, Page Range.

ISBN 978-3-7258-3315-3 (Hbk)
ISBN 978-3-7258-3316-0 (PDF)
https://doi.org/10.3390/books978-3-7258-3316-0

# Contents

# About the Editors

**Bing Yan**

Dr. Bing Yan is currently an assistant professor in the Department of Electrical and Microelectronic Engineering at Rochester Institute of Technology. She received a B.S. degree in information management and information system from Renmin University of China in 2010, an M.S. degree in electrical engineering and statistics from University of Connecticut in 2012 and 2017, and a Ph.D. in electrical engineering from University of Connecticut in 2016. Before joining Rochester Institute of Technology, she was an assistant research professor in the Department of Electrical and Computer Engineering, University of Connecticut. Dr. Yan's research interests include operation optimization of smart power and energy systems, the planning and scheduling of intelligent manufacturing systems, self-optimizing factories, and the mathematical optimization of large-scale mixed-integer linear programming problems.

**Marialaura Di Somma**

Prof. Marialaura Di Somma received a Ph.D in Mechanical Systems Engineering from the University of Naples Federico II (Italy) in 2016. In 2014, she held a position as a visiting research assistant at the Department of Electrical and Computer Engineering of the University of Connecticut in the U.S., whereas from 2015 to 2023, she served as a research scientist and project manager at ENEA in Italy. In this context, she has managed and coordinated several European and national projects on the topic of multi-carrier energy system optimization and control. As of December 2023, she is an associate professor in the area of Applied Thermodynamics in the Department of Industrial Engineering at the University of Naples Federico II (Italy). Her main research interests mainly focus on design and operation optimization of distributed energy resources in the context of local multi-carrier energy systems, and energy communities through a multi-objective approach . She is involved in EERA JP Energy Systems Integration in UNINA and she is an expert member of ETIP SNET working group flexible generation. She is also an editor of three books on the topic of integrated energy systems published by Elsevier and Wiley, and the chair in scientific international conferences. Prof. Marialaura Di Somma is the author of more than 80 scientific contributions, most of them published in international journals, with proceedings of international conferences, and awarded as highly cited papers.

**Jianxiao Wang**

Dr. Jianxiao Wang received his B.S. and Ph.D. degrees in Electrical Engineering from Tsinghua University, Beijing, China, in 2014 and 2019. He was a visiting student researcher at Stanford University, CA, USA. He is currently an associate professor with Peking University, Beijing, China. His research interests include data analytics and optimization for smart grid and energy storage. He has published in top journals including Nature Sustainability, Nature Communications, The Innovation, Patterns, etc. He received the IEEE PES Outstanding Volunteer Award in 2021. He is leading the Excellent Young Scholar Fund of National Science Foundation of China, and Young Scientist Program of National Key Research and Development Program of China. He is a Youth Editor of Cell The Innovation, an Associate Editor of IET Renewable Power Generation, IET Energy Conversion and Economics, etc.

# Preface

We are pleased to present the reprint of the Special Issue "Power System Analysis, Control, and Operation," a comprehensive exploration of the emerging challenges and opportunities in modern power systems. This collection brings together cutting-edge research to address the increasing complexity of integrating renewable energy resources (RERs), energy storage systems (ESRs), and electric vehicles (EVs) into power grids. With a growing emphasis on demand response and sector coupling, these topics are vital to improving efficiency, reliability, and flexibility in energy systems.

The scope of this reprint spans a wide range of topics, including optimal power flows, the integration of distributed energy resources, data analytics for load and renewable generation forecasting, and cross-sector coordination between power and transportation systems. It also highlights strategies for enhancing grid sustainability, resiliency, and adaptability, offering insights into innovative solutions for advancing energy system operations.

This Special Issue was made possible through the contributions of leading researchers, whose efforts we deeply appreciate. We would also like to express our gratitude to the reviewers for their constructive feedback, as well as to the editorial team at energies for their unwavering support in bringing this reprint to fruition.

The reprint is intended for a diverse audience, including researchers, industry professionals, and policymakers, all of whom play a pivotal role in shaping the future of sustainable energy systems. We hope this collection will serve as an invaluable resource and inspire continued advancements in the field.

**Bing Yan, Marialaura Di Somma, and Jianxiao Wang**
*Guest Editors*

*Article*

# A Comprehensive Tool for Scenario Generation of Solar Irradiance Profiles

Amedeo Buonanno [1], Martina Caliano [1], Marialaura Di Somma [1,*], Giorgio Graditi [2] and Maria Valenti [1]

[1]  Department of Energy Technologies and Renewable Energy Sources, ENEA, 80055 Portici, NA, Italy
[2]  Department of Energy Technologies and Renewable Energy Sources, ENEA, 00123 Rome, Italy
*  Correspondence: marialaura.disomma@enea.it

**Abstract:** Despite their positive effects on the decarbonization of energy systems, renewable energy sources can dramatically influence the short-term scheduling of distributed energy resources (DER) in smart grids due to their intermittent and non-programmable nature. Renewables' uncertainties need to be properly considered in order to avoid DER operation strategies that may deviate from the optimal ones. This paper presents a comprehensive tool for the scenario generation of solar irradiance profiles by using historical data for a specific location. The tool is particularly useful for creating scenarios in the context of the stochastic operation optimization of DER systems. Making use of the Roulette Wheel mechanism for generating an initial set of scenarios, the tool applies a reduction process based on the Fast-Forward method, which allows the preservation of the most representative ones while reducing the computational efforts in the next potential stochastic optimization phase. From the application of the proposed tool to a numerical case study, it emerged that plausible scenarios are generated for solar irradiance profiles to be used as input for DER stochastic optimization purposes. Moreover, the high flexibility of the proposed tool allows the estimation of the behavior of the stochastic operation optimization of DER in the presence of more fluctuating but plausible solar irradiance patterns. A sensitivity analysis has also been carried out to evaluate the impact of key parameters, such as the number of regions, a metric, and a specific parameter used for the outlier removal process on the generated solar irradiance profiles, by showing their influence on their smoothness and variability. The results of this analysis are found to be particularly suitable to guide users in the definition of scenarios with specific characteristics.

**Keywords:** scenario generation; scenario reduction; solar irradiance profiles; smart grid

**Citation:** Buonanno, A.; Caliano, M.; Di Somma, M.; Graditi, G.; Valenti, M. A Comprehensive Tool for Scenario Generation of Solar Irradiance Profiles. *Energies* 2022, 15, 8830. https://doi.org/10.3390/en15238830

Academic Editor: Adalgisa Sinicropi

Received: 31 October 2022
Accepted: 20 November 2022
Published: 23 November 2022

**Publisher's Note:** MDPI stays neutral with regard to jurisdictional claims in published maps and institutional affiliations.

## 1. Introduction

The sustainability objectives set by the European Green Deal require the increasing use of generation systems based on renewable energy sources (RES), as well as the use of electricity as the main energy vector. The pathway to reducing carbon emissions by 2030 will require efforts across society and sectors. With the European Green Deal as the main plan to implement for promoting this change, the European Union (EU) finalized its master program to fight carbon emissions, namely, the "Fit for 55" package [1]. Released in two batches in July and December 2021, the package drafts of EU climate and energy legislation underpin the bloc's political pledge to cut emissions by at least 55% by 2030 compared with 1990 levels. This target is more ambitious than the previous one of a 40% reduction for 2030 and is the key to achieving carbon neutrality in the EU by 2050. In view of this scenario, solar energy will become one of the key players in the electricity generation sector, thanks to its viability in combating global warming and its effectiveness in reducing pollution caused by fossil-fuel-based generation and diversifying the energy mix to ensure energy security. In particular, the installed capacity of solar photovoltaics (PV) has grown rapidly over the past decade due to the great improvements in PV technology performance, reductions in cost, and the development of efficient business models that have fostered new investments in this technology. This trend is expected to continue in the future, affecting not only large-scale centralized solar farms but, above all, small-/medium-scale PV at the distribution level, where the number of PV applications owned by residential and industrial prosumers (power producers and consumers) will also expand, driven by environmental policies and economic incentives.

From a range of studies, solar PV is expected to contribute 36% to 69% of European electricity consumption by 2050 [2], and its role is predominant in Energy Communities [3]. Due to the variability and intermittent nature of the solar PV output, such a high share of solar PV will impact the overall system costs due to the increase in operating costs and the infrastructure needed. This problem is aggravated by the inaccuracy of the methodologies in modeling renewables' uncertainties, which are related to the uncertainty of weather conditions for RES [4], which represent a key factor to be properly handled in the smart grid environment. In fact, they may influence how distributed energy resources (DER) are scheduled in the short term to provide the available flexibility for system balancing at all times [5–8]. If such uncertainties are not identified and handled properly in the operation scheduling of DER, their operation strategies may deviate from the optimal ones by causing a number of issues, such as an increase in operational costs or system stability and security. Modeling RES uncertainties in the stochastic operation optimization of DER is thus extremely important [9–12].

Several works in the literature deal with different sources of uncertainties, such as renewables, electricity consumption, electric vehicles, etc. [13–17]. The objective of uncertainty-modeling methods is to evaluate the impact of uncertain input parameters on system output parameters. These methods can be subdivided into several groups, as suggested in [18,19]:

- Probabilistic: the probability density functions (PDFs) of the input parameters are used;
- Possibilistic (fuzzy): the uncertainty of the input parameters is modeled with a membership function (MF);
- Hybrid probabilistic and possibilistic: both probabilistic and possibilistic approaches are used;
- Based on Information Gap Decision Theory (IGDT): it measures the deviation of the estimation error;
- Robust optimization: the uncertainty of the input parameters is described using uncertainty sets;
- Interval analysis: the uncertain inputs can assume values in a known interval (similar to the probabilistic approach with uniform PDFs).

The approaches to estimating solar irradiance can be grouped into linear, nonlinear, Artificial Neural Network (ANN)-based, and Fuzzy Logic (FL) techniques [20]. For linear and nonlinear models, the authors have created associations between solar irradiance and other variables, such as meteorological ones [21–23]. In ANN-based approaches [24,25], the usual inputs used are geographical coordinates, meteorological data, and information related to the current date and time. In FL approaches, the input to the estimation model is the classified sky condition. Moreover, other authors have applied statistical methods to study the hourly variation in solar irradiance data considering different climatic locations [26] or empirical models to estimate solar irradiance on a monthly basis for different locations [27,28]. In [29], the authors state that FL can yield better estimation results when the data available for estimation are ambiguous and vague.

Among the probability distributions, beta is considered one of the most effective for modeling solar irradiance [30–33] and is often employed in planning studies related to PV systems [34–36]. Other works propose the Weibull distribution for modeling solar irradiance [37].

The contribution of this paper is the presentation of a comprehensive tool to generate solar irradiance profiles using a scenario generation approach and historical data for a specific location. The methodology is general and highly replicable and can thus be applied in several contexts for generating 24 h solar irradiance scenarios, which are useful for the stochastic operation optimization of DER. The tool has been completely implemented in Python, and it is suitable for transformation into a Web Service to generate solar irradiance profiles related to a particular geographical region and a time period of interest.

The historical hourly solar irradiance data were fitted using the beta distribution, and the Roulette Wheel mechanism [38] was used to generate an initial set of scenarios; then, a reduction process based on the Fast-Forward method [39,40] was applied in order to preserve the most representative ones while reducing the computational efforts in the next potential stochastic optimization phase. The generation and reduction phases are ruled by certain parameters, such as the number of regions, a metric, and a specific parameter used for an optional process for outlier removal that can be modified. Moreover, a sensitivity analysis was performed to evaluate the impacts of the variations in these parameters on the solar irradiance scenarios generated. The numerical results of the analysis show that these parameters have a visible effect on the smoothness and variability of the generated scenarios. Based on the current scientific literature, there are no previous works that examined these aspects, thereby highlighting the importance of this study, which could be useful as a guide for tuning the scenario generation process with the aim of obtaining scenarios with certain characteristics. In the case study, the proposed tool is found to be efficient in generating plausible scenarios for daily solar irradiance profiles with 1 h as the time-step to be used as input for DER stochastic operation optimization purposes. Moreover, the high flexibility of the proposed tool allows the estimation of the behavior of DER stochastic operation optimization in the presence of more fluctuating—but plausible—solar irradiance profiles. The current paper extends the results presented in [41] by including additional results in the case study, as well as introducing a new verification system of the plausibility of the reduced scenarios.

In the following, the dataset, data preprocessing, data fitting, and methods for scenario generation and reduction are discussed in Section 2. The results of varying certain key parameters in the scenario generation and reduction processes are presented in Section 3. In Section 4, the sensitivity analysis is discussed, along with the obtained results.

## 2. Materials and Methods

The tool proposed in this paper is based on a statistical approach used to model solar irradiance based on historical data. By using the Roulette Wheel method [38], an initial set of scenarios is first generated, and then, through a reduction process, the most representative ones are preserved.

A scheme for describing the proposed tool is shown in Figure 1. In particular, the dataset retrieved from the Photovoltaic Geographical Information System (PVGIS) [42] (Section 2.1) is preprocessed by using an optional process for outlier removal and min–max scaling, as described in Section 2.2. For each hour, a data-fitting process (Section 2.3) is performed in order to obtain a probability distribution for each hour. From each hourly probability distribution, a sample is extracted to obtain 24 randomly sampled values, the so-called 24 h solar irradiance scenario (Section 2.4). This process is performed $n_S$ times in order to obtain $n_S$ scenarios. In order to reduce the computational complexity of the following optimization task, the generated scenarios have to be reduced by using an approach that preserves the "information content" present in the original set of scenarios. This is performed by using the procedure described in Section 2.5.

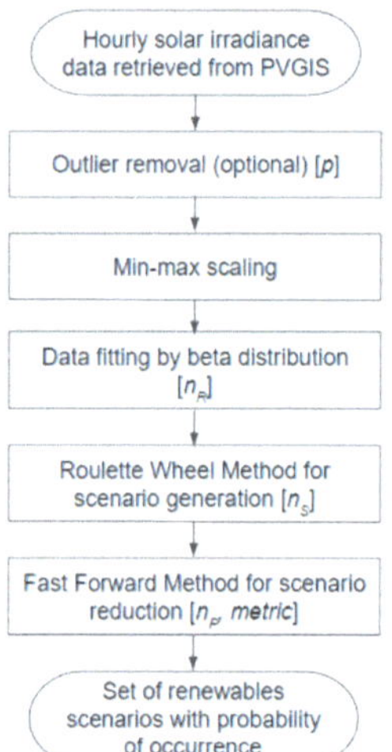


**Figure 1.** Overall description of the proposed tool. In square parentheses, the parameters that affect the considered step are reported.

## 2.1. Dataset Description

The hourly solar irradiance data from 2005 to 2016 for the city of Turin (Italy) were gathered using PVGIS [42]. In order to model both the winter and summer seasons, January and July were selected as months of interest.

The daily patterns of solar irradiance for the days in July and January from 2005 to 2016 are shown in Figures 2 and 3, respectively.

From the figures, it is possible to observe the great variability in each hour, which will be modeled by means of a probability distribution that fits the data.

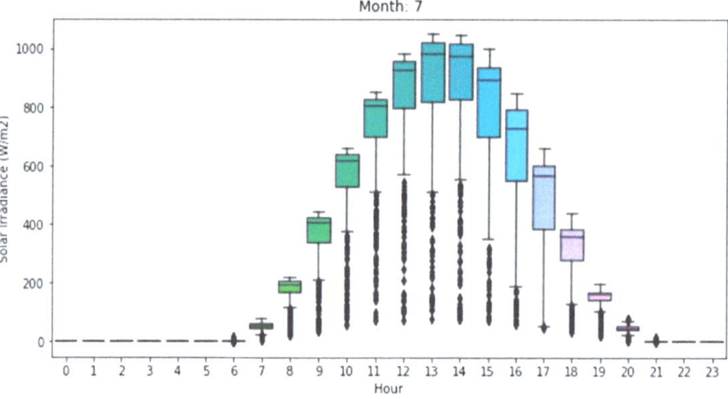


**Figure 2.** Box-whisker plot of solar irradiance for the days in July from 2005 to 2016. The bold points represent values identified as outliers following Equation (1) with $p = 1.5$. To each hour is associated a different color of box.

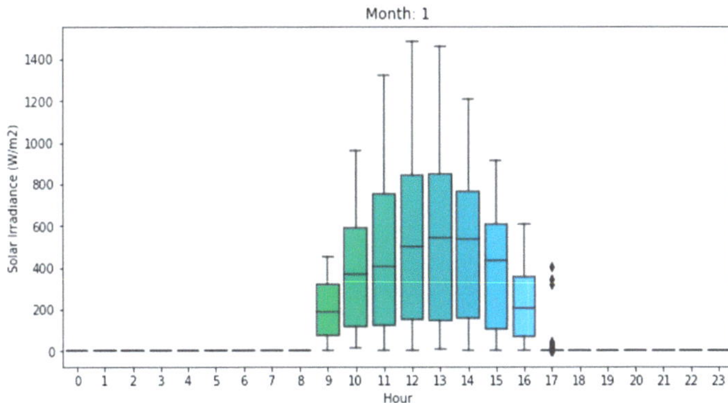

**Figure 3.** Box-whisker plot of solar irradiance for the days in January from 2005 to 2016. The bold points represent values identified as outliers following Equation (1) with $p = 1.5$. To each hour is associated a different color of box.

### 2.2. Data Preprocessing

The hourly data for each considered month were normalized using min–max scaling in order to map the values observed in the range [0, 1].

To reduce the variability in the observations, the outliers can be removed using the Interquartile Range (IQR) method [43], according to which the values are considered outliers—and hence removed—when they are outside of the following range:

$$[Q_1 - p \cdot IQR, Q_3 + p \cdot IQR] \tag{1}$$

where $Q_1$ and $Q_3$ are the 1st and 3rd quartiles (25th and 75th percentiles), respectively; $IQR = Q_3 - Q_1$; and $p$ is a value that permits the expansion or restriction of the range and hence the consideration of fewer or more values as outliers.

However, it is the user's choice to enable (or not) the outlier removal process, as well as the value of $p$.

### 2.3. Data Fitting

The normalized hourly irradiance data were fitted using several probability distributions, namely, Weibull, beta, logistic, and arcsine. Among the tested distributions, beta was found to be the best fit for most hours in January and July, confirming what was observed in the relevant literature for solar irradiance data [32,33].

Beta is a continuous probability distribution with support in [0, 1], and its PDF is defined as in [44]:

$$f_S(x, a, b) = \frac{x^{a-1}(1 - x)^{b-1}}{B(a, b)} \tag{2}$$

where $B(a,b)$ is the beta function, formulated as:

$$B(a, b) = \frac{\Gamma(a)\Gamma(b)}{\Gamma(a + b)} \tag{3}$$

where $\Gamma(x)$ is the gamma function, formulated as:

$$\Gamma(x) = \int_0^\infty u^{x-1}e^{-u}du \tag{4}$$

Different values of $a$ and $b$ allow uniform ($a = 1$, $b = 1$), bimodal ($a < 1$, $b < 1$), or unimodal ($a > 1$, $b > 1$) distributions to be obtained. $n_R$ regions, or bins, are defined to divide the support of the distribution.

Figure 4 shows the fitting results of data related to the hour 12:00 p.m. in July using the beta distribution and dividing its support into 7 regions.

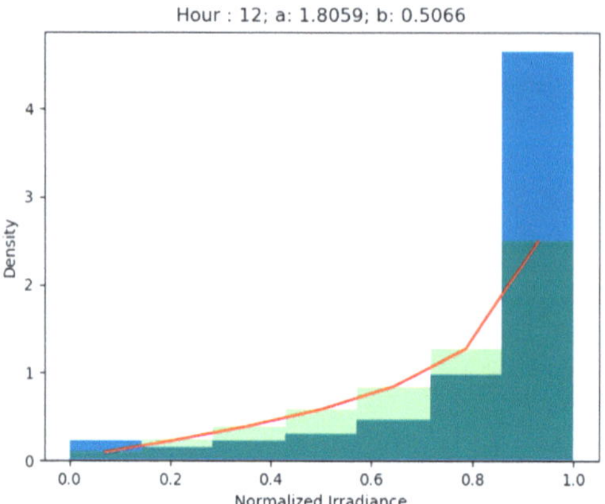


**Figure 4.** Data fitting using beta distribution with 7 regions: (blue bars) histograms of the empirical distribution; (green bars overlaid with transparency) histograms of the fitted beta distribution; (red line) the line connecting the central points of all regions.

## 2.4. Roulette Wheel Method

The Roulette Wheel method [38] was used to extract a set of samples from the fitted probability density functions with their supports divided into $n_R$ regions [45].

The probability of the occurrence of a particular region $r$ ($r \in \{1, \ldots, n_R\}$) at time $t$ (the hour of the day, $t \in \{0, \ldots, 23\}$) can be computed as the product of width $w_{t,r}$ (width of region $r$ at time $t$) and height $h_{t,r}$ (height of region $r$ at time $t$): $\alpha_{t,r} = w_{t,r} \cdot h_{t,r}$, appropriately normalized as reported in Equation (5).

$$\hat{\alpha}_{t,r} = \frac{\alpha_{t,r}}{\sum_{\rho=1}^{n_R} \alpha_{t,\rho}} \tag{5}$$

The probabilities of the occurrence for all regions of a particular hour were sorted in descending order and cumulated.

In order to sample from the beta distribution with its support divided into $n_R$ regions, it is possible to sample a value from a uniform distribution ($v \sim U(0,1)$) and, comparing it with the cumulative probability of occurrence, select one of $n_R$ possible regions that $v$ belongs to (this method is also known as the Inverse Transform [46]).

The central value of the selected region is the value sampled from the beta probability distribution with support divided into $n_R$ regions. The procedure described above was performed for each hour, obtaining 24 sampled values that compose the scenario. To be precise, not all 24 values were sampled in this way because, in some hours, the observed solar irradiance was constantly zero, and this behavior was maintained in the scenario generation process.

The binary variable $W_{k,t,r}$ is used to contain the information on whether region $r$ is selected for a scenario $s^k$ at hour $t$ ($W_{k,t,r} = 1$) or not ($W_{k,t,r} = 0$).

The main assumption of the proposed model is that, at a specific hour, the solar irradiance is independent of the values observed in the previous hours. With this assumption, the probability of the occurrence of scenario $s^k$, $\pi_k$, is the product of the probabilities of the regions that compose it [45]:

$$\pi_k = \frac{\prod_{t=0}^{23} \sum_{r=1}^{n_R} \left\{ W_{k,t,r} \cdot \hat{\alpha}_{t,r} \right\}}{\sum_{k=1}^{n_S} \prod_{t=0}^{23} \sum_{r=1}^{n_R} \left\{ W_{k,t,r} \cdot \hat{\alpha}_{t,r} \right\}}, \quad k = 1, \ldots, n_S \tag{6}$$

The independence assumption can be relaxed, considering that the solar irradiance assumed at a particular hour depends on the solar irradiance values of previous hours. The methodology described here is, however, still valid, and this extension is reported elsewhere.

*2.5. Scenarios' Reduction Process*

From an initial set of $n_S$ scenarios with their probabilities $\pi_k$ (with $k = 1, \ldots, n_S$), it is necessary to obtain a reduced set of $n_P$ scenarios to use in the successive potential stochastic optimization phase. The considered reduction method is the Fast-Forward method [39,40]. The algorithm creates a subset of scenarios with the minimum Kantorovich distance from the initial set. After computing the distance (with respect to a *metric* defined as the mean absolute distance, Euclidean distance, etc.) between all pairs of scenarios, the scenario with the minimum weighted distance (the weights are the probabilities of the occurrence of each scenario) from all of the other ones is selected. The probability of the occurrence of all removed scenarios is absorbed by the preserved scenario that is nearest to them.

The main steps of the Fast-Forward algorithm are described in the Algorithm 1 box [39].

---

**Algorithm 1.** Fast-Forward

---

**Step 1**

1. For each pair of scenarios ($s^k$ and $s^u$), the distance is computed by using the metric $c_T$. The generic element $C_{k,u}$ of matrix $C$ in step 1 is:

2.
$$C_{k,u}^{[1]} = c_T(s^k, s^u),\ k, u = 1, \ldots, n_S \tag{7}$$

3. The metric usually used is the $\ell_q$-Norm of $\mathbb{R}^T$, which can be defined as:

4.
$$c_T(s^k, s^u) = \left( \sum_{t=1}^{T} \left| s_t^k - s_t^u \right|^q \right)^{\frac{1}{q}} \tag{8}$$

5. Each scenario $s^u$ is associated with the weighted distance to any other scenario $s^k$, where the weights are the probabilities of occurrence $\pi_k$:

6.
$$z_u^{[1]} = \sum_{\substack{k=1 \\ k \neq u}}^{n_S} \pi_k C_{k,u}^{[1]},\ u = 1, \ldots, n_S \tag{9}$$

For example, the $z$ values for scenarios $s^1$ and $s^2$ are the following:

7.
$$z_1^{[1]} = \pi_2 C_{21}^{[1]} + \pi_3 C_{31}^{[1]} + \pi_4 C_{41}^{[1]} + \ldots$$

$$z_2^{[1]} = \pi_1 C_{12}^{[1]} + \pi_3 C_{32}^{[1]} + \pi_4 C_{42}^{[1]} + \ldots$$

8. Among the results, the index of the scenario with the minimum value of $z$ is selected ($u_1$):

9.
$$u_1 \in argmin_{u \in \{1,\ldots,n_S\}}\ z_u^{[1]} \tag{10}$$

---

---

**Algorithm 1.** *cont.*

---

10.    Then, $s^{u_1}$ is preserved (operatively, $u_1$ is removed from the indexes of scenarios to delete in step 1, $J^{[1]}$):

11.

$$J^{[1]} = \{1, \ldots, n_S\} \backslash \{u_1\} \tag{11}$$

**Step i**

12.    Using the information from previous steps, the distance matrix is updated using Equation (12), new values of $z$ are computed using Equation (13), and a new scenario is selected to be preserved ($s^{u_i}$) using Equations (14) and (15):

13.

$$C_{ku}^{[i]} = \min\left\{ C_{ku}^{[i-1]}, \ C_{ku_{i-1}}^{[i-1]} \right\}, \ k, u \in J^{[i-1]} \tag{12}$$

14.

$$z_u^{[i]} = \sum_{k \in J^{[i-1]} \backslash \{u\}} \pi_k C_{ku}^{[i]}, \ u \in J^{[i-1]} \tag{13}$$

15.

$$u_i \in argmin_{u \in J^{[i-1]}} z_u^{[i]} \tag{14}$$

16.

$$J^{[i]} = J^{[i-1]} \backslash \{u_i\} \tag{15}$$

**Step $n_P + 1$**

17.    In the final step, the list of scenarios to remove $J = J^{[n_P]}$ is completed. Each scenario to be removed will be linked to a preserved scenario that will "substitute" it. In fact, $j(i)$ is the index of the preserved scenario nearest to the removed scenario $s^i$:

18.

$$j(i) \in argmin_{j \notin J} c_T(s^i, s^j), \ \forall i \in J \tag{16}$$

19.    The set of indexes of the removed scenarios that have $s^j$ as the nearest preserved scenario can be defined as follows:

20.

$$J(j) = \{i \in J : j = j(i)\} \tag{17}$$

21.    Using the *optimal redistribution rule* [40], the probability of the occurrence $\pi_j$ of the preserved scenario $s^j$ is computed:

22.

$$\pi_j = \pi_j + \sum_{i \in J(j)} \pi_i \tag{18}$$

23.    The probabilities of the occurrence of the removed scenarios nearest to $s^j$ are added to the initial value of $\pi_j$.

---

## 3. Numerical Results

In this section, the numerical results of the scenario generation process carried out considering different values of $n_R$, *metric*, and $p$ are presented.

Due to the randomness of sampling from uniform random variables in the Roulette Wheel method, two successive executions of the scenario generation process with the same parameters could return different scenarios unless fixing the *seed* for sampling from the uniform distribution. In order to compare the several generated scenarios with different parameters, the same *seed* was used for all trials related to the same month.

### 3.1. Outlier Removal

Figures 5 and 6 show the results obtained by using $n_S = 1000$, $n_P = 10$, $n_R = 7$, and *metric* = $\ell_2$-Norm and by varying $p$.

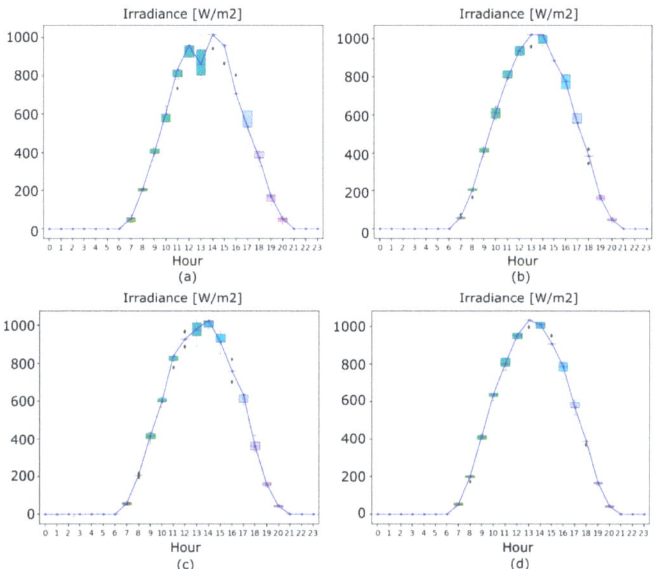

**Figure 5.** Median scenarios (blue) with box-whisker plot for scenarios generated for July using $n_S = 1000$, $n_P = 10$, $n_R = 7$, and *metric* = $\ell_2$ -Norm and varying *p*: 1.5 (**a**), 1 (**b**), 0.5 (**c**), and 0.15 (**d**).

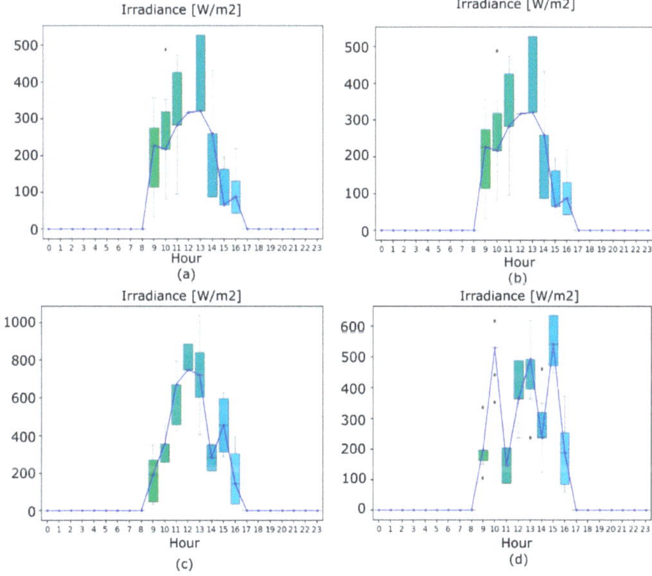

**Figure 6.** Median scenarios (blue) with box-whisker plot for scenarios generated for January using $n_S = 1000$, $n_P = 10$, $n_R = 7$, and *metric* = $\ell_2$ -Norm and varying *p*: 1.5 (**a**), 1 (**b**), 0.5 (**c**), and 0.15 (**d**).

*3.2. Number of Regions*

Figures 7 and 8 show the results obtained by considering $n_S = 1000$, $n_P = 10$, and *metric* = $\ell_2$-Norm without outlier removal and by varying $n_R$.

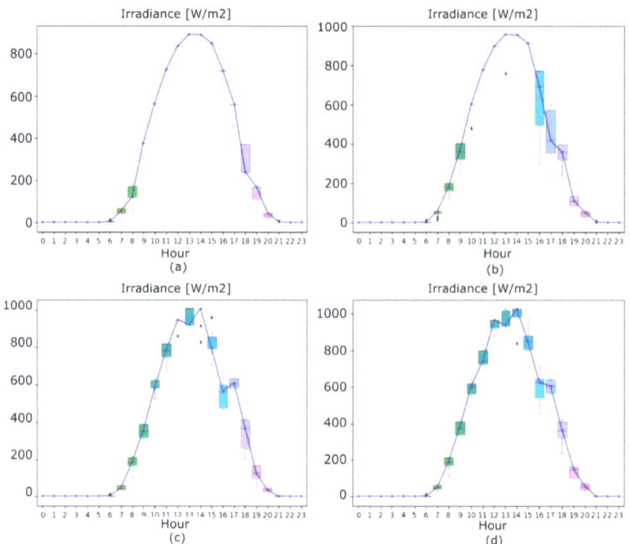

**Figure 7.** Median scenarios (blue) with box-whisker plot for scenarios generated for July using $n_S = 1000$, $n_P = 10$, and *metric* = $l_2$-Norm without outlier removal and varying the number of regions ($n_R$): 3 (**a**), 5 (**b**), 11 (**c**), and 21 (**d**).

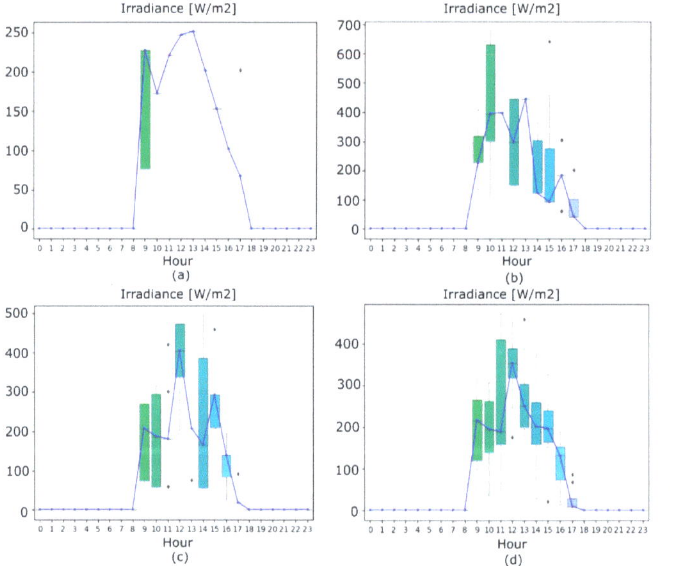

**Figure 8.** Median scenarios (blue) with box-whisker plot for scenarios generated for January using $n_S = 1000$, $n_P = 10$, and *metric* = $\ell_2$-Norm without outlier removal and varying the number of regions ($n_R$): 3 (**a**), 5 (**b**), 11 (**c**), and 21 (**d**).

### 3.3. Metric

Figures 9 and 10 show the results obtained by using $n_S = 1000$, $n_P = 10$, and $n_R = 7$ without outlier removal and varying the metric used to compare two scenarios. For that, several $\ell_q$-Norms were tested (Equation (8)): $\ell_1$-Norm ($q = 1$), $\ell_2$-Norm ($q = 2$), $\ell_4$-Norm ($q = 4$), and $\ell_\infty$-Norm (Equation (19)).

$$c_T(s^k, s^u) = \max_t \left| s_t^k - s_t^u \right| \tag{19}$$

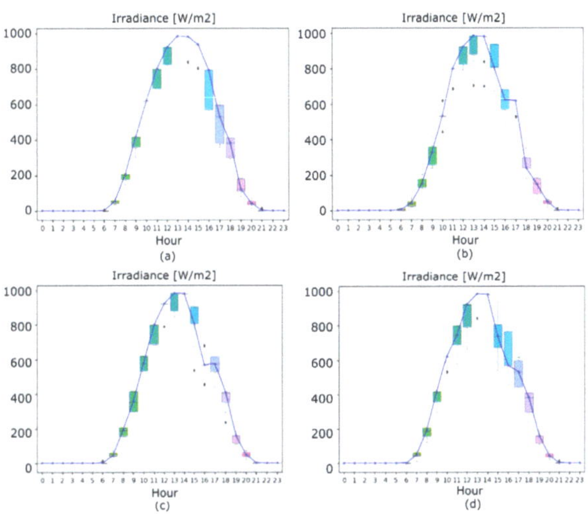

**Figure 9.** Median scenarios (blue) with box-whisker plot for scenarios generated for July using $n_S = 1000$, $n_P = 10$, and $n_R = 7$ without outlier removal and varying the type of *metric*: $\ell_1$ -Norm (**a**), $\ell_2$ -Norm (**b**), $\ell_4$ -Norm (**c**), and $\ell_\infty$ -Norm (**d**).

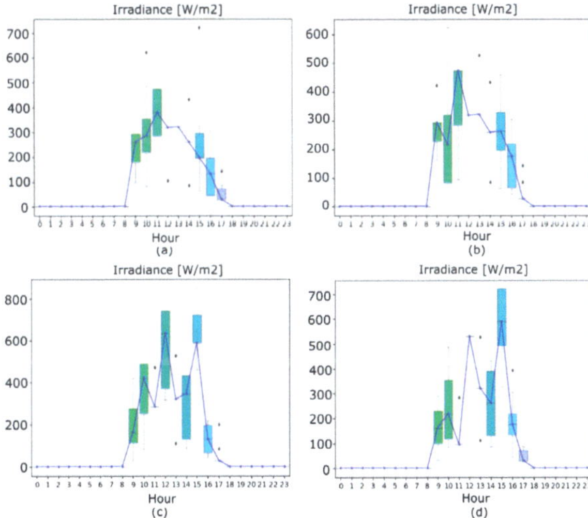

**Figure 10.** Median scenarios (blue) with box-whisker plot for scenarios generated for January using $n_S = 1000$, $n_P = 10$, and $n_R = 7$ without outlier removal and varying the type of *metric*: $\ell_1$ -Norm (**a**), $\ell_2$ -Norm (**b**), $\ell_4$ -Norm (**c**), and $\ell_\infty$ -Norm (**d**).

## 4. Discussion

In order to show the impact of the parameters $p$, $n_R$, and *metric* on the preserved scenarios (in terms of smoothness and variability in the same hour), a sensitivity analysis was performed.

For the estimation of the variability in the preserved scenarios, the average (for all hours excluding those in which irradiance is always zero) of the difference between the 97.5th and 2.5th percentiles of the preserved scenarios was considered. The trends of the average of the difference between the 97.5th and 2.5th percentiles for the different trials described in Sections 3.1–3.3 for July (summer) and January (winter) are shown in Figures 11 and 12, respectively.

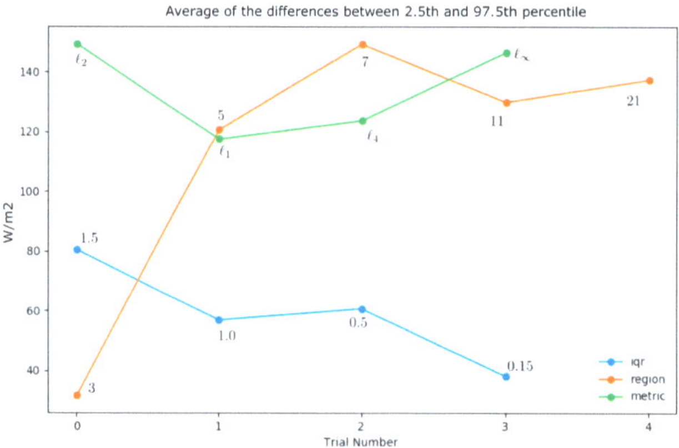

**Figure 11.** Trend of average of the difference between 97.5th and 2.5th percentiles for preserved scenarios for the summer: (blue) $n_R = 7$ and *metric* = $\ell_2$-Norm and varying $p$: 1.5, 1, 0.5, and 0.15; (green) $n_R = 7$, without outlier removal and varying *metric*: $\ell_2$-Norm, $\ell_1$-Norm, $\ell_4$-Norm, and $\ell_\infty$-Norm; (orange) *metric* = $\ell_2$-Norm, without outlier removal and varying $n_R$: 3, 5, 7, 11, and 21. The text in the graph indicates the values assumed by the varied parameters.

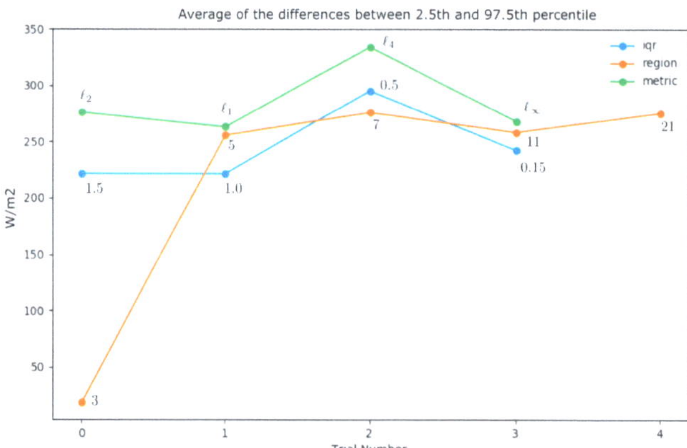

**Figure 12.** Trend of average of the difference between 97.5th and 2.5th percentiles for preserved scenarios for the winter: (blue) $n_R = 7$ and *metric* = $\ell_2$-Norm and varying $p$: 1.5, 1, 0.5, and 0.15; (green) $n_R = 7$, without outlier removal and varying *metric*: $\ell_2$-Norm, $\ell_1$-Norm, $\ell_4$-Norm, and $\ell_\infty$-Norm; (orange) *metric* = $\ell_2$-Norm, without outlier removal and varying $n_R$: 3, 5, 7, 11, and 21. The text in the graph indicates the values assumed by the varied parameters.

Regarding the outlier removal process, when reducing $p$, more samples for each hour are considered outliers (and hence removed). For the solar irradiance distributions with small hourly IQRs and many values outside IQRs, as in July, where the average IQR is about 111 W/m$^2$ (Figure 2), removing outliers and reducing the value of $p$ lead to generated scenarios with small variability (Figure 5 and blue line in Figure 11). For the solar irradiance distributions where hourly IQRs are high and few observations are outside IQRs, as in January, where the average IQR is about 460 W/m$^2$ (Figure 3), the outlier removal process does not have, in general, a high impact (Figure 6 and the blue line in Figure 12).

Regarding the impact of the number of regions ($n_R$), the variability is increased for both seasons when $n_R$ is increased from 3 to 7 (Figures 7 and 8 and orange lines in Figures 11 and 12). With a value of $n_R$ that is higher than 7, the variability is quite stable. This can be explained by the fact that, even though the support of the beta distribution has been divided using more regions, more of them have a low probability of occurrence and, hence, do not impact the resulting variability.

Regarding the impact of several metrics for comparing two scenarios during the reduction process (Figures 9 and 10 and green lines in Figures 11 and 12), it is possible to observe that very different results are obtained from different metrics in the two considered seasons (e.g., $\ell_2$-Norm produces scenarios with high variability in summer and low variability in winter, $\ell_4$-Norm produces scenarios with lower variability than $\ell_2$-Norm in summer and higher variability than $\ell_2$-Norm in winter, etc.), and the variability obtained with $\ell_\infty$-Norm is close to that obtained with $\ell_2$-Norm.

The user of the method can select the best combination of the presented parameters in order to obtain the best trade-off between the variability among preserved scenarios and their plausibility. To verify the plausibility of the preserved scenarios, the user can plot them in the box-whisker plot obtained from the real solar irradiance data (Figures 1 and 2) and verify that the generated values of hourly solar irradiance for the particular scenario do not deviate too much from the boxes and whiskers. (In the box-whisker plot, the box describes the range between $Q_1$ and $Q_3$, the upper whisker is on $Q_3 + 1.5 \cdot IQR$, and the lower whisker is on $Q_1 - 1.5 \cdot IQR$. The values outside the range defined by the whiskers can be considered outliers). In this case, this means that the generated values of hourly solar irradiance for that scenario can be considered inliers with respect to the observed data.

The scenarios obtained with $n_S = 1000$, $n_P = 10$, and $n_R = 7$, $\ell_2$-Norm metric and without the outlier removal step are shown in Figure 13 for summer and Figure 14 for winter. Figures 15 and 16 show the preserved scenarios plotted over the box-whisker plot of the observed solar irradiance. In these images, it is possible to see that the preserved scenarios are almost completely contained in the boxes (representing the IQR of the observed solar irradiance), and almost all values that are outside the boxes are confined to the variability range observed in the real data (the range between the two whiskers) and hence can be considered plausible.

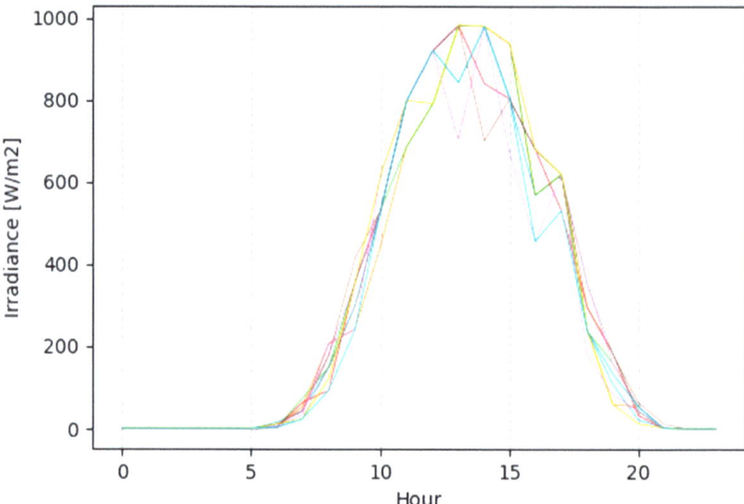

**Figure 13.** Ten generated scenarios reduced from one thousand initial scenarios using Fast-Forward algorithm with 7 regions, $\ell_2$-Norm metric, and without outlier removal. The historical data are related to July for the city of Turin (Italy) from 2005 to 2016. Different line colors represent different scenarios.

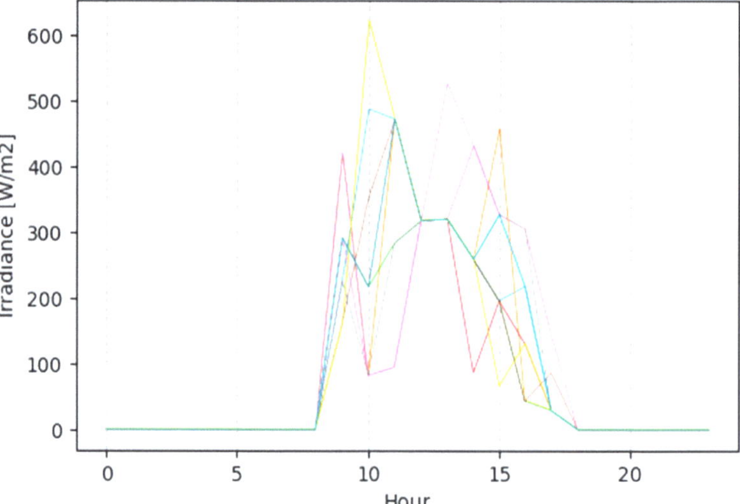

**Figure 14.** Ten generated scenarios reduced from one thousand initial scenarios using Fast-Forward algorithm with 7 regions, $\ell_2$-Norm metric, and without outlier removal. The historical data are related to January for the city of Turin (Italy) from 2005 to 2016. Different line colors represent different scenarios.

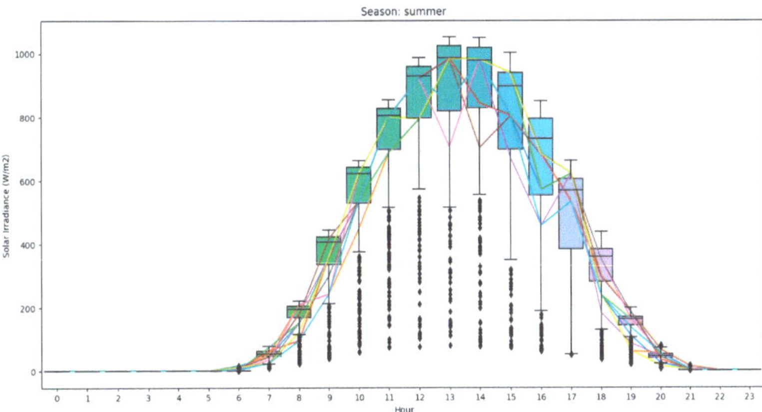

**Figure 15.** Ten preserved scenarios plotted over the box-whisker plot of the observed solar irradiance for the days in July from 2005 to 2016. Different line colors represent different scenarios.

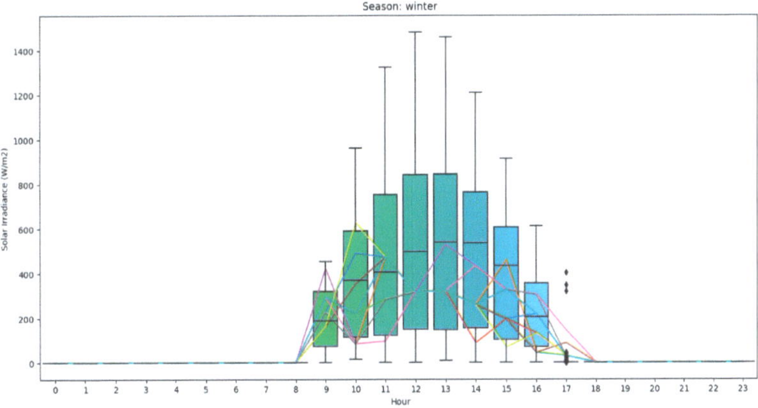

**Figure 16.** Ten preserved scenarios plotted over the box-whisker plot of the observed solar irradiance for the days in January from 2005 to 2016. Different line colors represent different scenarios.

## 5. Conclusions

In this work, a comprehensive tool to generate solar irradiance profiles is presented. The proposed approach is based on a scenario generation process aimed at generating 24 h solar irradiance scenarios using the historical data of solar irradiance for a specific location.

In the case study, the proposed method was applied to generate a set of daily solar irradiance scenarios for the months of January and July for the city of Turin (Italy). The Roulette Wheel mechanism was used to generate the initial set of scenarios, and the Fast-Forward method for the reduction process was applied to preserve the most representative scenarios and reduce the computational efforts associated with the potential stochastic operation optimization phase. The results demonstrate the flexibility of the method in generating scenarios for solar irradiance and in assessing their plausibility. These characteristics make the proposed approach an effective tool to be used for the stochastic operation optimization of DER.

Moreover, the results of the sensitivity analysis show the influence of the variation in the key parameters on the results in terms of increasing the variability and/or the smoothness of the generated scenarios, which could be very effective in estimating the behavior of the stochastic operation optimization of DER in the presence of more fluctuating but plausible solar irradiance patterns.

Given the generality of the proposed method, it can be easily adapted to model solar irradiance profiles for different locations and use cases, and hence, it can serve as a guide to users for the definition of scenarios with specific characteristics. Moreover, the proposed pipeline can be implemented as a Web Service queryable by users in order to generate solar irradiance scenarios with their probability of occurrence, which is fundamental for the stochastic optimization of DER.

**Author Contributions:** Conceptualization, A.B., M.C. and M.D.S.; methodology, A.B.; software, A.B.; data curation, A.B.; writing, review and editing, A.B., M.C., M.D.S., G.G. and M.V. All authors have read and agreed to the published version of the manuscript.

**Funding:** This research was funded under Project 1.7 "Tecnologie per la penetrazione efficiente del vettore elettrico negli usi finali" within the Italian Research Program ENEA-MISE 2022–2024 "Ricerca di Sistema Elettrico".

**Data Availability Statement:** The data used for this study were gathered using PVGIS as described in the Section 2.1.

**Conflicts of Interest:** The authors declare no conflict of interest.

## Abbreviations

| | |
|---|---|
| AI | Artificial Intelligence |
| ANN | Artificial Neural Networks |
| DER | Distributed energy resources |
| FL | Fuzzy Logic |
| IGDT | Information Gap Decision Theory |
| IQR | Interquartile Range |
| MF | Membership function |
| PDF | Probability density function |
| PV | Photovoltaic |
| PVGIS | Photovoltaic Geographical Information System |
| RES | Renewable energy sources |

**Nomenclature**

| | |
|---|---|
| $\alpha_{t,r}$ | Probability of occurrence of a particular region $r$ at time $t$ |
| $\hat{\alpha}_{t,r}$ | Normalized probability of occurrence of a particular region $r$ at time $t$ |
| $\pi_k$ | Probability of occurrence of scenario $s^k$ |
| $\Gamma(x)$ | Gamma function |
| $c_T$ | Metric used to compute the distance between two scenarios |
| $h_{t,r}$ | Height of region $r$ at time $t$ |
| $n_P$ | Number of preserved scenarios |
| $n_R$ | Number of regions (bins) used to divide the support of the distribution |
| $n_S$ | Number of generated scenarios |
| $p$ | Parameter used to define outliers |
| $r$ | Number of considered regions ($r \in \{1, \ldots, n_R\}$) |
| $s^k$ | $k$-th scenario (signal containing 24 irradiance values) |
| $t$ | Hour of the day ($t \in \{0, \ldots, 23\}$) |
| $w_{t,r}$ | Width of region $r$ at time $t$ |
| $z_u^{[m]}$ | Weighted distance of scenario $s^u$ from all other scenarios in step $m$ |
| $B(a,b)$ | Beta function with parameters $a$ and $b$ |
| $C$ | Matrix containing the distances between all pairs of scenarios |
| $C_{k,u}^{[m]}$ | $(k,u)$th entry of matrix $C$, representing the distance between scenarios $s^k$ and $s^u$ in step $m$ |
| $J(j)$ | Set of indexes of the removed scenarios that have $s^j$ as the nearest preserved scenario |
| $J^{[m]}$ | List of indexes of deleted scenarios in step $m$ |
| $Q_1$ | 1st quartile, 25th percentile of observed values |
| $Q_3$ | 3rd quartile, 75th percentile of observed values |
| $W_{k,t,r}$ | Binary variable that describes whether region $r$ of scenario $s^k$ is selected at time $t$ |

## References


1. 'Fi1. Fit for 55': Delivering the EU's 2030 Climate Target on the Way to Climate neutralitycom(2021) 550 Final, Brussels, 14.7.2021. Available online: https://ec.europa.eu/info/sites/default/files/chapeau_communication.pdf (accessed on 30 October 2022).
2. Breyer, C.; Khalili, S.; Bogdanov, D. Solar photovoltaic capacity demand for a sustainable transport sector to fulfil the Paris Agreement by 2050. *Prog. Photovolt. Res. Appl.* **2019**, *27*, 978–989. [CrossRef]
3. Sæle, H.; Morch, A.; Buonanno, A.; Caliano, M.; di Somma, M.; Papadimitriou, C. Development of Energy Communities in Europe. In Proceedings of the 18th International Conference on the European Energy Market (EEM), Ljubljana, Slovenia, 13–15 September 2022.
4. Fateh, D.; Eldoromi, M.; Birjandi, A.A.M. Uncertainty Modeling of Renewable Energy Sources. In *Scheduling and Operation of Virtual Power Plants*; Elsevier: Amsterdam, The Netherlands, 2022; pp. 193–208.
5. Brenna, M.; Foiadelli, F.; Zaninelli, D.; Graditi, G.; Di Somma, M. The Integration of Electric Vehicles in Smart Distribution Grids with Other Distributed Resources. In *Distributed Energy Resources in Local Integrated Energy Systems*; Elsevier: Amsterdam, The Netherlands, 2021; pp. 315–345. [CrossRef]
6. Papadimitriou, C.; Patsalides, M.; Venizelos, V.; Therapontos, P.; Efthymiou, V. Assessing Renewables Uncertainties in the Short-Term (Day-Ahead) Scheduling of DER. In *Distributed Energy Resources in Local Integrated Energy Systems*; Elsevier: Amsterdam, The Netherlands, 2021; pp. 347–387.
7. Zakaria, A.; Ismail, F.B.; Lipu, M.H.; Hannan, M. Uncertainty models for stochastic optimization in renewable energy applications. *Renew. Energy* **2019**, *145*, 1543–1571. [CrossRef]
8. Li, Y.; Han, M.; Yang, Z.; Li, G. Coordinating Flexible Demand Response and Renewable Uncertainties for Scheduling of Community Integrated Energy Systems with an Electric Vehicle Charging Station: A Bi-level Approach. *IEEE Trans. Sustain. Energy* **2021**, *12*, 2321–2331. [CrossRef]
9. di Somma, M.; Graditi, G.; Heydarian-Forushani, G.; Shafie-khah, M.; Siano, P. Stochastic optimal scheduling of distributed energy resources with renewables considering economic and environrmental aspects. *Renew. Energy* **2018**, *116*, 272–287. [CrossRef]
10. Zhou, Z.; Zhang, J.; Liu, P.; Li, Z.; Georgiadis, M.C.; Pistikopoulos, E.N. A two-stage stochastic programming model for the optimal design of distributed energy systems. *Appl. Energy* **2012**, *103*, 135–144. [CrossRef]
11. Hong, F.; Cuiying, W.; Lu, L.; Xuan, L. Review of Uncertainty Modeling for Optimal Operation of Integrated Energy System. *Front. Energy Res.* **2022**, *9*, 641337. [CrossRef]
12. Di Somma, M.; Buonanno, A.; Caliano, M.; Graditi, G.; Piazza, G.; Bracco, S.; Delfino, F. Stochastic Operation Optimization of the Smart Savona Campus as an Integrated Local Energy Community Considering Energy Costs and Carbon Emissions. *Energies* **2022**, *15*, 8418. [CrossRef]
13. Ashtari, A.; Bibeau, E.; Shahidinejad, S.; Molinski, T. PEV Charging Profile Prediction and Analysis Based on Vehicle Usage Data. *IEEE Trans. Smart Grid* **2011**, *3*, 341–350. [CrossRef]
14. Ali, S.; Lee, S.-M.; Jang, C.-M. Statistical analysis of wind characteristics using Weibull and Rayleigh distributions in Deokjeok-do Island—Incheon, South Korea. *Renew. Energy* **2018**, *123*, 652–663. [CrossRef]
15. Shoaib, M.; Siddiqui, I.; Amir, Y.M.; Rehman, S.U. Evaluation of wind power potential in Baburband (Pakistan) using Weibull distribution function. *Renew. Sustain. Energy Rev.* **2017**, *70*, 1343–1351. [CrossRef]
16. Labeeuw, W.; Deconinck, G. Customer sampling in a smart grid pilot. In Proceedings of the 2012 IEEE Power and Energy Society General Meeting, San Diego, CA, USA, 22–26 July 2012.
17. Li, J.; Zhou, J.; Chen, B. Review of wind power scenario generation methods for optimal operation of renewable energy systems. *Appl. Energy* **2020**, *280*, 115992. [CrossRef]
18. Aien, M.; Hajebrahimi, A.; Fotuhi-Firuzabad, M. A comprehensive review on uncertainty modeling techniques in power system studies. *Renew. Sustain. Energy Rev.* **2016**, *57*, 1077–1089. [CrossRef]
19. Alipour, M.; Jalali, M.; Abapour, M.; Tohidi, S. Uncertainty Modeling in Operation of Multi-carrier Energy Networks. In *Planning and Operation of Multi-Carrier Energy Networks. Power Systems*; Spinger: Berlin/Heidelberg, Germany, 2021.
20. Teke, A.; Yıldırım, B.H.; Çelik, Ö. Evaluation and performance comparison of different models for the estimation of solar radiation. *Renew. Sustain. Energy Rev.* **2015**, *50*, 1097–1107. [CrossRef]
21. Khatib, T.; Mohamed, A.; Sopian, K. A review of solar energy modeling techniques. *Renew. Sustain. Energy Rev.* **2012**, *16*, 2864–2869. [CrossRef]
22. Yorukoglu, M.; Celik, A.N. A critical review on the estimation of daily global solar radiation from sunshine duration. *Energy Convers. Manag.* **2006**, *47*, 2441–2450. [CrossRef]
23. Besharat, F.; Dehghan, A.A.; Faghih, A.R. Empirical models for estimating global solar radiation: A review and case study. *Renew. Sustain. Energy Rev.* **2013**, *21*, 798–821. [CrossRef]
24. Elminir, H.; Azzam, Y.; Younes, F. Prediction of hourly and daily diffuse fraction using neural network, as compared to linear regression models. *Energy* **2007**, *32*, 1513–1523. [CrossRef]
25. Rahimikhoob, A. Estimating global solar radiation using artificial neural network and air temperature data in a semi-arid environment. *Renew. Energy* **2010**, *35*, 2131–2135. [CrossRef]
26. Aguiar, R.; Collares-Pereira, M. Statistical properties of hourly global radiation. *Sol. Energy* **1992**, *48*, 157–167. [CrossRef]
27. Duzen, H.; Aydin, H. Sunshine-based estimation of global solar radiation on horizontal surface at Lake Van region (Turkey). *Energy Convers. Manag.* **2012**, *58*, 35–46. [CrossRef]

28.  Teke, A.; Yıldırım, H.B. Estimating the monthly global solar radiation for Eastern Mediterranean Region. *Energy Convers. Manag.* **2014**, *87*, 628–635. [CrossRef]
29.  Korachagaon, I. Estimating Global Solar Radiation Potential for Brazil by Iranna-Bapat's Regression Models. *Int. J. Emerg. Technol. Adv.* **2012**, *2*, 178–186.
30.  Khatod, D.K.; Pant, V.; Sharma, J. Evolutionary programming based optimal placement of renewable distributed generators. *IEEE Trans. Power Syst.* **2012**, *28*, 683–695. [CrossRef]
31.  Salameh, Z.; Borowy, B.; Amin, A. Photovoltaic module-site matching based on the capacity factors. *IEEE Trans. Energy Convers.* **1995**, *10*, 326–332. [CrossRef]
32.  Li, Y.; Zio, E. Uncertainty analysis of the adequacy assessment model of a distributed generation system. *Renew. Energy* **2012**, *41*, 235–244. [CrossRef]
33.  Karaki, S.; Chedid, R.; Ramadan, R. Probabilistic performance assessment of autonomous solar-wind energy conversion systems. *IEEE Trans. Energy Convers.* **1999**, *14*, 766–772. [CrossRef]
34.  Khan, M.F.N.; Malik, T.N. Probablistic generation model for optimal allocation of PV DG in distribution system with time-varying load models. *J. Renew. Sustain. Energy* **2017**, *9*, 065503. [CrossRef]
35.  Hung, D.Q.; Mithulananthan, N.; Lee, K.Y. Determining PV Penetration for Distribution Systems With Time-Varying Load Models. *IEEE Trans. Power Syst.* **2014**, *29*, 3048–3057. [CrossRef]
36.  Hagan, K.E.; Oyebanjo, O.O.; Masaud, T.M.; Challoo, R. A probabilistic forecasting model for accurate estimation of PV solar and wind power generation. In Proceedings of the 2016 IEEE Power and Energy Conference at Illinois (PECI), Urbana, IL, USA, 19–20 February 2016; pp. 1–5. [CrossRef]
37.  Afzaal, M.U.; Sajjad, I.A.; Awan, A.B.; Paracha, K.N.; Khan, M.F.N.; Bhatti, A.R.; Zubair, M.; Rehman, W.U.; Amin, S.; Haroon, S.S.; et al. Probabilistic Generation Model of Solar Irradiance for Grid Connected Photovoltaic Systems Using Weibull Distribution. *Sustainability* **2020**, *12*, 2241. [CrossRef]
38.  Rocke, D.M.; Michalewicz, Z. Genetic Algorithms + Data Structures = Evolution Programs. *J. Am. Stat. Assoc.* **2000**, *95*, 347. [CrossRef]
39.  Growe-Kuska, N.; Heitsch, H.; Romisch, W. Scenario Reduction and Scenario Tree Construction for Power Management Problems. In Proceedings of the 2003 IEEE Bologna Power Tech Conference Proceedings, Bologna, Italy, 23–26 June 2003; pp. 1–7.
40.  Heitsch, H.; Römisch, W. Scenario Reduction Algorithms in Stochastic Programming. *Comput. Optim. Appl.* **2003**, *24*, 187–206. [CrossRef]
41.  Buonanno, A.; Caliano, M.; Di Somma, M.; Graditi, G.; Valenti, M. Comprehensive method for modeling uncertainties of solar irradiance for PV power generation in smart grids. In Proceedings of the 2021 International Conference on Smart Energy Systems and Technologies (SEST), Vaasa, Finland, 6–8 September 2021.
42.  PVGIS. Available online: https://re.jrc.ec.europa.eu/pvg_tools/en/tools.html (accessed on 30 October 2022).
43.  Tukey, J.W. *Exploratory Data Analysis*; Addison-Wesley: Boston, MA, USA, 1977.
44.  Murphy, K.P. *Probabilistic Machine Learning—An Introduction*; The MIT Press: Cambridge, MA, USA, 2022.
45.  Pourghasem, P.; Sohrabi, F.; Abapour, M.; Mohammadi-Ivatloo, B. Stochastic multi-objective dynamic dispatch of renewable and CHP-based islanded microgrids. *Electr. Power Syst. Res.* **2019**, *173*, 193–201. [CrossRef]
46.  Rubinstein, R.Y.; Kroese, D.P. *Simulation and the Monte Carlo Method*, 3rd ed.; Wiley: Hoboken, NJ, USA, 2016.

*Article*

# Static Voltage Stability Zoning Analysis Based on a Sensitivity Index Reflecting the Influence Degree of Photovoltaic Power Output on Voltage Stability

Sheng Li *, Yuting Lu and Yulin Ge

School of Electric Power Engineering, Nanjing Institute of Technology, Nanjing 211167, China
* Correspondence: lisheng_njit@126.com; Tel.: +86-25-8611-8400

**Abstract:** The large-scale integration of photovoltaic (PV) power can bring a greatly negative influence on the grid-connected system's voltage stability. To study the static voltage stability (SVS) of PV grid-connected systems, the traditional SVS index, L-index, was re-examined. It was firstly derived and proved that the PV active output $P_{pv}$ is proportional to the voltage phase angle of the PV station's POI (Point of Interconnection), based on a simplified two-node system integrated with a PV station operating in *PV* (active power—voltage) mode or *PQ* (active power—reactive power) mode with unit power factor. Then a novel voltage stability sensitivity index LPAS-index was proposed that takes the derivative of the L-index with respect to the POI's voltage phase angle, so as to reflect the influence degree of $P_{pv}$ on the SVS of each load node. A SVS zoning analysis method for the PV grid-connected system was designed according to the classification results of load nodes based on the proposed LPAS-index, the power grid can be zoned into three kinds of areas that reflect different correlations between the SVS and $P_{pv}$: strong correlation, moderate correlation and weak correlation. Since the LPAS-index is less impacted by $P_{pv}$, the SVS zoning results are relatively unchanged. On the basis of a classic 14-node system with PV, the practicability of the zoning analysis method was verified. The simulation results show that the PV access point in general falls within the strongly or moderately associated area with $P_{pv}$. When most of the load nodes fall within the weakly associated area with $P_{pv}$, it is not necessary to consider the impact of $P_{pv}$ and load power is still the main influencing factor on the SVS. In the multi-PV case, owing to the expansion of areas more affected by $P_{pv}$, an excessive $P_{pv}$ can cause adverse influence on the SVS of the whole power grid; and an effective PV power-shedding measure is proposed to solve this problem. The proposed SVS zoning analysis method can be used for reference by power grid dispatchers.

**Keywords:** static voltage stability (SVS); photovoltaic (PV) active output; Point of Interconnection (POI); sensitivity index; zoning analysis

**Citation:** Li, S.; Lu, Y.; Ge, Y. Static Voltage Stability Zoning Analysis Based on a Sensitivity Index Reflecting the Influence Degree of Photovoltaic Power Output on Voltage Stability. *Energies* **2023**, *16*, 2808. https://doi.org/10.3390/en16062808

Academic Editors: Marialaura Di Somma, Jianxiao Wang and Bing Yan

Received: 14 February 2023
Revised: 9 March 2023
Accepted: 15 March 2023
Published: 17 March 2023

## 1. Introduction

In order to alleviate problems such as energy shortage and environmental pollution and achieve the goal of 'dual carbon', China has made great efforts to develop photo-voltaic (PV) power generation technology in recent years [1,2]. With the maturity of PV grid-connected technology and the continuous reduction of installation costs, PV power generation shows a development trend of large-scale grid-connected operation and becomes the main new-energy in the construction of modern power systems [3]. In China, according to the latest data released by the National Energy Administration, the installed capacity of grid-connected PV power generation reached 392.04 GW by the end of 2022, and centralized PV accounts for about 60 percent of the total installed capacity. The PV power output is greatly affected by natural and meteorological factors and has a strong random fluctuation and intermittence [4,5]. Meanwhile, the integration of large-scale PV intensifies the power electronic characteristics and reduces the power system inertia, and brings more effects and challenges to the static and dynamic stability of the power system [6–8], in particular the static-state (steady-state), small-disturbance, transient-state and long-term voltage stability [9–11].

Static voltage stability (SVS) refers to the voltage stability when the power system sustains various small disturbances (such as small changes in load power) without considering the dynamic characteristics of each electric element [12]. In recent years, the studies on the SVS of traditional AC power systems have been still a hot spot and various machine learning algorithms are applied to the SVS prediction issue [13–15].

In the context of the integration of large-scale PV power stations, what are the main factors affecting the SVS of the power grid? How to determine the fluctuation range of SVS critical point, SVS domain and SVS margin; how to analyze the impact of PV power fluctuation on the weakest SVS area; how to design a novel SVS index (or criterion) that can apply to the PV power fluctuation and how to use certain appropriate control strategies to improve the system's SVS are all the main concerns.

The research on the SVS issue of the large-scale PV station grid-connected system has been reported in relevant references. In Reference [16], the impacts of multiple factors were studied on the SVS, including PV transmission line length, PV active output, and PV topological structure, etc. In Reference [17], the impacts of several factors such as solar irradiance, PV generation power factor, PV installed capacity and PV transmission line impedance, were further studied.

In Reference [18], the SVS margin of the IEEE 14-node system with PV under the two modes of unity power factor operation and constant voltage operation was calculated, and it is considered that the integration of large-scale PV is conducive to improving the SVS margin. However, it was pointed out that too much PV active output can decrease the SVS margin in Reference [9]. In Reference [19], an assessment for the probabilistic SVS margin was studied, by using the probabilistic model of PV/wind power and the Monte-Carlo simulation method.

Several novel indexes that are suited for the SVS analysis in the system with large-scale PV stations were proposed. In Reference [20], the influence of the PCC (Point of Common Coupling) of the PV station on the SVS was studied by defining a sensitivity index of load node voltage—PCC active power that can reflect the impact of PV penetration rate on the SVS. In Reference [21], an improved NVSI-index based on the traditional IVSI-index was used to measure the SVS of PCC, which directly considers the impact of PC/wind injection power. A short-circuit ratio index was proposed in Reference [22], which is more related to the SVS of a weak sending-end system with PV. In Reference [23], a synthetical application framework was proposed to measure the SVS of a PV grid-connected system, considering the critical eigenvalue by modal analysis, the reactive power margin by $QV$ (reactive power—voltage) analysis and the line-loss by power flow analysis.

The SVS control for the system with a large-scale PV station was also explored. In Reference [24], a SVS fuzzy controller was designed, synthetically adopting the load node voltage and the load-margin index considering PV power fluctuation as the fuzzy input variables. In Reference [25], SSSC (Static Synchronous Series Compensator) was used to improve the SVS of weak nodes considering the integration of high penetration PV/wind.

In the above references, the influence law of PV power size and fluctuation on the SVS of load nodes, load areas and the whole power grid is not studied. In the SVS analysis of large-scale PV grid-connected systems, in addition to the frequently-used small disturbance condition of gradually increasing load power, the impact of PV power output should also be considered when the system load is at a certain level.

From the above concerns, a traditional SVS index, L-index [26], is re-examined in this paper. Firstly, the relationship between the L-index and PV active output $P_{pv}$ is derived. Then a novel sensitivity index of L-index of each load node relative to $P_{pv}$ is proposed. According to the numerical level of the proposed sensitivity index value, a zoning method for the SVS analysis is proposed and verified. Compared with previous studies, the proposed zoning analysis method can reveal the influence degree on the SVS of different areas in the power grid by $P_{pv}$, and some effective measures can be suggested to improve the SVS according to the zoning results.

## 2. A Novel Sensitivity Index of L-Index Relative to PV Active Output

### 2.1. The Traditional L-Index

The traditional L-index is a local SVS index and was first proposed by Kessel [26], which is used to monitor and evaluate the SVS of load nodes in the traditional power system.

For a large-scale PV grid-connected system, the node-voltage equation of the system is given below:

$$\begin{bmatrix} \dot{I}_L \\ \dot{I}_G \\ 0 \end{bmatrix} = \begin{bmatrix} Y_{LL} & Y_{LG} & Y_{LC} \\ Y_{GL} & Y_{GG} & Y_{GC} \\ Y_{CL} & Y_{CG} & Y_{CC} \end{bmatrix} \begin{bmatrix} \dot{V}_L \\ \dot{V}_G \\ \dot{V}_C \end{bmatrix} \tag{1}$$

In Equation (1):

$Y_{LL}$, $Y_{LG}$, $Y_{LC}$, $Y_{GL}$, $Y_{GG}$, $Y_{GC}$, $Y_{CL}$, $Y_{CG}$ and $Y_{CC}$ correspond to the sub-matrices of the grid-connected system's node admittance matrix, respectively.

$\dot{V}_L$ and $\dot{I}_L$ are the load nodes' voltage and current vectors, respectively.

$\dot{V}_G$ and $\dot{I}_G$ are the power nodes' voltage and current vectors, respectively, including slack bus, *PV* (active power—voltage) bus of synchronous generator, and POI (Point of Interconnection) of PV or PV converter outlet bus.

$\dot{V}_C$ and $\dot{I}_C$ are the contact nodes' voltage and current vectors respectively, and the contact node is the node with neither power supply nor load demand.

By eliminating the contact nodes in Equation (1), we can get:

$$\begin{bmatrix} \dot{V}_L \\ \dot{I}_G \end{bmatrix} = H \cdot \begin{bmatrix} \dot{I}_L \\ \dot{V}_G \end{bmatrix} = \begin{bmatrix} Z_{LL} & F_{LG} \\ K_{GL} & Y_{GG} \end{bmatrix} \cdot \begin{bmatrix} \dot{I}_L \\ \dot{V}_G \end{bmatrix} \tag{2}$$

where $Z_{LL}$, $F_{LG}$, $K_{GL}$ and $Y_{GG}$ are the block sub-matrices of the H-matrix, thereinto $F_{LG}$ is the load participation factor sub-matrix.

Then the L-index of each load node can be given below:

$$L_j = \left| \widetilde{L}_j \right| = \left| 1 - \frac{\sum\limits_{i \in a_G} \widetilde{F}_{ji} \cdot \dot{V}_i}{\dot{V}_j} \right| \tag{3}$$

where $i$ and $j$ are the number of power nodes and load nodes, respectively; $\dot{V}_i$ and $\dot{V}_j$ are the voltage phasors of node $i$ and node $j$ respectively; $a_G$ is the set of power nodes, $\widetilde{F}_{ji}$ is the load participation factor (complex form); $\widetilde{L}_j$ is the complex expression of L-index.

Virtually, the L-index is in the complex form $\widetilde{L}_j$, and its modulus is taken as the practical index in order to measure the SVS of each load node. The range of $L_j$ is between 0 and 1. When the value of $L_j$ trends to 0, the SVS of load node $j$ trends to more stability.

In general, the voltage fluctuations of load nodes and power nodes caused by the PV power fluctuation are very small, so the L-index value is less affected by the PV active output $P_{pv}$. The synchronous generator (set as *PV* bus) is the same.

### 2.2. The Relationship between PV Active Output and POI's Voltage Phase Angle

In the SVS analysis, the PV power can be regarded as a negative load power, so that the Thevenin equivalent can be carried out from the POI of the PV station. Then a simplified PV station grid-connected two-node system is formed, as shown in Figure 1.

In Figure 1, $P_{pv}$ and $Q_{pv}$ are the active output and reactive output of PV, respectively; $V_{pv}$ and $\delta$ are the voltage amplitude and phase angle of POI, respectively. $Z$ is the impedance modulus of the equivalent power grid line and $\theta$ is the impedance angle. $E$ is the potential of an equivalent electric source.

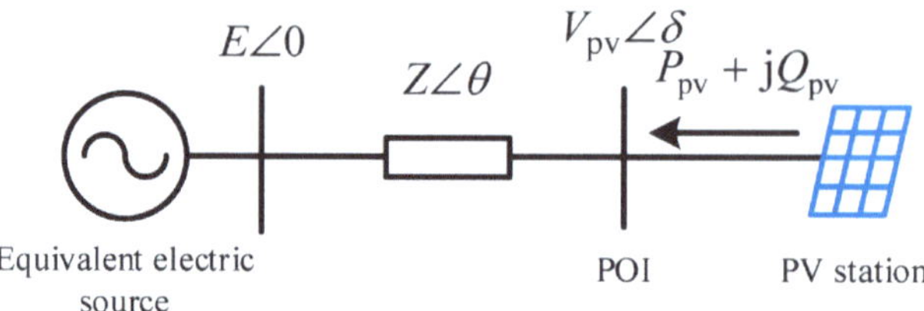


**Figure 1.** The simplified PV grid-connected two-node system.

In this paper, all electrical quantities are in per-unit value (pu) and the phase angle's unit is rad.

Then we can get:

$$
\begin{aligned}
-P_{pv} - jQ_{pv} &= V_{pv} \angle \delta \left( \frac{E \angle 0 - V_{pv} \angle \delta}{Z \angle \theta} \right)^{*} \\
&= \frac{EV_{pv}}{Z} \cos(\delta + \theta) - \frac{V_{pv}^2}{Z} \cos\theta + j\left[ \frac{EV_{pv}}{Z} \sin(\delta + \theta) - \frac{V_{pv}^2}{Z} \sin\theta \right]
\end{aligned}
\tag{4}
$$

The followings can be obtained by arranging Equation (4):

$$
\cos(\delta + \theta) = \frac{V_{pv}}{E} \cos\theta - \frac{Z}{EV_{pv}} P_{pv}
\tag{5}
$$

$$
\sin(\delta + \theta) = \frac{V_{pv}}{E} \sin\theta - \frac{Z}{EV_{pv}} Q_{pv}
\tag{6}
$$

In normal operation of the grid-connected system, $\delta$ is a positive or negative value, and its absolute value is generally small (around 0 rad); while $\theta$ is generally large (close to $\pi/2$ rad), therefore $(\delta + \theta)$ is greater than 0 rad and less than $\pi$ rad.

The PV station's operation mode can generally be divided into *PV* (active power—voltage) mode and *PQ* (active power—reactive power) mode [18,27]. *PV* mode is the constant voltage operation mode and the voltage amplitude of the POI or PV inverter outlet bus is set as a constant value. When $P_{pv}$ changes, the POI's voltage phase angle will change. *PQ* mode is the constant generation power factor operation mode. When $P_{pv}$ changes, the voltage amplitude and voltage phase angle of POI will change.

### 2.2.1. *PV* Mode

In Figure 1, if the PV station operates in *PV* mode, $V_{pv}$ is constant. According to Equation (5), with the increase of $P_{pv}$, $\cos(\delta + \theta)$ gradually decreases, $(\delta + \theta)$ gradually increases. It can be known that $\delta$ gradually increases since $\theta$ is a constant, that is, $\delta$ is proportional to $P_{pv}$.

### 2.2.2. *PQ* Mode

If the PV station operates in *PQ* mode, the situation is more complicated. Now, we only consider that $P_{pv}$ takes the unit power factor, that is, $Q_{pv} = 0$ pu. Considering the effect of equivalent line resistance, then $\theta$ is greater than 0 rad and less than $\pi/2$ rad.

From Equation (6), we can get:

$$
\frac{d[\sin(\delta + \theta)]}{dP_{pv}} = \frac{\sin\theta}{E} \frac{dV_{pv}}{dP_{pv}}
\tag{7}
$$

Since $\frac{dV_{pv}}{dP_{pv}}$ decreases monotonically from a positive value to a negative value (over 0) with the increase of $P_{pv}$ (the derivation process is given in Appendix A), $\sin(\delta + \theta)$ first increases and then decreases monotonically. Whereas $(\delta + \theta)$ is greater than 0 rad and less than $\pi$ rad, $\delta$ changes from small to large, that is, $\delta$ is proportional to $P_{pv}$.

By summarizing the above results, it can be concluded that the active output $P_{pv}$ of the PV station operating in *PV* mode or *PQ* mode with unit power factor is proportional to the voltage phase angle $\delta$ of POI.

### 2.3. A Novel Voltage Stability Sensitivity Index LPAS

Since the PV active output, $P_{pv}$ is directly proportional to the POI's voltage phase angle $\delta$, we can take the derivative of L-index with respect to $\delta$ to reflect the influence degree of $P_{pv}$ on the SVS of each load node.

Set $\tilde{F}_{ji} = F_{ji}\angle\alpha_{ji}$, $\dot{V}_i = V_i\angle\delta_i$ and $\dot{V}_j = V_j\angle\delta_j$, then Equation (3) can be expressed below:

$$L_j = \left|1 - \frac{1}{V_j}\sum_{i\in a_G} F_{ji}V_i[\cos(\alpha_{ji} + \delta_i - \delta_j) + j\sin(\alpha_{ji} + \delta_i - \delta_j)]\right| \tag{8}$$

The sensitivity of the L-index of load node $j$ relative to the voltage phase angle of power node $i$ can be obtained by taking the partial derivative of the complex expression of L-index ($\tilde{L}_j$) with respect to the voltage phase angle $\delta_i$ and taking the modulus value. According to the verification, the partial derivative of the modulus expression of L-index ($L_j$) with respect to $\delta_i$ is the same as it, so we can obtain:

$$\frac{\partial L_j}{\partial\delta_i} = \left|\frac{\partial\tilde{L}_j}{\partial\delta_i}\right| = \frac{F_{ji}V_i}{V_j}\left|\sin(\alpha_{ji} + \delta_i - \delta_j) - j\cos(\alpha_{ji} + \delta_i - \delta_j)\right| = \frac{F_{ji}V_i}{V_j} \tag{9}$$

Equation (9) represents the coupling degree between the L-index of each load node and the voltage phase angle of each power node (including synchronous generators and PV stations). The influence of voltage phase angle change $\Delta\delta_i$ of each power node on the L-index of each load node can be expressed below:

$$\begin{bmatrix} \Delta L_1 \\ \Delta L_2 \\ \vdots \\ \Delta L_m \end{bmatrix} = \begin{bmatrix} \frac{\partial L_1}{\partial\delta_1} & \frac{\partial L_1}{\partial\delta_2} & \cdots & \frac{\partial L_1}{\partial\delta_n} \\ \frac{\partial L_2}{\partial\delta_1} & \frac{\partial L_2}{\partial\delta_2} & \cdots & \frac{\partial L_2}{\partial\delta_n} \\ \vdots & \vdots & \cdots & \vdots \\ \frac{\partial L_m}{\partial\delta_1} & \frac{\partial L_m}{\partial\delta_2} & \cdots & \frac{\partial L_m}{\partial\delta_n} \end{bmatrix} \begin{bmatrix} \Delta\delta_1 \\ \Delta\delta_2 \\ \vdots \\ \Delta\delta_n \end{bmatrix} \tag{10}$$

where $m$ is the maximum number of load nodes and $n$ is the maximum number of power nodes.

For a PV station in the system, set the number of POI as $k$ and its voltage amplitude as $V_{pvk}$. Then, a voltage stability sensitivity index (named LPAS-index), namely the derivative of L-index with respect to POI's voltage phase angle can be defined below:

$$LPAS_j = \frac{F_{jk}V_{pvk}}{V_j} \tag{11}$$

where $j$ is the number of load nodes.

Equation (11) reflects the sensitivity of the L-index of each load node relative to POI's voltage phase angle of each PV station, that is, the sensitivity of the L-index relative to $P_{pv}$. By calculating the LPAS-index, the relationship between the SVS of each load node and $P_{pv}$ can be explored.

In a known power grid, if the power grid's structure does not change, the load participation factor $F_{jk}$ will remain unchanged. For a PV station operating in *PV* mode, since $V_{pvk}$ is constant, the LPAS-index value is mainly affected by the load node voltage $V_j$. For a PV station operating in *PQ* mode with a unit power factor, the LPAS-index value is affected by the POI's voltage $V_{pvk}$ and the load node voltage $V_j$ at the same time. It is noteworthy that the voltage fluctuations of load nodes and POI caused by the PV power fluctuation are very small normally, so the LPAS-index value is less affected by $P_{pv}$. The same is true for the synchronous generator (set as *PV* bus).

## 3. Static Voltage Stability Zoning Analysis Method

In the system with a large-scale PV station, if the PV station operates in *PV* mode or *PQ* mode with a unit power factor, for each load node, the sensitivity of the L-index relative to $P_{\mathrm{pv}}$ can be obtained by calculating the corresponding LPAS-index. According to the numerical level of the LPAS-index value, the whole system can be zoned into several areas which can reflect the correlations between the SVS and $P_{\mathrm{pv}}$. On the other hand, the weakest SVS area and the weakest node can be determined by calculating the L-index value. Then according to the zoning results, the SVS analysis and control aiming at the weakest area and the whole power grid can be conducted.

Figure 2 is the flow chart of the SVS zoning analysis method based on the LPAS-index.

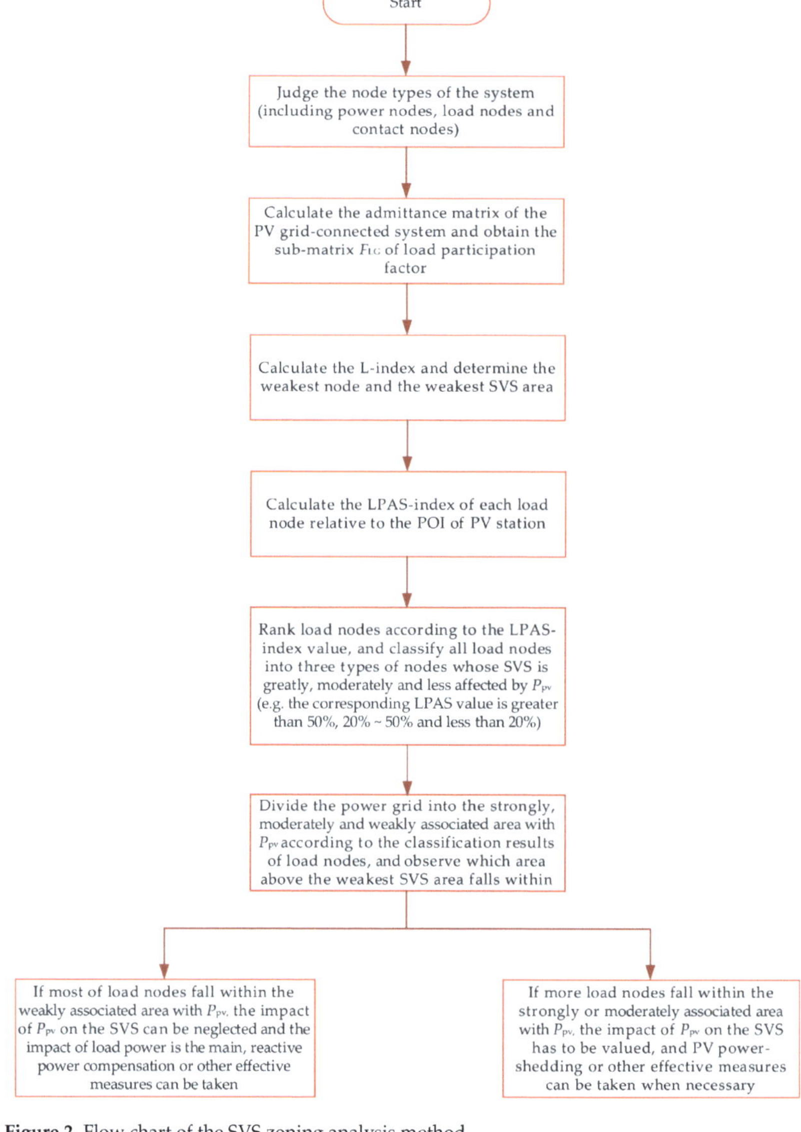


**Figure 2.** Flow chart of the SVS zoning analysis method.

The main steps of the SVS zoning analysis method are as follows:

(1) Get the sets of power nodes, load nodes and contact nodes by assessing the node types, and calculate the admittance matrix of the grid-connected system, find the sub-matrix $F_{LG}$ of load participation factor and obtain the participation factor of each load node relative to the POI of each PV station.

(2) According to Equations (3) and (11), the L-index and LPAS-index of each load node can be calculated. By ranking the L-index values of all load nodes, determine the weakest SVS area and the weakest node.

(3) Load nodes are classified according to the numerical level of LPAS-index value, which can be generally classified into three types of nodes whose SVS is greatly, moderately and less affected by $P_{pv}$. According to the classification results for load nodes, the PV grid-connected system can be zoned into three kinds of SVS areas that are strongly, moderately and weakly associated with $P_{pv}$.

The numerical level of LPAS-index value can be classified according to the standard: greater than 50% (corresponding to the strong association area), from 20% to 50% (corresponding to the moderate association area), and less than 20% (corresponding to the weak association area). However, the classification standard is not unchangeable and should be determined according to the actual situation of the PV grid-connected system.

(4) According to the classification results for load nodes, the PV grid-connected system can be zoned into three kinds of SVS areas that are strongly, moderately and weakly associated with $P_{pv}$. As the LPAS-index value of the load node is less affected by $P_{pv}$ and other generators, the zoning results are relatively unchanged.

(5) Analyze the SVS of the whole power grid and the weakest SVS area on the basis of the zoning results, and find out the rules that are affected by $P_{pv}$. Based on the analysis results, some useful control strategies can be proposed to improve the SVS of the weakest area and the whole system.

If most load nodes fall within the weakly associated area with $P_{pv}$, the impact of $P_{pv}$ on the SVS can be neglected, and the change of load power is the main influencing factor on the SVS of the whole power grid. Similar to the traditional power grid, reactive power compensation or other effective measures can be used to improve the SVS.

If more load nodes fall within the strongly or moderately associated area with $P_{pv}$, the impact of $P_{pv}$ on the SVS has to be valued. Load-shedding or other effective measures can be taken when necessary, and we will use a new PV power-shedding measure.

## 4. Simulation Verification for the Zoning Analysis Method

In order to verify the rationality and practicability of the above zoning analysis method, we take the IEEE 14-node system as the test example, as shown in Figure 3. The 14-node system is a sub-transmission system [18], and its voltage classes are 69 kV (including Node 1~Node 5), 13.8 kV (including Node 6, Node 7, and Node 9~Node 14), and 18 kV (Node 8).

See Figure 3, Node 1 is the slack bus (swing bus), Node 2 is a *PV* bus. SC1~SC3 are synchronous condensers. Set the base power as 100 MVA and the initial power of the total load as 2.59 + j0.814 pu. Set the voltage amplitude of Node 2 as 1.045 pu and the initial active output of generator Gen 2 ($P_{G2}$) as 1 pu.

See the dotted line in Figure 3, three plans about the centralized PV station integrated into the 14-node system will be adopted:

Plan A: PV station A is integrated into Node 5 by a PV transmission line, and the corresponding POI is POI_A;

Plan B: PV station B is integrated into Node 14 by a PV transmission line, and the corresponding POI is POI_B;

Plan C: PV station B is integrated into Node 14, the generator Gen 2 is replaced with PV station C (operating in *PV* mode), and the corresponding POI is Node 2. This is a multi-PV case.

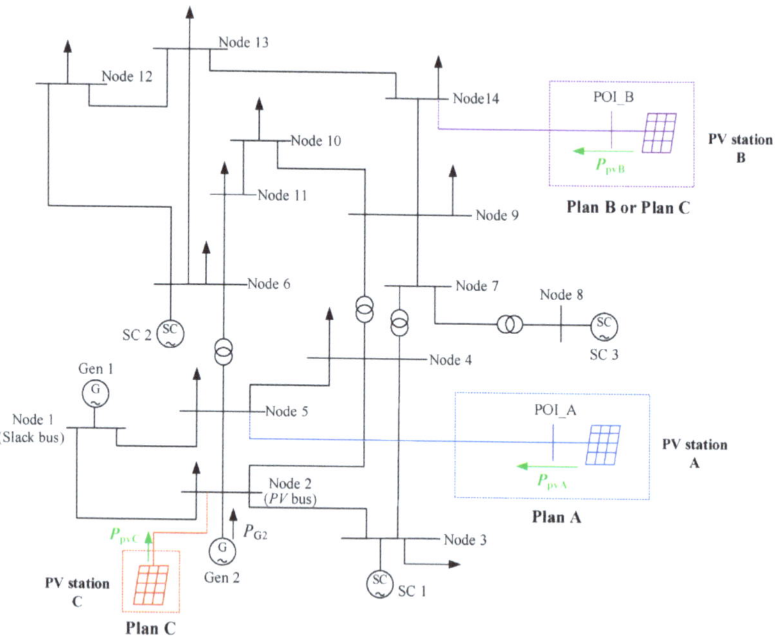



**Figure 3.** The classic IEEE 14-node system with PV station.

$P_{pvA}$, $P_{pvB}$ and $P_{pvC}$ represent the active output of PV stations A, B and C, respectively, and the maximum active output of them are 1 pu, 0.3 pu and 1 pu, respectively.

### 4.1. Verification for the Relationship between PV Active Output and Voltage Phase Angle of POI

Section 2.2 proved that the PV active output $P_{pv}$ is proportional to the POI's voltage phase angle $\delta$, now we take Plan A as an example to verify it. When PV station A operates in *PV* mode, the PV inverter outlet bus voltage amplitude is set as 1.05 pu. When PV station A operates in *PQ* mode with a unit power factor, its reactive power is set as 0 pu. The $P_{pvA}$—$\delta_A$ ($\delta_A$ is the voltage phase angle of POI_A.) curve is shown in Figure 4, we can see that $P_{pvA}$ is proportional to $\delta_A$ under the two operation modes.

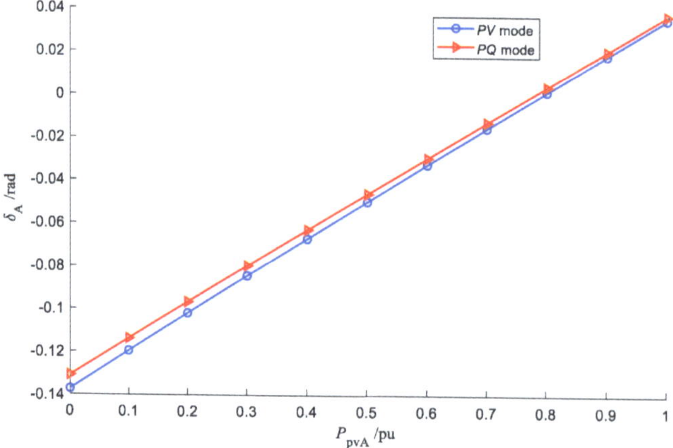



**Figure 4.** $P_{pvA}$—$\delta_A$ (active output of PV station A—voltage phase angle of POI_A) curve.

## 4.2. Index Calculation

### 4.2.1. Plan A and Plan B

Firstly, Plan A and Plan B are investigated. Now take PV stations A and B operating in *PV* mode as an example. Set $P_{pvA}$ = 1 pu, $P_{pvB}$ = 0.3 pu (The capacity of PV station integrated into 13.8 kV node can't be too large). Set the current load multiple of the whole system *lm* = 1 pu. Calculate the L-index values of all load nodes when the PV station adopts Plan A and Plan B, respectively, and does not integrate into the 14-node system, the results are listed in Table 1.

**Table 1.** L-index values of load nodes under Plan A and Plan B (Load multiple *lm* = 1 pu).

| Node Number | L-Index Value | | |
|---|---|---|---|
| | Plan A ($P_{pvA}$ = 1 pu) | Plan B ($P_{pvB}$ = 0.3 pu) | Without PV |
| 14 | 0.0760 | 0.0297 | 0.0784 |
| 9 | 0.0646 | 0.0532 | 0.0681 |
| 10 | 0.0619 | 0.0526 | 0.0649 |
| 11 | 0.0351 | 0.0303 | 0.0366 |
| 13 | 0.0316 | 0.0209 | 0.0321 |
| 4 | 0.0293 | 0.0283 | 0.0307 |
| 12 | 0.0238 | 0.0185 | 0.0241 |
| 5 | 0.0197 | 0.0194 | 0.0209 |

It can be seen from Table 1 that when the PV station is not integrated into the system, Node 14 is the weakest node and Node 5 is the strongest node. Node 9, Node 10 and Node 14 compose the weakest SVS area.

When Plan A is carried out, the weakest SVS area is still composed of Node 9, Node 10 and Node 14, Node 14 and Node 5 are still the weakest node and the strongest node.

When Plan B is adopted, the weakest SVS area is composed of Node 9 and Node 10. The weakest node becomes Node 9, this is because the SVS of Node 14 is observably enhanced after integrating with PV station B. The strongest node is still Node 5.

It can be seen that the change of $P_{pv}$ has a minor impact on the L-index from Figure 5 (Node 5 in Plan A and Node 14 in Plan B are taken as examples). According to the computing results, the variable amplitude of the L-index is less than 0.1%. Therefore, when $P_{pv}$ changes, the determined weakest SVS area can remain consistent.

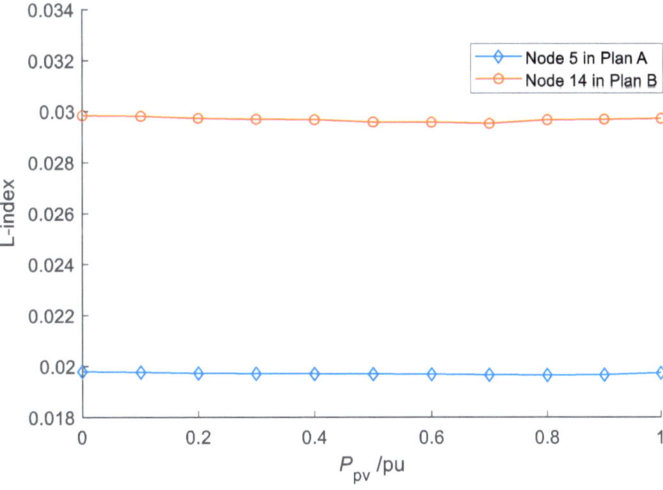


**Figure 5.** PV active output $P_{pv}$—L-index curves (Load multiple *lm* = 1 pu).

See Table 1, when a PV station is integrated into the system, compared with the system without a PV station, the L-index values of all load nodes are relatively reduced, indicating that the SVS of load nodes is improved to a certain extent. Moreover, when the PV station is integrated into different load nodes, the weakest SVS area and the weakest node can be changed.

As shown in Table 2, the LPAS-index values of all load nodes, respectively, relative to POI_A and POI_B are sorted from large to small under Plan A and Plan B.

**Table 2.** LPAS-index values of load nodes under Plan A and Plan B (Load multiple *lm* = 1 pu).

| Plan A ($P_{pvA}$ = 1 pu) | | Plan B ($P_{pvB}$ = 0.3 pu) | |
|---|---|---|---|
| **Node Number** | **LPAS-Index** | **Node Number** | **LPAS-Index** |
| 5 | 23.17% | 14 | 61.23% |
| 4 | 14.18% | 9 | 18.71% |
| 9 | 5.95% | 10 | 15.48% |
| 10 | 4.92% | 13 | 14.05% |
| 14 | 3.78% | 11 | 7.85% |
| 11 | 2.49% | 12 | 6.97% |
| 13 | 0.85% | 4 | 3.15% |
| 12 | 0.42% | 5 | 1.93% |

Figure 6 shows the $P_{pv}$—LPAS-index curves (Node 5 in Plan A and Node 14 in Plan B are taken for examples). To the same load node, the change of $P_{pv}$ has a minor impact on the LPAS-index value. Other active power supplies are similar such as a generator. In a general way, the variable amplitude of the LPAS-index is less than 1% when $P_{pv}$ changes. Therefore, aiming at the changes of $P_{pv}$, the classification of load nodes and the SVS zoning results can remain unchanged.

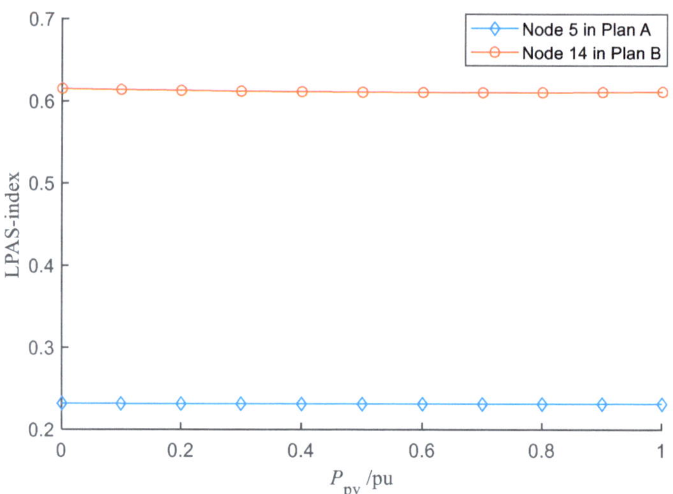


**Figure 6.** PV active output $P_{pv}$—LPAS-index curves (Load multiple *lm* = 1 pu).

In summary, the maximum active output of the PV station can be directly selected to calculate the LPAS-index and L-index of load nodes, so as to carry out the classification of load nodes and the determination of the weakest SVS area.

### 4.2.2. Plan C

In Plan C, PV station B has remained, and Gen 2 is replaced with PV station C. PV station C operates in *PV* mode, and the voltage of Node 2 remains as 1.045 pu. When

$P_{pvB} = 0.3$ pu, $P_{pvC} = 1$ pu, the values of L-index and LPAS-index relative to Node 2 (namely POI of PV station C) and POI_B are listed in Table 3. Node 9 is the weakest node and Node 12 is the strongest node. Node 9 and Node 10 compose the weakest SVS area.

**Table 3.** The values of L-index and LPAS-index of load nodes under Plan C (Load multiple $lm = 1$ pu).

| Plan C ($P_{pvB} = 0.3$ pu and $P_{pvC} = 1$ pu) | | | |
|---|---|---|---|
| **Node Number** | **L-Index** | **LPAS-Index (to Node 2)** | **LPAS-Index (to POI_B)** |
| 5 | 0.0194 | 38.58% | 1.93% |
| 4 | 0.0283 | 38.18% | 3.15% |
| 9 | 0.0532 | 14.33% | 18.71% |
| 10 | 0.0526 | 11.86% | 15.48% |
| 11 | 0.0303 | 6.01% | 7.85% |
| 14 | 0.0297 | 3.99% | 61.23% |
| 13 | 0.0209 | 0.92% | 14.05% |
| 12 | 0.0185 | 0.45% | 6.97% |

Similarly, the active output of PV station C ($P_{pvC}$) has a minor impact on the values of the L-index and LPAS-index. For example, in Plan C, the LPAS-index values of Node 5 relative to Node 2 and Node POI_B are 0.3859 and 0.0193, respectively, when $P_{pvC} = 0.3$ pu and $P_{pvB} = 0.3$ pu, and are approximately equal to the corresponding LPAS-index values when $P_{pvC} = 1$ pu and $P_{pvB} = 0.3$ pu, as shown in Table 3.

### 4.3. SVS Zoning and Analysis

According to the data in Tables 2 and 3, the load nodes can be classified and the 14-node system with PV can be zoned.

#### 4.3.1. Plan A

See Table 2, in Plan A, only the LPAS-index value of Node 5 is greater than 20% and less than 50%, illustrating that the SVS of Node 5 is moderately affected by $P_{pvA}$. The LPAS-index values of other load nodes (including Node 4, Node 9, Node 10, Node 11, Node 12, Node 13 and Node 14) are all less than 20%, illustrating that the SVS of these nodes are all less affected by $P_{pvA}$.

Furthermore, the LPAS-index value of Node 5, namely the access point of PV station A, is not too big, indicating that the PV access point is not necessarily greatly affected by $P_{pv}$.

Figure 7 shows the zoning chart of Plan A. The system can be zoned as a moderately associated area (including Node 5) and a weakly associated area (including other load nodes) with $P_{pvA}$. Owing to the LPAS-index and L-index are less affected by $P_{pv}$ and $P_{G2}$, the zoning results of Plan A are relatively unchanged.

See Figure 7, the weakest SVS area (composed of Node 9, Node 10 and Node 14) falls within the area that is weakly associated with $P_{pvA}$, so this area's SVS is less affected by $P_{pvA}$. In fact, since the overwhelming majority of load nodes (except Node 5) are falls with the weakly associated area with $P_{pvA}$, the SVS of the whole power grid is less affected by $P_{pvA}$, and more attention should focus on the impact of load power change on the SVS.

As the impact of $P_{pv}$ on the L-index is minimal, we select the system load-margin index $I_{LM}$ to measure the SVS and verify the above conclusion. The load-margin index $I_{LM}$ can be calculated by Equation (12):

$$I_{LM} = 1 - \frac{1}{\lambda_{MAX}} \tag{12}$$

where $\lambda_{MAX}$ is the maximum load margin parameter that corresponds to the critical point of the SVS.

The value range of $I_{LM}$ is 0~1. 0 means voltage collapse, and 1 means absolute voltage stability.

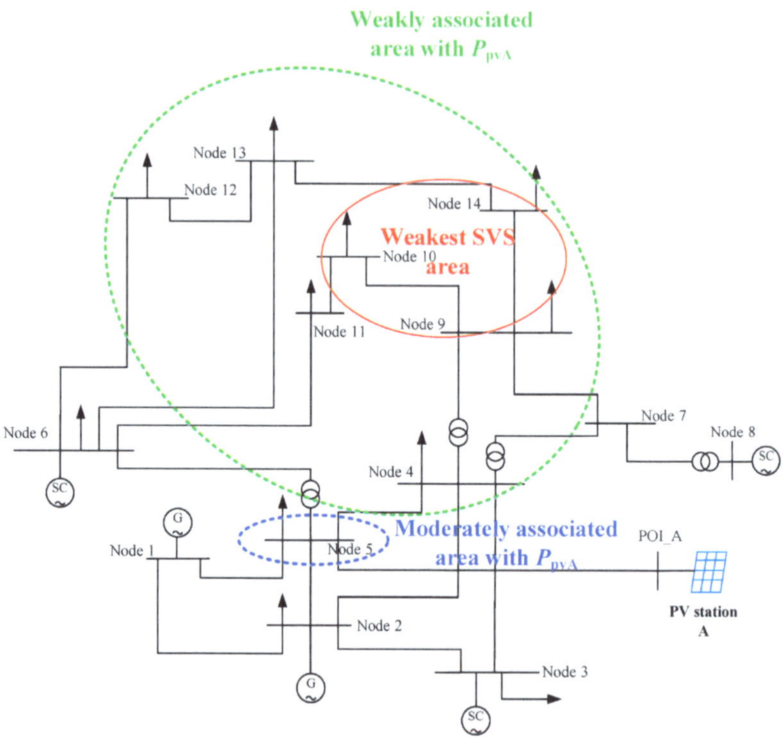


**Figure 7.** Static voltage stability zoning chart of Plan A.

Now set $P_{G2} = 0.4$ pu. The values of $I_{LM}$ are given in Table 4. When the system load multiple $lm$ is 1 pu, the system operates in a normal state (namely a low load level). When $P_{pvA}$ changes from 0 to 1 pu, the $I_{LM}$ values keep almost unchanged (Adjoining variation amplitude is less than 1%).

**Table 4.** Values of system load-margin index (Plan A, active output of Gen 2 $P_{G2} = 0.4$ pu).

| $P_{pvA}$/pu | Load Multiple $lm$ = 1 pu | | Load Multiple $lm$ = 1.5 pu | |
|---|---|---|---|---|
| | $\lambda_{MAX}$/pu | $I_{LM}$ | $\lambda_{MAX}$/pu | $I_{LM}$ |
| 0 | 2.5846 | 0.6131 | 1.7346 | 0.4235 |
| 0.1 | 2.5957 | 0.6147 | 1.7444 | 0.4267 |
| 0.2 | 2.6068 | 0.6164 | 1.7536 | 0.4297 |
| 0.3 | 2.6163 | 0.6178 | 1.7625 | 0.4326 |
| 0.4 | 2.6251 | 0.6191 | 1.7708 | 0.4353 |
| 0.5 | 2.6330 | 0.6203 | 1.7787 | 0.4378 |
| 0.6 | 2.6400 | 0.6212 | 1.7861 | 0.4401 |
| 0.7 | 2.6460 | 0.6221 | 1.7921 | 0.4420 |
| 0.8 | 2.6508 | 0.6228 | 1.7976 | 0.4437 |
| 0.9 | 2.6542 | 0.6232 | 1.8032 | 0.4454 |
| 1 | 2.6560 | 0.6235 | 1.8085 | 0.4471 |

When the load multiple $lm$ is 1.5 pu, the system operates in a heavy load (namely a high load level). See Table 4, with the change of $P_{pvA}$, the $I_{LM}$ values still keep almost unchanged. However, compared to the normal operation state ($lm$ = 1 pu), the $I_{LM}$ values and the SVS margin significantly reduce, indicating that the SVS of the whole system (including the weakest area) is mainly affected by the load power. The load-shedding method can be used to improve the SVS [24].

4.3.2. Plan B

See Table 2, in Plan B, the LPAS-index value of Node 14 (the access point of PV station B) is greater than 50%, so the SVS of Node 14 is greatly affected by $P_{pvB}$. However, the LPAS-index values of other load nodes (including Node 4, Node 5, Node 9, Node 10, Node 11, Node 12 and Node 13) are all less than 20%, so the SVS of these nodes is less affected by $P_{pvB}$.

Figure 8 shows the zoning chart of Plan B. Node 14 can be set as a strongly associated area with $P_{pvB}$, and other load nodes can be set as a weakly associated area with $P_{pvB}$. The weakest SVS area (composed of Node 9 and Node 10) falls within the area that is weakly associated with $P_{pvB}$.

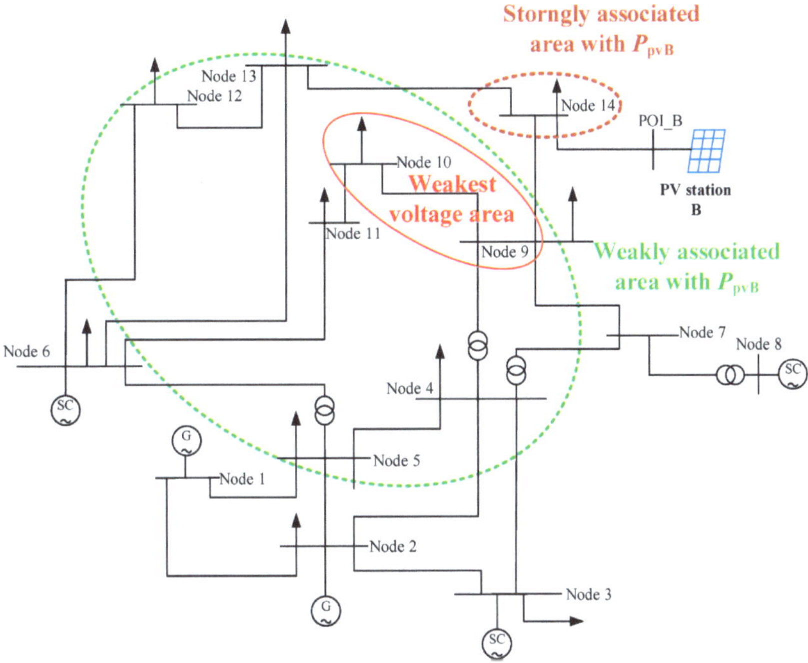


**Figure 8.** Static voltage stability zoning chart of Plan B.

See Table 5, with the change of $P_{pvB}$, the variation amplitudes of $I_{LM}$ value are still very small under normal state and heavy load, and the SVS of the whole power grid is mainly affected by the load power. It is similar to the case in Plan A, more attention should focus on the impact of load power change on the SVS.

**Table 5.** Values of system load-margin index (Plan B, active output of Gen 2 $P_{G2} = 0.4$ pu).

| $P_{pvB}$/pu | Load Multiple *lm* = 1 pu | | Load Multiple *lm* = 1.5 pu | |
|---|---|---|---|---|
| | $\lambda_{MAX}$/pu | $I_{LM}$ | $\lambda_{MAX}$/pu | $I_{LM}$ |
| 0 | 2.5272 | 0.6043 | 1.6961 | 0.4104 |
| 0.1 | 2.5695 | 0.6108 | 1.7239 | 0.4199 |
| 0.2 | 2.6089 | 0.6167 | 1.7504 | 0.4287 |
| 0.3 | 2.6459 | 0.6221 | 1.7751 | 0.4367 |

### 4.3.3. Plan C

See Table 3, in Plan C, the LPAS-index values of Node 4 and Node 5 relative to $P_{\text{pvC}}$ is greater than 20% and less than 50%, so the two nodes can compose a moderately associated area with $P_{\text{pvC}}$. The LPAS-index values of other load nodes relative to $P_{\text{pvC}}$ are all less than 20%, so they can compose a weakly associated area with $P_{\text{pvC}}$. However, the LPAS-index value of Node 14 relative to $P_{\text{pvB}}$ is greater than 50%, so we can neglect the impact of $P_{\text{pvC}}$ and zone Node 14 as a strongly associated area with $P_{\text{pvB}}$. Ultimately, Node 9~Node 13 can be determined to compose a weakly associated area with $P_{\text{pvB}}$ and $P_{\text{pvC}}$. The zoning chart of Plan C is shown in Figure 9.

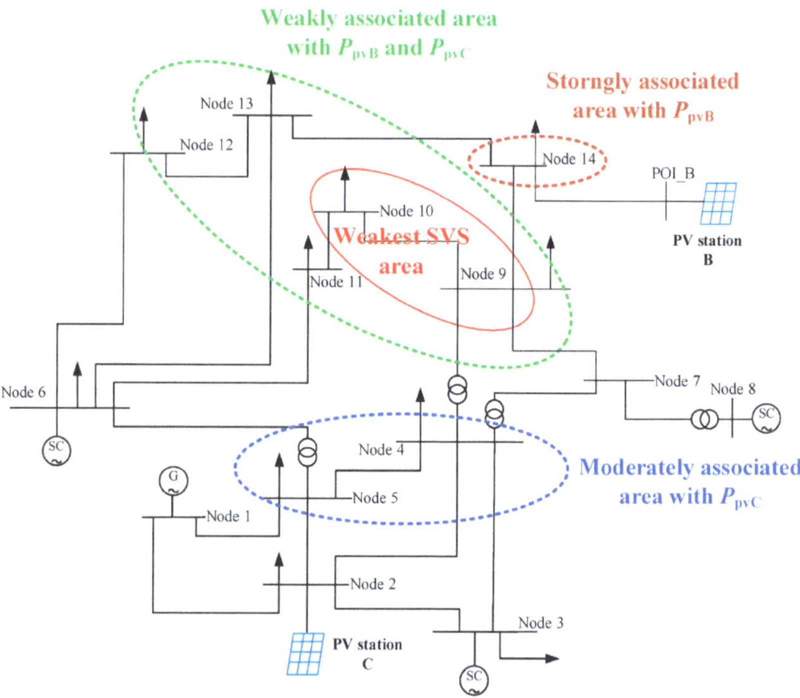


**Figure 9.** Static voltage stability zoning chart of Plan C.

The weakest SVS area (composed of Node 9 and Node 10) falls within the weakly associated area with $P_{\text{pvB}}$ and $P_{\text{pvC}}$, so the impact of $P_{\text{pv}}$ on the weakest area can be neglected.

See Figure 9, what is different from Plan A and Plan B is that the areas more affected by $P_{\text{pv}}$ are significantly enlarged in Plan C. Especially, Node 4 and Node 5 (69 kV voltage class) fall within a moderately associated area with $P_{\text{pvC}}$, so the impact of $P_{\text{pv}}$ on the SVS should be further investigated.

Now we set $P_{\text{pvB}} = 0.3$ pu, and $P_{\text{pvC}}$ changes from 0 to 1 pu.

The $I_{\text{LM}}$ values are listed in Table 6. It can be seen that when the system operates in a normal state (*lm* = 1 pu), with the gradual increase of $P_{\text{pvC}}$ from 0 to 0.5 pu, the $I_{\text{LM}}$ value gradually increases too, indicating that the SVS is gradual enhanced with the increase of $P_{\text{pvC}}$. However, with the gradual increase of $P_{\text{pvC}}$ from 0.6 pu to 1 pu, the $I_{\text{LM}}$ value gradually decreases and the SVS is gradually weakened. When $P_{\text{pvC}}$ reaches the maximum (1 pu), the impact of $P_{\text{pv}}$ on the $I_{\text{LM}}$ value is very obvious (decreases to 0.4597), and the SVS becomes as weak as the status that the system operates in heavy load.

**Table 6.** Values of system load-margin index (Plan C, active output of PV station B $P_{\text{pvB}}$ = 0.3 pu).

| $P_{\text{pvC}}$/pu | Load Multiple *lm* = 1 pu | | Load Multiple *lm* = 1.5 pu | |
|---|---|---|---|---|
| | $\lambda_{\text{MAX}}$/pu | $I_{\text{LM}}$ | $\lambda_{\text{MAX}}$/pu | $I_{\text{LM}}$ |
| 0 | 2.5846 | 0.6131 | 1.7378 | 0.4246 |
| 0.1 | 2.6138 | 0.6174 | 1.7505 | 0.4287 |
| 0.2 | 2.6343 | 0.6204 | 1.7608 | 0.4321 |
| 0.3 | 2.6451 | 0.6219 | 1.7692 | 0.4348 |
| 0.4 | 2.6459 | 0.6221 | 1.7751 | 0.4367 |
| 0.5 | 2.6321 | 0.6201 | 1.7787 | 0.4378 |
| 0.6 | 2.5960 | 0.6148 | 1.7791 | 0.4379 |
| 0.7 | 2.5171 | 0.6027 | 1.7761 | 0.4370 |
| 0.8 | 2.3401 | 0.5727 | 1.7685 | 0.4345 |
| 0.9 | 2.1311 | 0.5308 | 1.7557 | 0.4304 |
| 1 | 1.8507 | 0.4597 | 1.7343 | 0.4234 |

See Table 6, when the system operates under heavy load (*lm* = 1.5 pu), the $I_{\text{LM}}$ value gradually increases with the increase of $P_{\text{pvC}}$ from 0 to 0.6 pu, and gradually decreases with the increase of $P_{\text{pvC}}$ from 0.7 pu to 1 pu, but the variation amplitude is very small.

Now we set $P_{\text{pvC}}$ = 1 pu, and $P_{\text{pvB}}$ changes from 0 to 0.3 pu. The $I_{\text{LM}}$ values are listed in Table 7. It can be seen that excessive PV power can evidently weaken the SVS when the system operates in a normal state (*lm* = 1 pu). When the system operates under heavy load (*lm* = 1.5 pu), the increase of $P_{\text{pvB}}$ only can slightly improve the SVS and has no negative impact on the SVS.

**Table 7.** Values of system load-margin index (Plan C, active output of PV station C $P_{\text{pvC}}$ = 1 pu).

| $P_{\text{pvB}}$/pu | Load Multiple *lm* = 1 pu | | Load Multiple *lm* = 1.5 pu | |
|---|---|---|---|---|
| | $\lambda_{\text{MAX}}$/pu | $I_{\text{LM}}$ | $\lambda_{\text{MAX}}$/pu | $I_{\text{LM}}$ |
| 0 | 2.1028 | 0.5244 | 1.6760 | 0.4033 |
| 0.1 | 2.0303 | 0.5075 | 1.6978 | 0.4110 |
| 0.2 | 1.9583 | 0.4894 | 1.7173 | 0.4177 |
| 0.3 | 1.8507 | 0.4597 | 1.7343 | 0.4234 |

From the above analyses, it is concluded that in a multi-PV case, excessive PV active power can evidently weaken the SVS when the system operates in a normal state, and only slightly impact the SVS when the system operates under heavy load.

A PV power-shedding method can be used in time to maintain the SVS when the system operates at a low load level and an excessive PV power. For example, when *lm* = 1 pu, $P_{\text{pvB}}$ = 0.3 pu and $P_{\text{pvC}}$ = 0.95 pu, the value of $I_{\text{LM}}$ is 0.4958 by calculation. Then the PV active power control system can execute a series of power-shedding operations and the shedding quantity is 0.1 pu each time, $P_{\text{pvC}}$ will be reduced to 0.55 pu, and the value of $I_{\text{LM}}$ will recover to 0.6180. The system's SVS is improved effectively after power-shedding.

Moreover, reducing the rated installation capacity of PV station is also a valid solution to this problem.

## 5. Conclusions

Firstly, this paper derives and proves that the POI's voltage phase angle of the PV station is proportional to the PV active output $P_{\text{pv}}$, based on a simplified two-node system with PV. From this, a novel sensitivity index LPAS-index of the traditional L-index relative to the voltage phase angle of POI is proposed, and the LPAS-index can be used to reflect the influence degree of $P_{\text{pv}}$ on the SVS of load nodes. A PV grid-connected system can be zoned into different areas: strongly, moderately and weakly associated with $P_{\text{pv}}$ for the SVS according to the numerical level of LPAS-index value. Based on the zoning results,

## References


1. Cheng, L.; Zang, H.; Wei, Z.; Sun, G. Ultra-short-term forecasting of regional photovoltaic power generation considering multispectral satellite remote sensing data. *Proc. CSEE* **2022**, *42*, 7451–7464.
2. Yao, S.; Zhang, C.; Liu, G.; Ma, J.; Wang, Y. Electromagnetic transient decoupling of photovoltaic power generation units and fast simulation method. *Autom. Electr. Power Syst.* **2022**, *46*, 170–178.
3. Ding, M.; Xu, Z.; Wang, W.; Wang, X.; Song, Y.; Chen, D. A review on China's large-scale PV integration: Progress, challenges and recommendations. *Renew. Sust. Energ. Rev.* **2016**, *53*, 639–652. [CrossRef]
4. Wang, F.; Li, J.; Zhen, Z.; Wang, C.; Ren, H.; Ma, H.; Zhang, W.; Huang, L. Cloud feature extraction and fluctuation pattern recognition based ultrashort-term regional PV power forecasting. *IEEE Trans. Ind. Appl.* **2022**, *58*, 6752–6767. [CrossRef]
5. Martins, C.C.; Sperandio, M.; Welfer, D. Cloud variance impact on the voltage and power flow of distribution networks with photovoltaic generators. *IEEE Lat. Am. Trans.* **2021**, *19*, 217–225. [CrossRef]
6. Tamimi, B.; Canizares, C.; Bhattacharya, K. System stability impact of large-scale and distributed solar photovoltaic generation: The case of Ontario, Canada. *IEEE Trans. Sustain. Energy* **2013**, *4*, 680–688. [CrossRef]
7. Eftekharnejad, S.; Vittal, V.; Heydt, G.T.; Keel, B.; Loehr, J. Impact of increased penetration of photovoltaic generation on power systems. *IEEE Trans. Power Syst.* **2013**, *28*, 893–901. [CrossRef]
8. Liang, Z.; Qin, F.; Cui, F. Impact analysis of annular solar eclipse on June 21, 2020 in China on photovoltaic power generation and power grid operation. *Autom. Electr. Power Syst.* **2021**, *45*, 1–7.
9. Li, S.; Wei, Z.; Sun, G.; Gao, P.; Xiao, J. Voltage stability bifurcation of large-scale grid-connected PV system. *Electr. Power Autom. Equip.* **2016**, *36*, 17–23.
10. Lammert, G.; Premm, D.; Ospina, L.D.P.; Boemer, J.C.; Braun, M.; Cutsem, T.V. Control of photovoltaic systems for enhanced short-term voltage stability and recovery. *IEEE Trans. Energy Conver.* **2019**, *34*, 243–254. [CrossRef]
11. Munkhchuluun, E.; Meegahapola, L. Impact of the solar photovoltaic (PV) generation on long-term voltage stability of a power network. In Proceedings of the 2017 IEEE Innovative Smart Grid Technologies—Asia (ISGT-Asia), Auckland, New Zealand, 4–7 December 2017; pp. 1–6.
12. Kundur, P. Voltage stability. In *Power System Stability and Control*; McGraw-Hill: New York, NY, USA, 1994; pp. 990–1006.
13. Tang, Y.; Dong, S.; Zhu, C.; Wu, J.; Song, Y. Online prediction method of static voltage stability margin based on Tri-Training-LASSO-BP network. *Proc. CSEE* **2020**, *40*, 3824–3834.
14. Li, J.; Liu, D.; An, J.; Li, Z.; Yang, H.; Zhao, G.; Yang, S.; Zheng, H. Static voltage stability margin assessment based on reinforcement learning theory. *Proc. CSEE* **2020**, *40*, 5136–5147.
15. Liu, Y.; Wang, T.; Qiu, G.; Wei, W.; Zhou, B.; Liu, T.; Liu, J.; Mei, S. Surrogate modeling method and its analytical algorithm for static voltage stability control embedded with input convex neural network. *Electr. Power Autom. Equip.* **2023**, *43*, 151–159.
16. Wang, D.; Yuan, X.; Zhao, M.; Qian, Y. Impact of large-scale photovoltaic generation integration structure on static voltage stability in China's Qinghai province network. *J. Eng.* **2017**, *2017*, 671–675. [CrossRef]
17. Du, X.; Zhou, L.; Guo, K.; Yang, M.; Liu, Q. Static voltage stability analysis of large-scale photovoltaic plants. *Power Syst. Technol.* **2015**, *39*, 3427–3434.
18. Kabir, S.; Nadarajah, M.; Bansal, R. Impact of large scale photovoltaic system on static voltage stability in sub-transmission network. In Proceedings of the 2013 IEEE ECCE Asia Downunder, Melbourne, VIC, Australia, 3–6 June 2013; pp. 468–473.
19. Rawat, M.S.; Vadhera, S. Probabilistic steady state voltage stability assessment method for correlated wind energy and solar photovoltaic integrated power systems. *Energy Technol.* **2020**, *9*, 2000732. [CrossRef]
20. Liu, Y.; Yao, L.; Liao, S.; Xu, J.; Dun, Y.; Cheng, F.; Cui, W.; Li, F. Study on the impact of photovoltaic penetration on power system static voltage stability. *Proc. CSEE* **2022**, *42*, 5484–5496.
21. Li, S.; Ma, Y.; Zhang, C.; Yu, J. Static voltage stability prediction for PCCs of new energy power stations in the grid-connected system. *J. Phys. Conf. Ser.* **2022**, *2369*, 012037. [CrossRef]
22. Xiao, F.; Han, M.; Tang, X.; Zhang, X. Voltage stability of weak sending-end system with large-scale grid-connected photovoltaic power plants. *Electr. Power* **2020**, *53*, 31–39.
23. Rahman, S.; Saha, S.; Islam, S.N.; Arif, M.T.; Mosadeghy, M.; Haque, M.E.; Oo, A.M.T. Analysis of power grid voltage stability with high penetration of solar PV systems. *IEEE Trans. Ind. Appl.* **2021**, *57*, 2245–2257. [CrossRef]
24. Li, S.; Wei, Z.; Ma, Y. Fuzzy load-shedding strategy considering photovoltaic output fluctuation characteristics and static voltage stability. *Energies* **2018**, *11*, 779. [CrossRef]
25. Zheng, L.; Ma, D.; Zhou, X.; Dai, J. Static voltage stability characteristics of high permeability new energy grid considering SSSC. *Power DSM* **2022**, *24*, 50–56.
26. Kessel, P.; Glavitsch, H. Estimating the voltage stability of a power system. *IEEE Trans. Power Deliv.* **1986**, *1*, 346–354. [CrossRef]
27. Li, S.; Wei, Z.; Sun, G.; Wang, Z. Stability research of transient voltage for multi-machine power systems integrated large-scale PV power plant. *Acta Energ. Sol. Sin.* **2018**, *39*, 3356–3362.

*Review*

# A Comprehensive Review of the Design and Operation Optimization of Energy Hubs and Their Interaction with the Markets and External Networks

Christina Papadimitriou [1], Marialaura Di Somma [2,*], Chrysanthos Charalambous [3], Martina Caliano [2], Valeria Palladino [2], Andrés Felipe Cortés Borray [4], Amaia González-Garrido [4], Nerea Ruiz [4] and Giorgio Graditi [2]

[1] Electrical Engineering Department, Eindhoven University of Technology, 5612 AZ Eindhoven, The Netherlands; c.papadimitriou@tue.nl

[2] Department of Energy Technologies and Renewable Sources, ENEA, 00184 Rome, Italy; martina.caliano@enea.it (M.C.); valeria.palladino@enea.it (V.P.); giorgio.graditi@enea.it (G.G.)

[3] FOSS Research Centre for Sustainable Energy, University of Cyprus, 2109 Nicosia, Cyprus

[4] TECNALIA, Basque Research and Technology Alliance (BRTA), Astondo Bidea, Building 700, 48160 Derio, Spain; acortes024@ikasle.ehu.eus (A.F.C.B.); amaia.gonzalez@tecnalia.com (A.G.-G.)

* Correspondence: marialaura.disomma@enea.it

**Abstract:** The European Union's vision for energy transition not only foresees decarbonization of the electricity sector, but also requires commitment across different sectors such as gas, heating, and cooling through an integrated approach. It also sets local energy communities at the center of the energy transition as a bottom-up approach to achieve these ambitious decarbonization goals. The energy hub is seen as a promising conceptual model to foster the optimization of multi-carrier energy systems and cross-sectoral interaction. Especially in the context of local energy communities, the energy hub concept can enable the optimal design, management, and control of future integrated and digitalized networks where multiple energy carriers operate seamlessly and in complementarity with each other. In that sense, the optimal design and operation of energy hubs are of critical importance, especially under the effect of multiple objectives taking on board not only technical, but also other aspects that would enable the sustainability of local energy communities, such as economic and environmental. This paper aims to provide an in-depth review of the literature surrounding the existing state-of-the-art approaches that are related to the design and operation optimization of energy hubs by also exploring their interaction with the external network and multiple markets. As the planning and operation of an energy hub is a multifaceted research topic, this paper covers issues such as the different optimization methods, optimization problems formulation including objective functions and constraints, and the hubs' optimal market participation, including flexibility mechanisms. By systematizing the existing literature, this paper highlights any limitations of the approaches so far and identifies the need for further research and enhancement of the existing approaches.

**Keywords:** energy hubs; energy markets; integrated energy system; optimal design and operation; sector coupling

**Citation:** Papadimitriou, C.; Di Somma, M.; Charalambous, C.; Caliano, M.; Palladino, V.; Cortés Borray, A.F.; González-Garrido, A.; Ruiz, N.; Graditi, G. A Comprehensive Review of the Design and Operation Optimization of Energy Hubs and Their Interaction with the Markets and External Networks. *Energies* 2023, 16, 4018. https://doi.org/10.3390/en16104018

Academic Editor: Alessia Arteconi

Received: 7 April 2023
Revised: 26 April 2023
Accepted: 5 May 2023
Published: 10 May 2023

## 1. Introduction

### 1.1. The Background and the Sector Coupling Need

The European Union (EU) set ambitious environmental and energy goals to design a low-carbon energy system by the middle of the 21st century. The EU Climate and Energy Framework aims to reduce greenhouse gas (GHG) emissions by 55% by 2030, improve the share of renewable electricity by 32%, and improve energy efficiency by 32.5%. These goals can be achieved through the development of integrated energy systems that foster the protection of the environment and the creation of market-oriented energy services, while guaranteeing security, reliability, and resilience of the energy supply [1].

The basic idea of an integrated energy system is to switch from a single energy carrier to multiple energy carriers to take advantage of the synergistic effects of interactions to increase the efficiency of the energy resources used [2]. The integrated energy systems concept was introduced in the European Technology & Innovation Platforms Smart Networks for Energy Transition (ETIP SNET) Vision 2050 [3]. Their main feature is the integrated management of various carriers beyond electricity including heating, cooling, hydrogen, and mobility. In detail, this system can be seen as a "system of systems," where various energy carriers coexist in an integrated infrastructure that is supported by the electrical system. Therefore, electrical networks are coupled with gas, heating, and cooling networks and this integration is made possible through energy conversion processes and storage of different types. This sector coupling is related to both the linking energy carriers with each other, and to end-use sectors including residential, tertiary, industry, and transportation. With this in mind, different types of energy technologies are managed with strong synergy to meet multi-energy needs, and the services can be supplied by the most convenient sector and related carrier. In addition, the increase in efficiency in the use of energy resources achieved thanks to the exploitation of synergies among energy carriers can lead to the reduction of renewable energy sources (RES) curtailment [4]. For instance, power-to-X technologies can act as a reservoir for excess electricity, using available energy in a cyclic and cost-effective manner. Integrating electricity and gas/hydrogen sectors (power-to-gas, power-to-hydrogen) will help exploit and retrofit existing gas infrastructures for renewable energy transport, thereby reducing the needs to expand power transmission through the storage of gas to deal with the seasonal changes in supply and demand of renewable energy. In fact, the gas generated with RES can be a resource as a low-carbon back-up capacity to supply electricity in power plants or fuel cells when there is no availability of other RES. Power-to-heat technologies such as heat pumps (HPs) allow achieving efficient heating and cooling processes of buildings from both energetic and economic points of view through the reduction of primary energy consumption. If thermal storage is coupled with HPs, it could allow for a change in thermal energy production in the event of an excess of renewable electricity.

Based on this premise, sector coupling is today considered a key to responding to the needs of the energy system, which is characterized by strong electrification of final consumption and high penetration of RES. This will bring several challenges for the operation of the power system, which in principle would need additional flexibility, reinforcement, and new investments for the transmission and distribution networks.

In the longer term, PRIMES (price-induced market equilibrium system) model results [5] show that 70% of the gas mixture could be renewable by 2050 [6], and this can be achieved only through sector coupling [7]. Further deployment of RES in collaboration and coordination with emerging technologies such as HPs or hydrogen production and storage from other carriers than electricity is considered more critical than ever for decarbonizing the whole energy system, as seen in the RePowerEU plan [8].

### 1.2. The Energy Hub Concept at the Service of Need

Due to the importance of sector coupling solutions, the energy hub (EH) concept, which allows the coupling/integration of different carriers at a local level, has been drawing researchers' attention in recent years. An EH is a conceptual unit where multiple energy carriers can be converted, conditioned, stored, and consumed. EHs can exchange energy at their interfaces, e.g., with electricity and natural gas external networks, and provide certain required energy services such as power, heating, cooling, mobility services, ancillary services, etc. Within the hub, energy is converted and conditioned by technologies such as combined heat and power (CHP), transformers, power electronic devices, compressors, heat exchangers, and storage, among others [9]. A series of technical, economic, and environmental advantages of the EH concept is summarized in Table 1, making EHs a perfect fit for the architectural concept basis of the future integrated grids.

**Table 1.** An overview of energy hubs' advantages.

| Categories | Main Advantages of EHs |
|---|---|
| Technical advantages | Enhanced efficiency for energy islands, i.e., systems with weak or no interconnections with the upstream grid; No size limitation. The size of an energy hub can vary from a building level (single house) to a community level (city—island); Increased system reliability; Increased load flexibility; |
| Economic advantages | Reduction of operating costs; Reduction of electrical grid congestion; |
| Environmental advantages | Reduction in GHG emissions; Reduction of fossil fuel use with the increased renewable energy penetration; Increase in energy efficiency. |

The energy technologies that are implementable and are building blocks of EHs can be classified as seen in [10]:

- Distributed generation technologies: renewable technologies to decarbonize energy supply systems;
- "End-user" sector coupling technologies: energy conversion technologies for the electrification of the end-uses that enable the flexibility of end-users/prosumers to be activated;
- "Cross-vector" sector coupling technologies: technologies that allow the integration of multiple energy carriers. The main technology that can be easily implemented in most energy hubs is the CHP, which can be installed both at the prosumer level (buildings, shopping centers, industries) and at the city/neighborhood level (district heating);
- Energy storage technologies of different energy carriers (electrical, thermal, mobility).

*1.3. The Need for This Review and Its Contribution*

The optimal design, operation, and interconnection of different energy carriers result in cost-efficient uses of local resources, aiming at maximizing the efficiency of the energy conversion processes. On top of that, EHs hosting storage facilities enable conversion to a greater extent between different energy carriers and thus offer higher demand-side flexibility potential. Therefore, optimal design and operation of the EHs are of primary importance to maximize their advantages. This especially holds true within the context of integrated local energy communities (ILECs). Within ILECs, a set of energy users need to agree on common choices in terms of satisfying their energy needs, maximizing the benefits derived from this collegial approach, thanks to the implementation of different multi-carrier technologies and the optimized management of energy flows. The optimal design and operation of such complex systems is a non-trivial task due to several aspects. For instance, for the design phase, the interest of developers in achieving a system configuration with the lowest costs might conflict with one of the EU energy legislations in public welfare in terms of sustainability of the energy supply, and this would require a multi-objective approach for guaranteeing the economic and environmental sustainability of such solutions. On the other hand, operation optimization is also challenging, not only for the need to consider multiple and conflicting objectives, but also to capture the interaction between energy carriers (e.g., electricity, heat, cooling, etc.) while satisfying the time-varying user demands. Last but not least, the interaction with external networks and multiple markets is also a key topic for EHs to optimize their benefits for larger systems.

So far, several reviews related to the EHs have been produced, but most of them are focused on the technical aspects of the configuration and the conversion technologies included in the EHs [11–13]. For instance, reference [14] focuses only on the aspects of the storage facilities and the flexibility that they have to offer within the EH, whereas reference [15] addresses the different demand response (DR) schemes served by EHs. Reference [16] touches upon the modeling of EHs based on multi-objective optimization for their design and operation without relating to an external framework such as the grid or the market, as it focuses on the uncertainty's impact. Reference [17] reviews the management techniques of the EHs, although it excludes heuristic methods. This means that the optimal operation of the EHs in relation to external factors such as the networks and the markets also under a multi-objective approach has not been systematized before or carefully reviewed.

The latter is significantly important to consider in the real context of an ILEC's planning and operation, where the multi-objective approach can foster different drivers such as technical, social, environmental, etc., along with the market interaction, on which the sustainability of the business cases relies. To overcome this existing review gap, the contribution of this paper is to present a comprehensive review related to the planning and operation of EHs holistically, considering a wide set of transversal aspects listed below:

- Analysis of the technologies and energy carriers in EHs;
- Analysis of the design and operation optimization of EHs, considering the full chain of relevant topics, i.e., problem formulation with constraints, objective functions overview, multi-objective approach and solution methodologies, solvers and modeling frameworks considering heuristic methods, uncertainty, and risk aversion, management of flexibility sources, and simulation methods for electric vehicles (EVs);
- Analysis of the EHs' interaction with multiple markets, from energy and balancing markets to peer-to-peer (P2P) markets, along with business models and interaction of EHs with the external network; and
- Analysis of collateral aspects such as temporal and spatial scopes.

For the analyzed aspects, limitations of the existing approaches and methodologies are also discussed, along with the need for further research and enhancement.

This paper is structured as follows. Section 2 presents the methodology used for this review. Section 3 presents a detailed analysis of EHs configuration in the literature. Section 4 analyzes the objective functions for optimal design and operation of EHs, while Sections 5 and 6 review the constraints and the optimization problems modeling for EHs, respectively. Section 7 presents the multi-objective optimization approach and methods. Section 8 focuses on heuristic models, while Section 9 provides an overview of optimization solvers and modeling environments. Section 10 focuses on uncertainties and risk aversion. Section 11 offers insights related to the interaction of EHs with external networks and multiple markets, with a key focus on P2P architectures and related markets. Section 12 presents the business landscape of EHs. Section 13 analyzes collateral concerns when setting up the operation frameworks of an EH, and Section 14 concludes the paper and summarizes limitations or gaps found in the current literature, while also providing insights on research pathways.

## 2. The Methodology Used for This Review Paper

In order to perform this systematic review, the following methodology with related steps has been adopted:

- Collect all the documents related to optimization problems, including multi-objective approaches, multi-carrier energy systems, and EHs configurations. In detail, 128 related documents were collected from the most popular and impactful research repositories of the research and innovation (R&I) community;
- Identify a list of topics of interest to focus on. The EH concept is a multi-faceted research question that entails different topics to be further investigated. Therefore, an exhaustive list of 18 topics of interest that are related to EHs has been developed. Within this list, the topics have been further categorized as "setting the background" topics or/and "research and innovation" topics. Background topics are the ones that formulate the state of the art of this review and establish the baseline knowledge of this effort, whereas "research and innovation" topics are classified as such to

formulate further innovation pathways and the research questions that are analyzed in detail. Of course, a topic can be characterized as both "background" and "research and innovation";

- An extensive review of the topics for each of the documents in order to capture the holistic approach of this review and the connection of EHs with the external framework, such as the networks, the market, and the business models; and
- Compilation of brief reports per topic for both state-of-the-art and innovation approaches before developing this review.

Figure 1 presents the methodology steps with the list of topics.

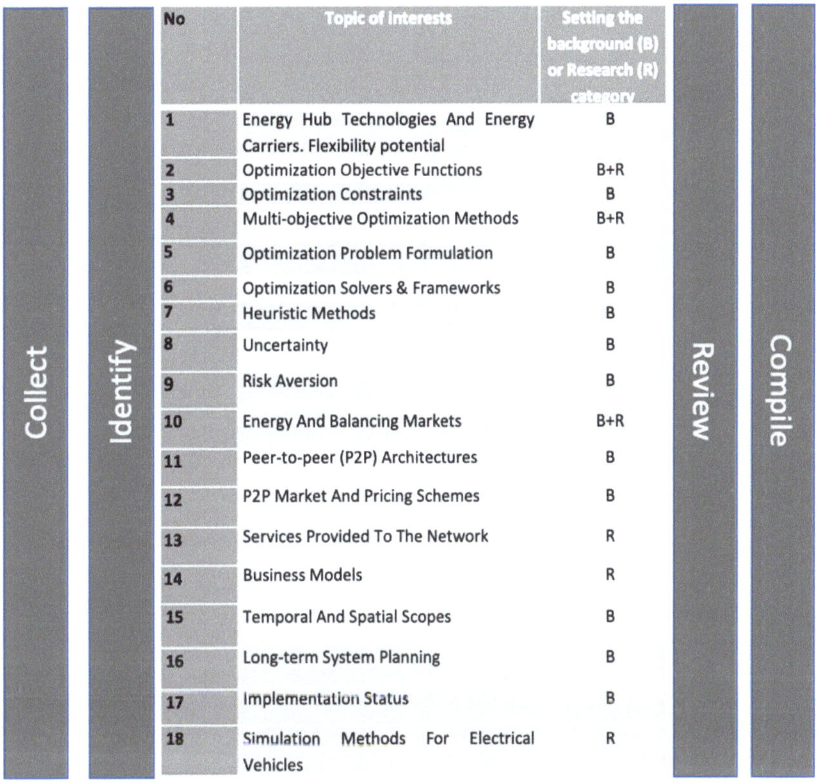

| Collect | Identify | No | Topic of Interests | Setting the background (B) or Research (R) category | Review | Compile |
|---|---|---|---|---|---|---|
| | | 1 | Energy Hub Technologies And Energy Carriers. Flexibility potential | B | | |
| | | 2 | Optimization Objective Functions | B+R | | |
| | | 3 | Optimization Constraints | B | | |
| | | 4 | Multi-objective Optimization Methods | B+R | | |
| | | 5 | Optimization Problem Formulation | B | | |
| | | 6 | Optimization Solvers & Frameworks | B | | |
| | | 7 | Heuristic Methods | B | | |
| | | 8 | Uncertainty | B | | |
| | | 9 | Risk Aversion | B | | |
| | | 10 | Energy And Balancing Markets | B+R | | |
| | | 11 | Peer-to-peer (P2P) Architectures | B | | |
| | | 12 | P2P Market And Pricing Schemes | B | | |
| | | 13 | Services Provided To The Network | R | | |
| | | 14 | Business Models | R | | |
| | | 15 | Temporal And Spatial Scopes | B | | |
| | | 16 | Long-term System Planning | B | | |
| | | 17 | Implementation Status | B | | |
| | | 18 | Simulation Methods For Electrical Vehicles | R | | |

**Figure 1.** The methodology of performing this review.

In addition, Figure 2 presents a graph capturing the distribution of the topics against the current works in the literature. From the blocks below, we can derive what has been mentioned earlier, i.e., the gaps in the literature in addressing the interaction of the EHs with the networks, markets, and business-model-related topics. It is also seen that no paper—research or review—has yet addressed all topics at the same time, thereby highlighting the main novelty of the current work.

**Figure 2.** Matrix of analyzed works in the literature vs. topics.

Table 2 shows the distribution of the related papers throughout the years. It is evident that in the last years, the multi-carrier energy approach has resulted in more scientific work around the topic of EHs.

**Table 2.** The distribution of the literature through the years.

| Year | Total Number of References | Percentage [%] |
|---|---|---|
| 2023 | 2 | 1.52 |
| 2022 | 11 | 8.33 |
| 2021 | 27 | 20.45 |
| 2020 | 23 | 17.42 |
| 2019 | 25 | 18.94 |
| 2018 | 11 | 8.33 |
| 2017 | 7 | 5.30 |
| 2016 | 6 | 4.55 |
| 2015 | 7 | 5.30 |
| 2014 | 4 | 3.03 |
| 2013 | 3 | 2.27 |
| 2012 | 3 | 2.27 |
| 2011 |  | 0.00 |
| 2010 | 2 | 1.52 |
| 2009 |  | 0.00 |
| 2008 |  | 0.00 |
| 2007 |  | 0.00 |
| 2006 | 1 | 0.76 |

## 3. The EH Configuration in the Literature

As it is important for the optimization of both design and operation, this section provides a detailed overview of the EH configuration that is mainly seen in the literature focusing on carriers and different technologies. Special mention is given to storage technologies and the demand flexibility potential under such configurations.

### 3.1. Energy Carriers of an EH

In all reviewed papers, electricity is considered the backbone carrier. Then, the electricity carrier is combined with other energy carriers such as natural gas, heating, cooling, hydrogen, and/or domestic hot water. Table 3 shows the distribution of the energy carrier combinations throughout the analyzed works in the literature. It can be observed that the electricity carrier is mostly combined with heating and cooling.

**Table 3.** Combination of considered energy carriers in the literature.

| References | Energy Carrier Combinations | | | | |
| --- | --- | --- | --- | --- | --- |
| | Electricity | Heating/Cooling | Hydrogen | Natural Gas | Domestic Hot Water |
| [18–31] | √ | √ | | | |
| [32] | √ | √ | √ | | |
| [33–38] | √ | √ | | √ | |
| [39–44] | √ | √ | | √ | √ |
| [45] | √ | √ | √ | √ | |

### 3.2. Cluster of EHs

Some papers in the literature [11,17,46,47] also discuss the benefits and challenges related to the integration of the energy carriers and the creation of a network of interconnected EHs identified as clusters of EHs.

The main benefits of such a clustering are as follows:

- Advanced security of energy supply;
- Increased provision of system services to neighboring systems, such as balancing and ancillary services;
- Reduced RES curtailment and therefore reduced GHG emissions;
- Increased system reliability;
- Increased load flexibility;
- Self-sufficiency and minimization of costs related to energy exchange with the upper grid.

On the other hand, the main challenges identified when integrating energy carriers and interconnecting of EHs that need to be addressed are specified below:

- Cost of the required infrastructure and the connecting technologies;
- The ownership of the interconnected networks has to be adequately defined;
- Advanced communication, data acquisition, and management infrastructure is needed for the optimum operation of the interconnected networks;
- The high initial investment with a long time for payback;
- Lack of cases and proper business models;
- Lack of regulations regarding functionalities and operation, including roles and responsibilities;
- Public acceptance of the interconnection and interaction between the EHs.

### 3.3. Energy Conversion Technologies

Different energy carrier conversion technologies are considered in the literature. Regarding heating and cooling carriers, the most considered technologies are CHPs [14,19,21–32] and gas-fired boilers [14,19–21,23–25,27,29,31,32]. As far as hydrogen technology is concerned, most papers consider hydrogen production through electrolysis [18,20]. For example, hydrogen produced by electrolysis can be used to produce methane via a methanation process, which can then be injected into the natural gas network [28]. Other thermal generation technologies considered are HPs [21,22,27,29,32] and chillers [21,26,29,32]. When it comes to the electricity carrier, both RES and conventional energy technologies may well be considered. In most cases, the largest amount of energy comes from RES [22,40,41,48], including photovoltaic systems (PV), wind turbines, solar thermal, and biomass. Energy generation technologies of the electricity carrier interact with the generation technologies of other energy carriers and can be combined with other generation technologies such as diesel generators, natural gas generators, and fuel cells. In addition, power-to-gas (P2G) is considered in [23], while gas-fired generation is also included in [49]. Hydrogen by means of fuel cells is used to produce not only electricity, but also thermal energy, as seen in [50]. In Table 4, the distribution of the different technologies per carrier is shown with a detailed overview of the existing literature.

**Table 4.** The distribution of technologies per carrier based on the existing literature.

<table>
<tr><td rowspan="2">Technology</td><td colspan="5" align="center">Energy Carrier</td></tr>
<tr><td align="center">Electricity</td><td align="center">Heating/Cooling</td><td align="center">Hydrogen</td><td align="center">Natural Gas</td><td align="center">Domestic Hot Water</td></tr>
<tr><td>PV systems [23,27,38,40,41,43–45,48,51–61]</td><td>CHP [12,14,17,19,21–31]</td><td>$H_2$ generator from fossil fuels [50]</td><td>P2G [23]</td><td>Heat recovery from CHP [33–37,39–44,52,59]</td></tr>
<tr><td>Wind turbines [48,51,53,61]</td><td>Gas boilers [14,19–21,23–25,27,29,31,32,38,45]</td><td>$H_2$ electrolyzer (P2G) [18,20,23]</td><td>Methanation processes and devices (biogas) [28,49]</td><td>Solar thermal [39–41,44,58,62]</td></tr>
<tr><td>Solar thermal [39–41,44,58,62,63]</td><td>Heat Pumps [19,21,22,25–27,29,31,32,45]</td><td></td><td></td><td>Gas boilers [38,61,64]</td></tr>
<tr><td>Biomass [33,34,39–41,52]</td><td>Absorption chillers [21,25,26,29,31,32,45]</td><td></td><td></td><td>Biomass boilers [39–41,43,44,59]</td></tr>
<tr><td>Diesel generators [65–68]</td><td></td><td></td><td></td><td>Electric boilers [69]</td></tr>
<tr><td>Fuel cells [58,70]</td><td></td><td></td><td></td><td></td></tr>
</table>

In Figure 3, a general configuration of an EH, including different energy carriers and their respective conversion technologies, is presented.

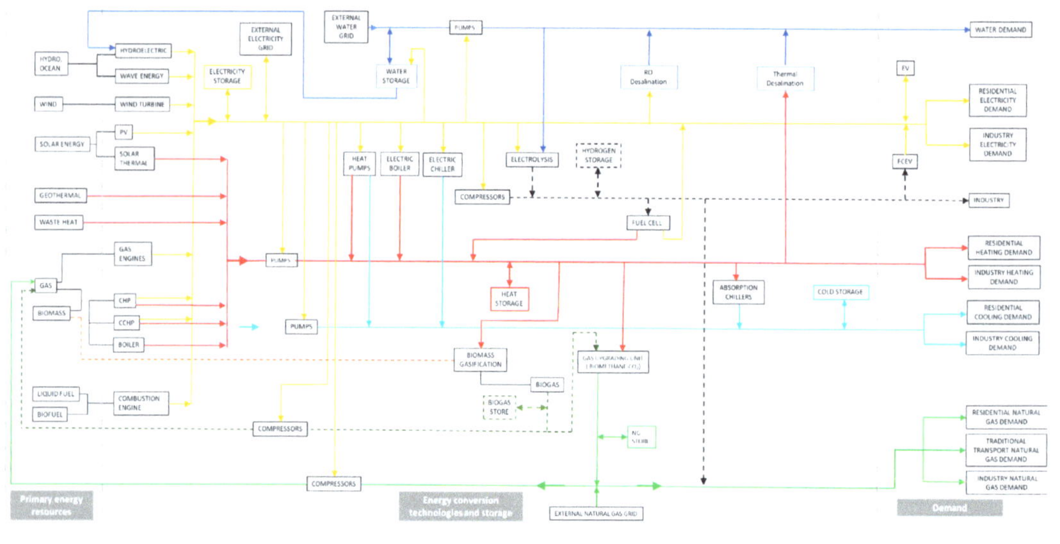

**Figure 3.** The general configuration of an EH.

### 3.4. Energy Storage and Flexibility Potential

As already mentioned in the previous sections, the EH concept is characterized by high flexibility potential that can be derived from storage or loads of different carriers. Regarding the storage facilities, in most of the reviewed papers [71–79], the configurations include at least electrical storage capabilities, combining other forms of storage in several cases, such as thermal storage and natural gas storage for providing better flexibility potential. Electrical storage may include any type of battery, EV, plug-in hybrid electric vehicle (PHEV), compressed air energy storage (CAES), or pumped hydro storage (PHS). In addition, fuel cells with hydrogen (H2) storage are considered for storage purposes in some cases, as seen in [58,70]. In other cases, hot water storage for heating purposes has been employed in [69,80].

Table 5 shows the combination of storage facilities per carrier for the complete list of the literature.

**Table 5.** Storage facilities per carrier based on the existing literature.

| | | Energy Carriers | | | | |
|---|---|---|---|---|---|---|
| | References | Electricity | Heating/Cooling | Hydrogen | Natural Gas | Domestic Hot Water |
| Storage facilities | [27,29,51,56,65,67,81] | Batteries | | | | |
| | [21,24–26,30,31,38,41,43,55,57, 61,71,75,78,79,82–85] | Batteries | Thermal Storage | | | |
| | [72] | Batteries | Thermal Storage | $H_2$ storage | | |
| | [20] | Batteries | Thermal Storage | $H_2$ storage | Natural gas storage | Thermal Storage |
| | [18,23] | Batteries | | $H_2$ storage | | |
| | [40,44,64] | Batteries | Thermal storage | | | Thermal storage |
| | [74] | Batteries | Thermal storage | | Natural gas storage | |
| | [80] | PHS | Thermal storage | | Natural gas storage | |
| | [54] | Batteries, EV | Thermal storage | | | |
| | [73] | Batteries, CAES | | | | |
| | [32] | Batteries, CAES, PHEVs | Thermal storage | $H_2$ storage | | |
| | [58] | Batteries, CAES | Thermal storage | $H_2$ storage | | Thermal storage |
| | [19] | PHEV | Thermal storage | | | |
| | [76] | Flywheel, batteries, CAES | | | | |
| | [42,86] | | Thermal storage | | | |
| | [69] | | Thermal storage | $H_2$ storage | | Thermal storage |
| | [70] | | | $H_2$ storage | | |
| | [87] | | | | Natural gas storage | |
| | [88] | | | | | Thermal storage |
| | [45] | V2G EVs | | | | |

From the reviewed papers, a specific focus is found on the concept of flexibility of different energy carriers of an EH. Very often, the supply-side flexibility is managed using only storage systems, such as a battery or thermal storage [41,43,51,54,55,58,79,89]. However, some of this research has been constrained in terms of supply-side flexibility because of the loose coupling of various energy carriers that were structured within their EH. For example, in some research, the CHP plant was the only technology that interacted with both the electricity and heating carrier [62].

On the other hand, in several works, demand-side flexibility is addressed by considering incentive-based programs and energy-price strategies [26,37] such as time-of-use electricity price or time-of-day unit price for electricity for demand response [19,25,32,42,43, 78,82,90]. In these cases, EVs and batteries are considered, but their collaboration strategy with other co-existing storage facilities in the EH is of primary importance.

The literature on EHs flexibility also focuses on flexibility coming from residential assets, i.e., appliances. Regarding the appliances (apart from the non-interruptible ones), three types that offer flexibility are analyzed: shiftable appliances, shapeable appliances, and thermostatically controlled appliances. Shiftable appliances are flexible and can shift consumption between time intervals. The operation of these appliances can be shifted to off-peak periods without affecting the consumer's comfort, such as washing machines and dishwashers [91,92]. Shapeable appliances are flexible appliances that can change their shape to store energy or minimize their consumption (the latter is known as interruptible load) [65]. Finally, thermostatically controlled appliances are the appliances that can be controlled by a thermostat, such as heating and cooling loads in buildings. The main objective of this control is the minimization of energy consumption [81].

As already seen, the EHs concept can cover much more than just a residential building and thus their flexibility assets. For example, an EH can cover a whole region under the ILEC concept exploiting flexibility from the different assets of the carriers that can be communally owned.

### 3.5. Flexibility Potential of EVs

Many different approaches are present in the literature dealing with the simulation and modeling of EVs within the EH environment for exploiting their flexibility potential. Functions foreseen, such as controllable charging/discharging and vehicle-to-X operation (V2X) representing the flexibility potential, and how they are modeled, may have a significant impact on the obtained results. Therefore, a major challenge when dealing with modeling plug-in EVs (PEVs) or PHEVs is the representation of the vehicles' availability and energy use while they are away. Indeed, those factors are dependent on the users' behavior and habits.

PEVs can be represented either as loads under the grid-to-vehicle (G2V) or as distributed storage when equipped with V2X technology. Therefore, in most cases, EVs and PHEVs are modeled as batteries, as in [45,75,82]; as a part of a large-size equivalent battery, as in [76]; as different peers in a bilateral trading system based on P2P, as in [48]; as a part of the end-users' load profile, as [89]; or as a flexible/shapeable load for electrical networks, as in [65,81].

Another major concern when modeling EVs is the availability of data and energy use, as well as their availability and status during the day. Some works assume that the vehicles depart and arrive at given times [32,60]. For example, in [60], EVs are assumed to be plugged out at 8 a.m. (approximately full battery) and plugged in at 4 p.m. (quite empty). From 4 p.m. to 8 a.m., the battery can store electric energy when it is more convenient (during low-price hours) and uses its energy during high-price hours. In order to ensure that at 8 a.m., the EV has a good charge level, a dissatisfaction term is considered that is directly proportional to the difference between the maximum level of charge and the level of charge at 8 a.m. Other works divide the PEVs into clusters [54], each with specific characteristics: (1) battery capacity, (2) arrival and departure times at/from the charging stations, (3) the state-of-charge (SOC) at the arrival time, and (4) the SOC desired at the departure time. Some works, such as [33], use a probabilistic approach representing the EVs' availability by a normal distribution and a probability function based on survey data. In addition, these data are used in a stochastic model predictive control (MPC) to control the local energy system, allowing for planning for the EV availability with refined scenarios for each hour. In [19], the uncertainty of the time intervals during which the EV owners are at home and can therefore charge or discharge the EV battery is considered by means of a Monte Carlo approach. The daily driving distance of each EV is modeled as a log-normal distribution function, whereas the home arrival and departure times of each EV are modeled as a normal distribution.

### 3.6. Key Challenges for the EH Configuration in the Literature

Through the analysis of the literature on the EH configuration, it is perceived that some key challenges arise which must be addressed to achieve their efficient and reliable operation. The most important key challenges are presented below:

- Limited connection between various energy sectors/carriers. Although may diverse carriers can be present in the different EHs structures, the interconnectivity among them is low in many cases. This means that the full potential of employing the advantages of the integration as described in the introduction remains unused.
- Limited economic incentives in order to encourage the use of flexibility, focusing only on the electricity carrier. This results in a limited number of technologies participat-

ing in flexibility management, such as batteries, whereas the thermal part and their flexibility potential are neglected in most cases;

- Management or/and pricing schemes of other energy carriers beyond electricity. This results in more complex and decentralized schemes for energy carriers other than electricity that are now not represented;
- The inclusion of EVs in ILECs can make their management more difficult. Even if there is some coordination, there will always be several more constraints than for a simple battery storage system;
- The stochastic nature of EV operation that intertwines with behavioral aspects can affect the stability of the system as well in case of high EV penetration.

## 4. Objective Functions for Optimal Design and Operation of EHs

The optimal configuration for a multi-energy system is a complex problem due to the wide variety of technology options, energy price variations, and significant daily and annual fluctuations in energy consumption. Additionally, as environmental issues are becoming increasingly important in the analysis of these systems, they should also be taken on board through the appropriate variables and constraints. However, conflicting objective functions may arise when it comes to such a complex problem.

The literature discusses different optimization approaches leading to single or multi-objective functions with mostly having a twofold focus, i.e., minimization of system cost and environmental emissions (mostly $CO_2$ reduction). In detail, a significant part of the analyzed papers aims to minimize several objective functions. In most cases, multi-objective optimization problems have at least two different objectives. The first one is cost-related, and the second one is usually environmental-related, e.g., $CO_2$ emissions reduction and the maximum integration of RES. Moreover, most EHs use electricity and gas as energy carriers, and their coordination is performed through an optimization model in which both economic and environmental objectives may be considered. Representative references are discussed below.

The following references explore multi-objective optimization approaches for energy systems. References [40,42,70] focus on minimizing energy costs and $CO_2$ emissions. References [44,54,58,59,62,93,94] also prioritize $CO_2$ emissions reduction as the primary objective, but with different secondary objectives such as minimizing total annual costs, including investment, operation, and maintenance costs, maximizing profits for the energy system operator, and minimizing energy costs. Reference [77] prioritizes economic targets, followed by a second objective related to system reliability. Reference [78] presents a non-dominated sorting genetic algorithm for optimizing both economic benefits and energy efficiency. Reference [79] details a fuzzy multi-objective decision and two-stage adaptive robust optimization method. Reference [82] proposes a fuzzy decision-making approach to minimize procurement costs and carbon emissions for interconnected energy systems. Reference [84] aims to strike a balance between operational costs and exergy efficiency using a three-stage optimization process. Reference [95] seeks to minimize the operational cost of interconnected energy systems, as well as the amount of freshwater used. Finally, reference [96] proposes a multi-objective optimization problem to minimize the operational cost and total emissions of an energy system. While using multiple objectives in problem formulation can lead to better outcomes, it can also increase complexity, as objectives may conflict and require trade-offs.

On the other hand, some papers present a single-objective optimization problem. When it comes to single-objective optimization functions, the minimization of cost, be it investment cost or operation cost, is dominant. The maximization of revenues for the different actors of the energy value chain is also a common pursuit. Therefore, works such as [23,32,41,57,81,92,97–100] aim to minimize the operational cost of the EH, while, for instance, Refs. [28,29] deal with the minimization of total investment and operating costs. These representative papers are included here for the sake of completeness.

In summary, Table 6 shows the complete list of references with multi-objective optimization functions, while Table 7 presents the full list of references with single-objective optimization functions.

**Table 6.** Objective functions in multi-objective optimization of EHs in the literature.

| Reference | Multi-Objective Optimization Functions |
| --- | --- |
| [88,95] | • Minimization of operational cost;<br>• Minimization of freshwater extracted from the reservoirs; |
| [94] | • Energy loss minimization;<br>• Minimization of operational cost;<br>• Minimization of $CO_2$ emissions; |
| [78] | • Minimization of operational cost;<br>• Maximization of the total energy efficiency; |
| [79] | • Minimization of investment, operation, and carbon emission costs;<br>• Maximization of the grid integration level, i.e., the interaction between the grid and the multi-carrier system; |
| [39,41,84] | • Minimization of energy costs;<br>• Maximization of exergy efficiency; |
| [40,42,70,82,96] | • Minimization of energy costs;<br>• Minimization of $CO_2$ emissions; |
| [54,93] | • Minimization of $CO_2$ emissions;<br>• Maximization of the EH/aggregator operator's profit; |
| [43] | • Minimization of the total annual cost as the sum of annualized investment costs of all technologies, energy costs, and O&M costs of all technologies in the EH;<br>• Minimization of total primary energy input to the EH; |
| [44,58,59,62] | • Minimization of $CO_2$ emissions;<br>• Minimization of the total annual cost as the sum of annualized investment costs of all technologies, energy costs, and O&M costs of all technologies in the EH; |
| [65] | • Two-phase algorithm for minimization of the generation cost and of the thermal losses by rescheduling users' shapeable loads and distributed energy resources (DER).; |
| [30] | • Minimization of the unit commitment cost (start-up and shutdown cost of the CHP units, HPs, and natural gas boilers), the cost of purchased electricity and natural gas, and the dispatch cost;<br>• Maximization of the robustness of the solutions, i.e., by finding the worst case in terms of the increase in the cost due to the uncertainty. |

**Table 7.** Objective functions in single-objective optimization of EHs in the literature.

| Reference | Single-Objective Optimization Function |
|---|---|
| [55] | Maximization of the EH/aggregator operator's profit; |
| [23,27,32,41,45,57,81,92,97–100] | Minimization of operational cost; |
| [51] | Maximization of the utilities of the customers in a P2P energy-sharing trading; |
| [52] | Maximization of the social welfare given by the sum of the profits of all participants in the P2P energy trading; |
| [56] | Minimization of the total energy expenditure of all individual customers in the EH; |
| [91] | Minimization of the total social energy cost to derive the optimal energy-sharing profiles for the building cluster; |
| [28] | Minimization of the cost of device operation, energy storage, energy transaction, and curtailment power of wind and PV; |
| [29] | Minimization of the total investment cost and total operating costs of energy technologies in the EH. |

The objective function is the key factor of any optimization problem and defines its general target and the influencing parameters and variables. In the analyzed works, the main objective is—in most cases—related to economic targets expressed as the maximization of revenues or profits or, similarly, the minimization of the costs (capital and operational) and the minimization of energy losses.

In a single-objective optimization problem, the space of solutions is usually easily identifiable since there is a unique optimal solution. The introduction of an additional objective function in an optimization problem (multi-objective optimization) and the requirement of simultaneous optimization between each other results in both an increase in the number of solutions (the best solution is not one but many) and difficulty in accurately determining the space of the solutions.

## 5. Optimization Problem Constraints

Constraints have a central role in optimization problems. Indeed, they limit the solution space while defining the choice and/or operation of the components of the modeled system. In energy system modeling, they are used to determine the performance characteristics and limitations of the considered energy conversion and storage technologies and of the elements through which the energy carriers are transported and distributed. They also regulate the interaction between the system and the considered commodity, grid services, and ancillary markets. Due to the complexity of the interaction of the active resources in the EHs and the surrounding energy systems, it is necessary to define a set of physical, technical, economic, and environmental constraints that allow delimiting the optimization problem to viable search space.

Therefore, this section aims to discuss the equality and inequality constraints that have been used in the analyzed optimization problems in the literature so far. The constraints can be classified into four categories: **technology constraints, network constraints, market constraints, and other constraints.**

Technology constraints can be divided into three different subcategories, as can be seen in Table 8. Specifically, these are operational constraints, design constraints, and selection of the technologies in the EH configuration. Most of the reviewed papers use operational constraints, while fewer papers use design constraints in conjunction with operational constraints. As seen in [41,43,44,85], it mainly depends on whether the optimization problem aim is focused on the operation or design of an EH. Some constraints simply describe technology modeling in terms of their efficiency and capacity limits, as seen in [85]. Other models are more detailed for the various technologies that they consider. Commonly modeled technologies include PV [23,27,57], wind turbine [23,57], electric chiller [32,57], solar thermal [59], absorption chiller [32,57], gas boiler [10,27,46], gas turbine [59], diesel generator [50], fuel cell [50], CHP [27,50,56], and HPs [27,56]. Storage technologies are also common, such as batteries [23,27,32,57,99], or other electricity storage, such as compressed air [32], heat storage [32], and hydrogen storage [56]. Other conversion technologies, such as P2G, can also be found in [32]. In addition to the technologies that convert or store energy, some works also add details to the modeling of loads. The EVs, which act either as load or storage, are often modeled as in [32,81]. A major reason to include detailed modeling of the loads is to consider their flexibility potential. Shiftable loads are modeled in [32,92], while [81] makes a distinction between non-interruptible and thermostatically controlled appliances.

**Table 8.** Technology constraints.

| Technology Constraints | References |
| --- | --- |
| Operational constraints (e.g., capacity constraints, ramp-rate constraints, storage constraints) | [19,21–33,35–39,41–44,47,49–51,54–62,65–72,74–86,88,90,92–94,96,99,101–116] |
| Design constraints (e.g., device availability and available sizes in the market) | [41,43,44,85] |
| Selection of the technologies in the configuration | [58,105,117,118] |

Network constraints are presented in Table 9 and are divided into three subcategories. The network constraints represent mainly import and export limits to the different grids or markets [32,57]. Other network-related restrictions, such as the difference in temperature between the inlet and outlet of heat pipes and gas or power flow equations, can also be found, though more rarely [22]. Some models use a constraint to explicitly forbid simultaneous import and export, as seen in [57]. Reference [92] includes a constraint representing the possibility for the grid operator to limit their export to the electric grid if there are grid security concerns. The network flow constraints include the limits of energy flow, limits of imported/exported energy, and use of the share of the network between the energy providers and users, etc., as in [19,21,23–25,27,32,36,39–44,47,50–52,54–59,62,65,67–72,74–76,78,79,82–85,88,90,91,93,94,101–105], the transmission limits, which mainly include boundaries of active and reactive power in transmission lines, and gas flow equations in active and passive pipelines [46,49,54,97,101], and finally, the nodal limitations at EH level [19,22,27,30,49,86,88,90,92,106].

**Table 9.** Network constraints.

| Network Constraints | Description | References |
| --- | --- | --- |
| Network flow constraints | Electricity (active and reactive power), gas, heating, cooling, domestic hot water, energy flows | [21,23–25,27,32,35,36,39–44,47,51,54–59,62,66–72,75,76,78–80,82–85,88,90,91,93,94,101–105] |
| | Network usage charge between the seller and the buyer | [52,101] |
| | Lower and upper limits of imported/exported energy and natural gas from/to utility companies | [19,32,35,39,56,65,83,90] |
| | Non-convex branch flow model to model the distribution network | [87] |

**Table 9.** *Cont.*

| Network Constraints | Description | References |
|---|---|---|
| Transmission limits (active and reactive power limits in electricity networks, maximum mass flow rate in gas and heating networks, power, gas, and thermal flow equations, single flow direction) | Boundaries of the active and reactive power in the transmission lines | [46,49,54] |
| | Gas flow equations in active and passive pipelines | [49,97] |
| | Single direction for the flow of energy in pipeline and pipeline capacity | [101] |
| Nodal limitations at the EH level | Mass balances for each node | [22,86] |
| | End-users constraints | [88,90] |
| | Active and reactive power balance equations in the hub | [27,106] |
| | Maximum and minimum nodal voltages, maximum and minimum gas pressure, maximum and minimum supply and return temperature | [19,30,49] |
| | Constraint on feed-in power when grid security concerns | [92] |

Market constraints are presented in Table 10, and they are less commonly encountered in the reviewed documents. They are sometimes used to represent the estimation of the internal price in models, including P2P energy sharing. Other types of market constraints include the constraint of not buying and selling energy in the same period [36,57,85], the revenue from selling electricity by-products on the spot market [86], and the internal price of electricity [81,92].

**Table 10.** Market constraints.

| Market Constraints | References |
|---|---|
| Energy trading balance between crowdsources | [65] |
| Prevent buying and selling electricity in the same time period | [36,57,85] |
| Selling electricity by-products on the spot market | [86] |
| Estimation of the internal price | [81,92] |
| Constraints related to mutual energy sharing | [19,41,50,91] |

Other types of constraints used in optimization models of EHs are presented in Table 11, and they include inequality constraints about the trading participants and the platform service charge, the established target for the RES penetration levels, and the $\varepsilon$-constraint (parametric optimization method) and integer cut constraints (ICC).

**Table 11.** Other constraints.

| Other Constraints | References |
|---|---|
| Inequality constraints about the trading and the platform service charge | [52] |
| Target the renewable penetration level | [49] |
| $\varepsilon$-constraint (parametric optimization method) and ICC | [104] |

The identification of constraints is a critical step when formulating the optimization problem of the EHs due to the complexity of the system itself. The literature shows that the identification of operational constraints is dominant for both the EH per se and for the efficient operation of the upper grid. In most optimization models, the objectives of the problem formulation are transformed into operational constraints. It needs to be highlighted that only a few papers combine both the operational and design constraints of the EHs. This is an important consideration to be taken, especially when setting up an EH or planning its expansion. Constraints should be carefully identified when considering the interaction of the EHs with other parties, such as the markets or other EHs as well. So far, this has been considered only for one type of interaction, i.e., one type of electricity market, thereby neglecting different types or gas markets. This makes the problem formulation more demanding, but it is needed for a realistic representation of the EHs' role.

## 6. Optimization Problems Modeling

Based on the reviewed works, it is found that some papers formulate the optimization problem as a linear programming (LP) model, but most of them formulate the optimization problems using a mixed-integer linear programming (MILP) or mixed-integer nonlinear programming (MINLP) approach. Others present both linear and nonlinear problems (MILP&MINLP) formulation. A detailed discussion of MILP, MINLP, and MILP&MINLP approaches follows.

### 6.1. MILP

MILP problems are LP problems that include integer variables but are much harder to solve. Binary variables are the most widely used type of integer variable in MILP problems. These types of problems usually employ solution techniques, such as branch- and bound or branch and cut, to obtain the optimal value. Binary variables can be used for several purposes. In some cases, they represent a yes/no investment decision for a particular technology, e.g., by easily considering fixed investment costs in the objective function. In addition, these variables can also be employed to decide on the status (on/off) of an energy conversion technology in each period of the problem's time horizon. Similarly, they can be used to consider start-up costs and minimum uptime or downtime of an energy conversion technology. Sometimes, binary variables are used to control investment in grid connection and ensure no simultaneous import and export of electricity from/to the upper grid. Restricting the number of binary variables permits limiting the computation time.

Several papers address the problem of EH optimization using MILP formulations, such as [19–23,26–28,30–35,47,50,63,67,68,72,74,81,82,90,95,104,105,111–115,117–121]. In particular, Ref. [19] presents a MILP model to minimize the EH's total cost. A similar approach -that also considers the optimal size of an EH- is seen in [20]. The authors of [21] formulate the optimal sizing problem of a multi-energy urban EH as a multi-objective MILP problem aimed at minimizing both costs and carbon emissions. Authors in [22] give an overview of the integrated electricity and heat systems (IEHS) modeling and solution methods for optimal operation and compare the main differences between the possible solutions. Authors in [27,32] propose a planning model for a multi-energy microgrid formulated as a MILP problem, while in [28], the day-ahead scheduling of an electricity–hydrogen–gas–heat integrated energy system (EHGHS) is formulated as a scenario-based MILP problem. In [33], the day-ahead scheduling of an EH is formulated as a MILP problem aimed at minimizing the system's operation cost. The MILP problem is solved on a rolling horizon basis under a model predictive control strategy so as to cope with uncertainty. In [34], a MILP formulation for the optimal design of an industrial manufacturer's EH is used. In [35], a MILP formulation for the optimal operation of an EH under different electricity, heating, and cooling scenarios is considered. In [68], authors use binary variables to control investment in a grid connection and ensure no simultaneous import and export of electricity through a MILP approach.

On the other hand, in [111], a MILP framework is proposed for the robust optimization of smart multi-energy districts under uncertainty with different energy conversion and storage technologies (e.g., PV, EHP, CHP, electric and thermal energy storage, and gas boilers) and detailed integrated electricity, heat, and gas network mathematical models. In [90], a MILP model of an EH considering CHPs, HPs, air conditioners (ACs), EVs, RES, and community energy storage (CES) is presented, with the main objective of minimizing the overall operating cost of the system. The authors of [114] use a MILP formulation for the optimal operation of an EH considering downside risk constraints. References [115, 117,118] use MILP formulations for the optimal design of EHs from scratch, i.e., with no predetermined system structure, carriers, or technologies mix. Reference [105] proposes a robust MILP (RMILP) model for optimizing the operation of an EH. In [63], the potential for improvement of a residential multi-carrier microgrid is analyzed via scenario-based simulations under different management strategies.

### 6.2. MINLP

MINLP problems are still very challenging today, especially if they are non-convex, i.e., the objective functions and/or constraints are defined by non-convex functions. Usually, MINLP problems are relaxed and reformulated as, e.g., mixed-integer second-order cone programming (MISOCP), mixed-integer quadratically constrained (MIQCP), and MILP problems and solved by appropriate decomposition techniques. Several works in the literature formulate the optimization problem as an MINLP problem due to the nonlinear nature of the objective functions and/or constraints, such as [23,25,36,46,49,70,76,96,97,102, 106–110,122]. In particular, Ref. [25] defines the network-constrained scheduling problem of an EH as an MINLP problem, which is reformulated as an MISOCP problem and solved by the use of the partial surrogate cuts method. In [36], an MINLP model for the day-ahead scheduling of electricity and natural gas networks is presented by considering three downward EHs and a P2G system. An MINLP model is developed in [49] with a multi-objective approach to finding trade-off solutions between maximizing RES penetration and minimizing costs in the optimal scheduling of an EH. The authors in [97] present a nonlinearly constrained optimization problem to optimize the couplings/connections among the different networks of an EH. The authors highlight the usual complications that may appear when using MINLP formulations, such as the solution dependency of the initial values of variables or the large number of suboptimal solutions that are not all technically reasonable or feasible. The necessary procedures to overcome the problems mentioned above usually require ad hoc solutions that cannot be implemented elsewhere. The authors in [102] formulate the day-ahead scheduling problem of four urban EHs as an MINLP problem, which is reformulated as an MISOCP problem, which in turn, is decomposed into a second-order cone programming (SOCP) problem and a mixed-integer quadratic programming (MIQP) problem. These problems are solved sequentially and iteratively, using the results of each program as input variables to the other. In [122], an MINLP model is developed for the multi-objective operation optimization of an EH consisting of district heating, cooling, and electricity networks interconnected among them. The presented methodology is general enough to be applied to several objective functions, and in the case study, operating costs and carbon emissions are considered objectives.

### 6.3. MILP & MINLP

In the literature, there are also several works formulating the optimization problem by using a mixed approach based on both MILP and MINLP. Among the analyzed works using this approach, in [24], the multi-objective optimization problem is solved in two stages, in the first stage through a MILP and in the second stage through an MINLP. In [65], the optimization problem is defined through a SOCP relaxation. In [83], all the mathematical expressions in the optimization problem are formulated as linear equations. However, the employed solution method uses a meta-heuristic model (bacterial foraging optimization) to solve the problem. In [84], the optimization problem for the day-ahead dispatch is formulated as a MILP model, whereas the intraday scheduling is formulated as a nonlinear model. In [88], the optimization problem is formulated as an MINLP model. In [92], a bi-level optimization model is employed. The upper level is not formulated via an objective function but is solved through an iterative algorithm. The authors in [123] analyze the optimal scheduling of a coupled electrical–natural gas network feeding a distributed electrical load in a combined day-ahead market and real-time operating conditions. The problem is formulated via a data-driven distributionally robust optimization (DDRO), considering wind data uncertainty. On the other hand, reference [124] reviews existing optimization techniques and their applications in power systems, focusing on multi-objective optimization in power system planning.

Table 12 summarizes all screened documents according to the approach used to formulate the optimization problems.

**Table 12.** Modeling approaches used for the optimization problems formulation in the existing literature.

| Approach | References |
|---|---|
| MILP | [19–23,26–28,30,32–35,39–41,43–45,47,50,54,56–59,62,63,66,67,69,72,74,75,79,81,82,86,88–90,93,95,101,105,111–115,117–121] |
| MINLP | [23,25,29,36,46,49,70,76,78,80,96–98,102,106–110,122] |
| MILP & MINLP | [24,65,83,84,88,92,123,124] |

*6.4. Key Challenges for the Optimization Problems Modeling*

From the reviewed papers, it emerges that MILP formulations are the most widely used for the design and operation optimization of EHs. Nevertheless, some variables that may be present in EHs respond to a nonlinear behavior, e.g., power flow variables, and to non-convexities, e.g., commitment status of energy conversion and storage units. The accurate representation of the physical phenomena taking place in the energy conversion and storage units of an EH and of their design and operational decisions may require the use of MINLP formulations. As already mentioned, the solution to MINLP problems is a challenging and computationally hard task. Usually, MINLP problems are relaxed and reformulated as MIQCP, MISOCP, or MILP problems and solved using proper decomposition techniques. The solutions of such approaches are not guaranteed to be optimal, nor do they represent with precision some physical phenomena taking place in the energy conversion and storage units of the multi-carrier energy system. To summarize, in the context of EHs, very detailed models of energy technologies and energy flows cannot be the best solution to adopt since they limit the effectiveness of the optimization models and methods employed to solve the problem. It is, therefore, necessary to find a good trade-off between model fidelity and the complexity of the optimization process.

## 7. Multi-Objective Optimization Problems and Methods

The design and management of EHs need to respond to different stakeholders' participation requirements and preferences. Therefore, the objectives are usually formulated from different perspectives and can be in conflict with each other. For instance, an economic optimization that serves developers' needs cannot ensure that the EH configuration or the operation strategies obtained are optimal from an environmental point of view that would serve the energy community's needs. This intrinsic conflict existing between different objectives calls for the need to establish a multi-objective approach when dealing with the optimal design and operation of EHs.

In practice, in the context of a multi-objective problem, obtaining a unique solution is only possible if the objectives do not conflict with each other, but in most real cases, this does not happen, and there are multiple aims represented by objective functions, which need to be traded off. Due to the multiple objectives, there is no single optimal solution but a set of non-inferior solutions (a 1D set for two objectives, a 2D set for three objectives, etc.). This set defines the so-called Pareto front, and it can be used to assess solutions by the relative meaning of the objectives. Non-inferior solutions are solutions where one objective cannot be improved without making another objective worse.

In order to identify a single solution for the multi-objective problem for the design or operation of an EH, two steps need to be followed, i.e., the optimization and the decision-making steps. According to the order used for these steps, two methods are derived as specified below [107]:

- The preference-based approach in which the decision-making is performed before the optimization. This approach requires a good knowledge of the preferences of decision-makers that need to be respected in the optimization problem formulation. Quantifying these preferences is a challenge;
- The second approach is considered ideal. The optimization is performed before the decision-making. This approach is more desirable than the previous one, as it is less subjective and leaves the final decision to the decision-makers.

In many cases, the multi-objective problem is simplified into a single-objective problem by modifying the objective function or combining specific constraints. In this case, a second objective is added to the first one thanks to a conversion factor. For example, in [34,97], where multi-energy systems are investigated, emission minimization is included in the cost minimization models. Another option is represented by the epsilon constraint method, as seen in [72,85]. In detail, a primary objective is employed, while the other objectives are converted into inequality constraints where the right-hand side is a factor that is wide-ranging to obtain various solutions. Meta-heuristic models can also be applied for multi-objective optimization [125], but they do not guarantee the optimality (non-inferiority for multi-objective problems) of the solutions found.

From the analysis of the relevant literature, the most used method for multi-objective optimization problems in EHs is the weighted-sum method, as in [66,86,101]. This method combines the different objectives into a weighted sum. The weighting factor can be varied from 0–1 to obtain the Pareto front with the best possible trade-off solutions between two objectives. In addition, objectives can be normalized, as seen in [86,101], to simplify the process and the knowledge of the weights. This method allows for the use of the second, ideal approach described earlier, with the identification of the Pareto front through the optimization process. Therefore, the final choice is left to decision-makers to make informed choices and select an appropriate solution on the Pareto front from a wider range of alternatives.

The various multi-objective optimization methodologies analyzed in the literature are categorized in Table 13.

**Table 13.** Multi-objective optimization methods used in the context of EHs in the literature.

| Method | Description | References |
|---|---|---|
| MO-MFEA-II | The multi-objective multifactorial evolutionary algorithm II (MO-MFEA-II) is a multitasking method where multiple multi-objective problems are optimized simultaneously. Each component of the multi-objective problems contributes to a unique factor affecting the evolution of a population of individuals. This algorithm uses the concept of non-dominated rank (NR) and crowding distance (CD) in the non-dominated sorting genetic algorithm (NSGA-II) to define the fitness of each individual. | [94] |
| Pareto-based multi-objective evolutionary algorithms (MOEAs) | A Pareto-based multi-objective solution uses evolutionary algorithms to find non-dominated solutions on the Pareto front, which considers multiple objective functions at the same time as trade-offs. This method guarantees good performance in numerous application areas. This algorithm is, in fact, easy to implement since it does not require detailed knowledge of the domain of the case under study. | [34,97,107,124] |
| Modified teaching–learning-based optimization algorithm (MTLBO) | In the MTLBO algorithm, the methods of the teaching phase and learning phase are, respectively, modified to enhance the disturbance potential of search space, and a new "self-learning" method is presented to enhance the innovation ability of the learner and the global exploration performance. | [96,108,109] |

**Table 13.** *Cont.*

| Method | Description | References |
| --- | --- | --- |
| NSGA-II algorithm | The NSGA-II is employed to guarantee the feasibility and accuracy of the model solution. In this methodology, elitism and a maintenance methodology are used to increase diversity. A classification of the solutions according to an order of dominance is used. The assignment of a level or a front of dominance to all the solutions of one population is the basis of the NSGA-II. This method is more appropriate for dealing with nonlinear problems, which are more complex to overcome with other multi-objective optimization methods. | [77,78,107,110,125] |
| VIKOR | The multi-criteria decision-making method, VlseKriterijumska Optimizacija I Kompromisno Resenj (Vikor), can be employed to select the optimal solution from Pareto solutions. This technique is specific to selecting alternatives with respect to conflicting criteria based on an aggregating function that measures the distance to the best solutions. | [78] |
| $\varepsilon$-constraint | The $\varepsilon$-constraint method optimizes the main objective while other objectives are assumed as constraints of the problem. This approach is influenced by constraints choice; in addition, it can solve non-convex optimization problems. | [14,46,58,72,79,85] |
| Weighted-sum method | A single-objective function is formulated as a weighted sum of the objective functions. This method is employed to find the Pareto front, consisting of the best feasible trade-offs between the objectives that can be discovered by varying the weight in the interval 0–1. | [39–44,54,66,86,93,96, 98,101,106,107] |
| Compromise programming method | The application of the compromise programming method aims to the modification of the decision model to include only one objective. The optimum solution can be identified as the one with the shortest distance to the optimum value. | [62,70] |

Based on the conducted analysis, the main limitations related to the multi-optimization methods described in the previous table are reported below.

The most important limitation of the $\varepsilon$-constraints method and compromise programming method lies within the need to transform multi-objective problems into single-objective problems based on good knowledge of the decision-makers. The most common approach, as shown in Table 12, is the weighted-sum method, thanks to the ease and straightforward way of obtaining multiple points on the Pareto-optimal front. In this case, the challenge is the selection of a weighting criterion ensuring that the points are spread evenly on the Pareto front. Taking into account the nature of different flexibility resources present in EHs, the method which transforms the multi-objective problem into a single-objective problem to be easily solved could be considered.

Definitively, choosing the most suitable method depends on the complexity and scale of the problem. In this case, the challenge in deciding the best methodology to solve a multi-objective optimization problem is to perform a detailed study of the algorithms and the characteristics of the problem before applying the optimization approach. In particular, the selected method must always find and provide all Pareto-optimal solutions, consider weights to express preferences, as well as employ the utopia point or its approximation.

## 8. Heuristic Methods

A heuristic method can be defined as a procedure for solving a well-defined mathematical problem by an intuitive approach in which the structure of the problem can be interpreted and exploited intelligently to obtain a reasonable solution. Heuristic methods, unlike optimization methods discussed in previous sections, are not able to guarantee the optimality of the decisions but can, if designed and tuned correctly, provide enough adequate solutions faster and/or with fewer computational resources than by using an optimization model. For example, Ref. [124] mentions its utilization in power systems contexts, with a special focus on multi-objective optimization for power system planning. The heuristic methods discussed in this literature review are categorized into simple heuristics and meta-heuristics, as shown in Table 14 below.

**Table 14.** Categorization of heuristic methods.

| Heuristic Method | Characteristics | Examples |
| --- | --- | --- |
| Simple heuristics | Faster calculation of solution; Prone to get stuck in local optima. | Local search; Greedy algorithms; Hill climbers. |
| Meta-heuristics | Attempts to obtain a better solution in a pre-defined neighborhood; Many methods are based on biological metaphors. | Evolutionary algorithms; Genetic algorithms Simulated annealing; Particle swarm; Tabu search; Ant colony; Hybrid algorithms. |

Simple heuristics come in the form of local search, greedy algorithms, and/or hill climbers. Meta-heuristics try to go beyond the local search for the best solution. Many of them are based on some biological metaphor, are bio-inspired, and are in the form of genetic algorithms (GA) and evolutionary algorithms (EA), simulated annealing, tabu search, ant colony, hybrid algorithms, fuzzy programming, neural networks, etc. A meta-heuristic creates a set of candidate solutions (population), checks the value of the objective function for each of them, and applies a heuristic for generating a second population. The heuristic varies between the methods but is often based on the most promising elements of the previous generation. Those steps are repeated until a stop criterion is satisfied. Meta-heuristics can be applied to planning problems, as seen in [31]. There, the robust planning of the energy system of an EH is tackled using quantum particle swarm optimization (QPSO). This meta-heuristic is also compared to the performances of PSO and GA approaches and shows the superiority of the proposed method in terms of convergence speed and global search ability. Similarly, Ref. [125] uses an elitist GA (a variant of NSGA-II) in a multi-objective planning problem. The objective considered is minimizing the primary energy demand and investment costs for RES installation in a building.

In [67], a fuzzy inference system has been used to solve the energy storage system (ESS) scheduling problem in the microgrid energy management (MGEM) system. Reference [78] developed a multi-strategy gravitational search algorithm (MSGA-II) for optimizing the operation of integrated energy systems with electro-thermal DR mechanisms. In [83], a modified bacterial foraging optimization (MBFO) is used to solve the ESS scheduling problem considering the economic and environmental objective functions simulating the trade-off between conflicting objectives. In [96,109], a variation of a fuzzy decision making method is proposed, merged with the well-known modified teaching–learning-based optimization algorithm.

In P2P systems, energy sharing has the potential to facilitate local energy balance and self-sufficiency. In [81], an evaluation of the performance of some P2P energy-sharing systems based on a multiagent-based simulation framework was performed. In order to facilitate the convergence of the algorithm, two heuristic techniques were considered: a step length control and a learning process involvement.

Other methods are used in the context of local energy markets, such as centralized/decentralized optimization, hybrid, continuous trading, auction-based, etc. [99]. They specifically occur when modeling multiple agents competing in a market or in decentralized or distributed approaches, such as in [65], where different optimization models such as the alternating direction method of multipliers (ADMM), Stackelberg game, and Nash game approaches were utilized.

The environmental/economic dispatch problem involves conflicting objectives, and it is known to be highly constrained, as already mentioned. In order to tackle this challenge, a method combining traditional optimization and a meta-heuristic method is presented in [103]. It combines convex optimization and meta-heuristics in a method named scenario-based branch and bound. This method is used to obtain reliable solutions in a model predictive-based MINLP with coupled timesteps. The meta-heuristic used is a modified version of the real coded genetic algorithm (RCGA) that is chosen over GA because it converges faster. The meta-heuristic is used to solve single steps independently, and their compatibility over the prediction horizon is checked afterward. In [106], a hybrid multi-objective optimization algorithm is presented based on particle swarm optimization (PSO) and differential evolution (DE). They showed the effectiveness and potential of the algorithm, comparing it with different techniques reported in the literature and by the application to the standard IEEE 30-bus test system.

Reference [107] mentions that these problems are usually difficult to solve using traditional mathematical methods, therefore EA is a good alternative. EA handles sets of possible solutions simultaneously and, as a result, permits the identification of several solutions of the Pareto front at once. Hence, EA is recognized as a natural way of solving multi-objective problems efficiently.

Heuristic/meta-heuristics methods can also be used in various aspects related to EHs, for example, for the operation/control of the EHs, as they are often able to handle problems in a shorter computational time. Reference [108] uses a modified teaching–learning-based optimization algorithm for solving the optimal power flow problem in multi-carrier energy systems, while reference [110] solves optimal energy flow problems (OEF) via a multi-agent genetic algorithm (MAGA) for decomposing the multi-carrier optimal power flow (OPF) problem into separate OPF problems.

A summary of the utilized heuristics methods per reference can be seen in Table 15.

**Table 15.** Heuristic methods used in the context of EHs in the literature.

| Reference | Purpose | Method |
|---|---|---|
| [16] | Operation and control of EH | Does not apply (literature review) |
| [31] | Planning | Quantum particle swarm optimization (QPSO) |
| [65] | P2P exchange | Alternating direction method of multipliers (ADMM) |
| [67] | Scheduling of ESS | Fuzzy inference system |
| [78] | Optimize electro-thermal DR | Multi-strategy gravitational search algorithm (MSGA-II) |
| [81] | P2P exchange | Step length control and learning process involvement |
| [83] | Optimize operative costs | Modified bacterial foraging optimization (MBFO) |
| [96] | Minimize costs and emissions | Fuzzy decision-making |
| [99] | Energy markets | Does not apply (literature review) |
| [103] | Scheduling of DER | Scenario-based branch and bound |
| [106] | Environmental/economic dispatch | Particle swarm optimization (PSO) and differential evolution (DE) |
| [107] | Planning | Does not apply (literature review) |
| [108] | MO-OPF | Modified teaching–learning-based optimization algorithm |
| [109] | MO-OPF | Fuzzy decision-making |
| [110] | MO-OPF | Multi-agent genetic algorithm (MAGA) |
| [124] | Planning | Does not apply (literature review) |
| [125] | Optimize RES mix | Elitist benetic algorithm |

As already mentioned in the limitations of the previous section, in a multi-carrier energy system, the operation management considering multi-objective functions is a large-size problem and, in general, is nonlinear, non-convex, non-smooth, and of high dimension. Employing different mathematical techniques in such problems could lead to being trapped in local minima. Hence, a well alternative to deal with large-size problems is to use evolutionary techniques. However, a significant limitation when using heuristics techniques to solve multi-objective optimization problems is the complexity of setting up their parameters to afford an efficient performance. Along with this, the fact that it is not possible to obtain the optimal solution with these techniques is the main issue to consider when employing them.

## 9. Optimization Solvers and Modeling Environments

Based on the literature review, the most used optimization modeling environments and solvers with related optimization algorithms in EH contexts are summarized in Table 16. It is found that researchers employ more proprietary software (MATLAB, GAMS, LINGO, X-press, IBM ILOG CPLEX) than open-source (CVXPY) tools to formulate, create, configure, and solve the proposed optimization problems. Table 15 shows that most works use MATLAB as a framework for modeling their approaches due to its capability to handle both linear and nonlinear problems through its proprietary optimization tools. However, other works combine MATLAB as an interface with other modeling environments such as GAMS or YALMIP to implement the optimization models, using commercial solvers such as CPLEX or GUROBI to solve them. A few papers also formulate their models in environments developed under Python language such as CVXPY or RLLab, the latter for developing and evaluating reinforcement learning algorithms, with the limitation that larger models are hardly handled by open-source solvers, e.g., GLPK, IPOPT, CBC, ECOS, SCIP, etc., leading to an infeasible convergence or getting stuck in local minima. Therefore, using a particular modeling environment strongly depends on the expertise of the researcher, development time, ease of model implementation, and maintenance. As the environments enable the user to express complex algebraic expressions concisely, they employ more memory than an optimization application programming interface (API), e.g., CPLEX or GUROBI object-oriented Python API, to create the model.

**Table 16.** Optimization solvers with related optimization algorithms and modeling environments used in EH contexts in the literature.

| Optimization Solver (Algorithm Used) | Modeling Environment | References |
|---|---|---|
| Not available * | MATLAB (Optimization toolbox) | [21,47,60,71,77,91,92,94,108,111] |
| Gurobi (barrier and simplex algorithms) | MATLAB | [27,59,84,101] |
| BMIBNB (branch and bound) | MATLAB (YALMIP toolbox) | [84] |
| Not available * | MATLAB | [79] |
| CPLEX (simplex and branch and bound algorithms) | MATLAB (YALMIP toolbox) | [50,56] |
| Not available * | MATLAB + GAMS: MATLAB was used to develop the system operation model, and GAMS was used for the optimization phase | [34,125] |
| CPLEX (simplex and branch and bound algorithms) | MATLAB | [19,28,72] |
| CPLEX (simplex and branch and bound algorithms) | GAMS | [23–26,30,45,57,58,64,67,74,75,82,90,102,112,113] |
| DICOPT (outer-approximations algorithm) | | [29,32,36,51,56,78,80,88] |
| BARON (branch and reduce algorithm) | | [102] |
| CPLEX (simplex and branch and bound algorithms) | IBM ILOG CPLEX | [33,39–44,54,55,81,93,114] |
| Not available * | X press | [62] |
| Not available * | LINGO | [70] |
| Not available * | MATPOWER TOOL | [52,110] |
| Not available * | CVXPY | [65] |
| Gurobi (barrier and simplex algorithms) | Python + GAMS/SCENRED tool for reduction of scenarios | [38,69,85] |
| Not available * | Python (RLLab) | [87] |

* Optimization solver was not mentioned in these works.

## 10. Uncertainties and Risk Aversion

### 10.1. Uncertainties

When considering uncertainties in the context of EHs, most of the analyzed research has considered the behavior of variable renewable energy (VRE), be it wind or PV, or the variations in consumption patterns among different users, be it by the representation of electrical and/or thermal loads. Added to this, some of the works have also considered variations in the market energy price. There is some research focusing also on thermal load uncertainties. In order to categorize these references according to the uncertainties that were tackled, a summary is provided in Table 17.

**Table 17.** Uncertainties in the context of EHs in the literature.

| Uncertainties | References |
|---|---|
| Renewable generation | [16,24,28,31,38,40,45,46,49,50,55,61,65,77,85,86,90,91,99,111,114,116,123] |
| Consumption | [11,13,20,28,30,36,38,40,41,45,46,60–62,65,72,76,77,79,85–87,91,99,100,111,114,115] |
| Storage and EVs | [19,31,33,67,123] |
| Energy price | [11,13,19,38–40,46,52,55,60,61,64,76,85–87,99,100,105,114,115,126] |
| Failure | [77] |
| Thermal load | [127,128] |

In order to consider uncertainties in the problem formulation, different methodologies have been proposed in the literature that can be divided into three categories, as explained in [108], which are as follows:

- Stochastic optimization discretizes the continuous stochastic parameters into a tree of scenarios, whose nodes of uncertainty are assumed to be known;
- Robust optimization defines the solution according to more adverse scenarios regardless of the probability of occurrence;
- Chance-constrained optimization introduces probabilistic constraints for obtaining a trade-off between the optimal value and the robustness of the solution.

Among the analyzed references, the most frequent methodology uses stochastic optimization models. From these, the utilization of the Monte Carlo simulations [19,20,36, 40,46,57,59,71,82,85] has stood out as the most common way of dealing with uncertainties, which randomly samples scenarios from historical data or probability distributions, albeit other sampling methods have also been utilized, as in [94] with Latin hypercube sampling or [111] with the average sampling approximation. In the latter, the authors propose a two-stage framework that combines a stochastic optimization model and robust techniques to identify solutions to the problem that are robust and flexible in terms of uncertainty. In the case of [33], a two-step approach is used for the operation of an EH, using stochastic optimization in the first step and an MPC strategy in the second step. The authors in [114] propose a stochastic optimization combined with a novel risk assessment approach called the downside risk constraints method for the modeling of the risk imposed by uncertain parameters.

Robust optimization techniques are found in the works [30,31]. The authors of [30] consider the uncertainty of each variable through a suitable uncertainty set or prediction interval that is defined as a function of the forecast value and the forecast error. The authors consider different degrees of robustness and different magnitudes of the forecast error. In [31], the authors use a robust method based on a QPSO approach for solving the optimization problem.

As for chance-constrained optimization, the authors of [50] solve the optimization problem using a distributionally robust chance-constrained model. Further, the authors of [67] made a forecast of uncertain parameters that were then used in a fuzzy inference system to make charging and discharging decisions for an ESS with the goal of simplifying the optimization of the energy system.

*10.2. Risk Aversion*

Risk-averse formulations interpolate between the classical expectation-based stochastic and minimax optimal control. This means that they are flexibly managing uncertainty from the worst case up to the expected (risk-neutral). In this way, risk-averse problems aim at hedging against extreme events of low probability without being overly conservative. There are two main approaches followed in dealing with this issue in the literature:

- The first one considers risk metrics that provide a grade of risk to moderate the decision;
- The second one is through distributionally robust optimization.

In general, the decision-maker may trade performance for safety by interpolating between the conventional stochastic and worst-case formulations looking forward to robustness to load and renewable power prediction errors.

In [71], the presence of uncertainties in the zero-carbon multi-energy system (ZCMES) influences its scheduling performance and brings some particular operational risks. Thus, in the operational-cost objective function, Markowitz's mean-variance theory is employed to simultaneously minimize and maintain a balance between economy and risk.

The information gap decision theory (IGDT) is a practical strategy with no need for a probability distribution function of the uncertain parameters that models the positive and negative aspects of uncertainty based on the known and unknown information. For example, in [84], positive and negative outcomes that may cause risk are modeled using two functions of information gap decision theory called robustness and opportunity functions. In [74], IGDT may be used either from risk-averse or risk-seeking perspectives. In risk-averse IGDT, the decision-maker would be satisfied if the cost is equal to or less than a pre-specified critical value.

Reference [114] presents a stochastic model for risk assessment that utilizes a flexible methodology to mitigate risks associated with uncertain environments. This is achieved by slightly increasing the operational costs, as demonstrated in various tables that compare the operation costs of the hybrid energy system (HES) under different scenarios with the risk control parameter.

In addition, data-driven solutions are utilized to coordinate the scheduling of multi-energy coupled systems (MECS) by taking into account the correlation and distribution characteristics of uncertainties. This approach reduces the conservativeness of decision-making and improves the operation reliability of coordination scheduling for MECS. Reference [123] employs the data-driven robust optimization (DDRO) method to address uncertainties while obtaining less conservative minimum cost solutions.

The table below (Table 18) presents the risks with the associated parameters, as seen in the literature.

**Table 18.** Risks and related parameters in the context of EHs in the literature.

| Risks | Related Parameters |
| --- | --- |
| Financial risks | Electrical loads;<br>Thermal loads;<br>Solar irradiation;<br>Electricity prices. |
| Reliability and power quality risks | Deviations of demands;<br>PV power;<br>Wind power;<br>Electricity prices. |

*10.3. Key Challenges in Handling Uncertainties and Risk Aversion*

To handle uncertainty in the context of EHs, scenario-based optimization (SO) and robust optimization (RO) methods are dominant. However, they provide certain challenges. SO needs to utilize scenarios while considering the probability of them occurring. Otherwise, the end-up solutions can be conservative and/or their costs suboptimal. RO assumes known uncertainties per node. However, treating the uncertainties from different sources independently leads to over-conservative strategies as well. By combining the advantages of SO and RO, the distributionally robust optimization (DRO) methods can be promising when solving optimization problems with uncertainties (renewable generation and loads). There are different approaches to solving these kinds of problems, typically by reducing the constraints formulation to be solved via MILP, where the distribution information is useful to obtain a less conservative solution. These data-driven approaches can be combined with risk-averse methods to identify realistic and operationally efficient solutions.

## 11. Interaction of EHs with Multiple Markets and Networks

The aim of this section is to analyze the interactions of EHs with external entities and markets, which include not only the existing power grid and gas utility networks, but also thermal networks and, in some cases, other neighboring EHs.

The exchanged energy and the involved external networks differ between the analyzed literature, depending—to a great extent—on the existing energy carriers and local resources of the EH. In the review papers [11,13,17], it is possible to gain a general idea of the entities with which different EHs can interact. EHs can exchange energy among them (electricity, gas, thermal, and hydrogen) through external networks, as seen in [58]. In contrast, EHs can also send and receive electrical and thermal energy between each other (named transactive energy) without considering any utility or third-party network, as seen in [88]. However, most of the revised literature considers the interaction of a unique EH with at least one external network and energy market. In [129,130], the EH through the aggregator role interacts with more than one different operator and provides multi-energy bids to participate in multiple energy markets.

These energy exchanges could be performed efficiently by means of the conventional wholesale markets, P2P markets, and/or other kinds of market mechanisms that allow the exchange of different forms of energy other than electricity (natural gas, fuel, thermal energy, and hydrogen). Due to the higher complexity of coupling and integration of the EH with the external grids, synergies with multiple markets, and novel operational schemes should be further investigated.

### 11.1. Involvement of EHs in Multiple Wholesale Markets

According to the literature, two approaches are mainly followed to consider the interaction of the EHs with the surroundings and can be evaluated from the market and the network perspective. The market approach refers to the monetary/financial energy transactions between the EH and the surrounding energy system or market of interest, while the network approach is limited to the physical interaction and constraints regarding the amount of energy exchange between those two main systems.

By focusing on the technical network-constrained approach, it is included in the strategy of optimal coordination of flexible resources to maintain reliability and stability in EHs' operation. Still, most of the literature within the EHs' scope has been focused on demonstrating an optimal operation of such resources, where market rules were scarcely designed or omitted altogether [67,71].

In this first network approach, the optimization models have to impose technical constraints associated with the electrical and gas networks, such as [19,23,30,36,49,96,97,116,123], and load balancing to guarantee the stability, operational security, and energy balance of the EHs. On top of these, in [129,130], technical constraints for heat networks are also considered. These constraints include power and energy balance at the electrical point of common coupling (PCC), an admissible voltage range, transmission line capacity, and operational limits related to the gas sub-network. For example, in [123], economic costs have been considered, such as generation cost, start-up/shutdown cost, upward/downward reserve cost, and real-time operation cost (regulation cost, curtailment/shedding penalties), where a co-optimization between the power system and the natural gas system is presented. Similarly, in [116], the electric power, natural gas, and district heating systems are coordinated to achieve the optimal economical operation of the whole system (energy and reserve scheduling) with minimum wind curtailment, both in the day-ahead and real-time stages. In [130], possible energy imbalances and reserve shortages due to network violations are minimized for the combined space of multi-carrier systems. In [49], the model for optimal scheduling of P2G and gas-fired generation (GfG) is evaluated on the IEEE standard power transmission grid integrated with a gas transmission grid that would form an EH. Reference [25] proposes a network-constrained optimal scheduling model for a power distribution network and district heating network (DHN) with consideration of the DR, in which gas is imported upstream for thermal needs covering. More details regarding the involved constraints are reported in previous Section 4.

There is another trend of research that assesses the impact of the interaction of the considered EHs with the electrical distribution grids in terms of grid stability or performance, but not from a market perspective. Grid integration level (GIL) is evaluated in [79], where a strong interaction via importing/exporting electrical power in large quantities may result in a destabilization and performance degradation of the grid, and it can reduce the economic performance of the EH with high GIL rates. In [49], the renewable penetration level into the EH and optimal scheduling of P2G and GfG is decided by the grid operator considering the cost that comes with it. Other works evaluate and solve these grid issues inside the EH and not under the upstream utility network perspective. In [116], simultaneous energy and reserve scheduling is performed to minimize the operation costs, while accommodation of wind power production is achieved through reserves in the second stage (real time). In [123], the power output and up/down reserve capacity of gas-fired units is determined in the day-ahead scheduling stage in the multi-energy coupled system for handling wind power uncertainty, which influences the operation feasibility of the gas system and the availability of reserve of gas-fired units in real time. In both works [116,123], the reserve availability is used to manage uncertainties inside EHs, but not as a network service to the grid. Reference [66] estimates the ancillary service fees and other chargers Inside the EHs, depending on the energy schedule, but does not apply the issue upstream in the utility grid.

Hereafter, the market approach is presented in-depth, covering the wholesale electricity markets consecutively, balancing/grid services, gas market/trading, and DR mechanisms.

As for the literature that has been studied, EHs participate in the day-ahead electricity market (DAM), considering variable costs and/or different timeframes. A single EH is usually connected to the utility grid at the PCC, assuming the existence of the wholesale market [25,26,31,39,50,57,58,63,66,68,70,72,74,78,79,82,83,85,86,88,94,97,104,114, 119,120,125]. The interaction with the DAM can be asymmetric by considering energy imports exclusively, as in [27,30,33,40,44], or the EH is enabled to import energy and excess export energy [100] to the wholesale market, as in [19,28,32,35,49,54,55,69,84,93]. In the latest case, the energy exports can be the result of surplus generation from a CHP in [86], the available electrical flexibility of DERs in [55], or EVs in vehicle-to-grid (V2G) operations [54]. Reference [88] addresses multi-EHs operation to minimize grids' cost of operation and carbon emission through electricity price signal. Reference [40] provides a stochastic model for the optimal operation scheduling of a DER system, including renewables, considering economic and environmental aspects. In [54], local multi-energy systems cover the electricity and thermal needs of a building cluster with a fleet of PHEVs, where DERs are dispatched (including optimized charging/discharging strategies of PHEVs) in order to maximize the operator's profit while also reducing $CO_2$ emissions. Reference [63] addresses hourly scheduling for a multi-apartment residential microgrid operation.

Other works also cover the design stage, for instance, in [32], where a planning model is presented for a multi-energy microgrid that supplies the electricity, heating, and cooling loads, including flexible demands and hydrogen energy storage. Reference [68] addresses the optimal microgrid design, including optimal technology portfolio, placement, and dispatch, for multi-energy microgrids, including annualized investment costs of discrete and continuous technologies, the total cost of electricity purchase inclusive of carbon taxation, demand charges, electricity export revenues, and generation cost for electrical, heating, or cooling technologies. The interaction of thermal or electricity carriers between multiple EHs is presented in [57], where a non-profit entity named the local market operator (LMO) trades various forms of energy in local P2P markets between EHs and manages the aggregated demand—supply curves in the wholesale and thermal district markets. Despite the fact that the majority of the models consider electricity markets to purchase or sell electricity, authors use historical day-ahead electricity prices to model the electricity market in its simplest version, as seen, for example, in [86,104]. Usually, the fluctuation of hourly wholesale electricity prices has a direct influence on the objective functions, in contrast to [79], where the electricity price is considered relatively stable.

After the DM, the intraday market (IDM) is represented as the second stage, where deviations from the DAM schedule and operation can be corrected in the IDM. Not many works consider electricity markets closer to operation, such as IDM or real-time markets [27, 50,66,84,116,123]. The market interaction is modeled in [19,103] through the day-ahead and real-time prices, but wholesale electricity markets are not addressed for each timeframe. In contrast, Ref. [103] aims to reduce energy utilization from the utility grid through a real-time price-based DR scheme. Similarly, in [19], the EH is allowed to purchase or sell electricity so as to meet its own electricity and heat demand and to minimize the cost, in which the uncertainties of "real-time prices" are highly correlated to the DAM prices. In [27], the intraday schedule is adjusted (such as increasing the purchase of electricity and gas) to reduce the energy differences between the supply and demand inside the multi-energy system.

IDM is incorporated in [66] via real-time pricing data for both electricity import and export (wholesale electricity prices, distribution and transmission charges, and other ancillary service fees), according to different projections of grid carbon intensity and cost. Reference [50] proposes a chance-constrained problem designed for DAM EH energy management with multiple uncertainties and taking account of risks in intraday real-time transactions. The electricity and gas purchases are optimized and presented at daily and intraday timeframes. Reference [84] considers a multi-stage optimization with three temporal horizons: day-ahead dispatch with 1 h resolution and intraday operation with 15 min and 5 min beforehand scheduling intervals to adjust short-term cost and efficiency based on forecast updates of local generation and load. This market design representation is closer to the realistic market dispatch. Reference [35] proposes a robust chance-constrained framework for optimal EH management in the presence of uncertainties, where the time resolutions of 60 min, 30 min, 15 min, and 1 min have been simulated to present a close-to-real operation model. Additionally, local electricity markets are presented as a mechanism to deal with uncertainty in prices, demand, and renewable production, which usually use an intraday timescale and only a few combined multiple timeframes [99]. Reference [130] optimizes an aggregator's portfolio by considering the technical constraints of the networks (electricity, gas, heat) and it enables trading in day-ahead electricity, natural gas, green hydrogen, and carbon markets, considering their intraday activation of the bidding as well.

Figure 4 illustrates the common timeline for energy scheduling of the wholesale electricity markets: day-ahead and intraday, including temporal horizon and discretization.

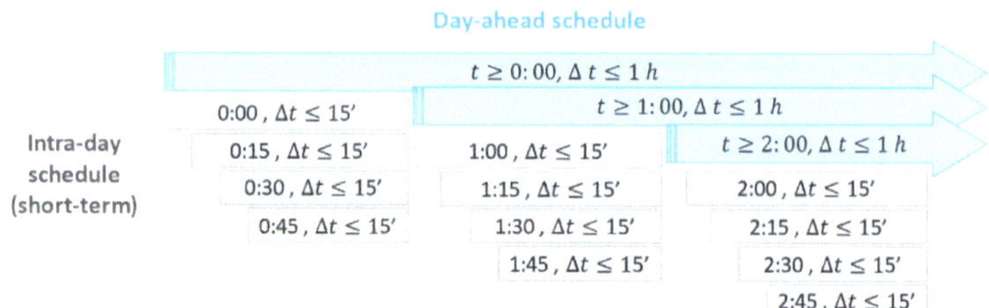


**Figure 4.** Day-ahead and intraday schedule.

Apart from the wholesale electricity markets, other ancillary services should be considered as well in favor of network operators, such as grid congestion, reserve, balancing markets, frequency control, or voltage regulation. In this regard, reference [99] reports different TSO-DSO coordination schemes to solve balancing problems and grid congestion at the local level. Nevertheless, ancillary or grid services are hardly considered in the literature under the scope of EHs [49,65,131]. In [131], the participation of a distributed multi-energy system in providing ancillary services for the reserve market is assessed as a function of the economic parameters (availability fee and exercise fee), while maintaining the end-use energy demand at a constant level and thus without affecting the consumers' comfort. The availability and exercise fees are established by the market operator rules. A P2G system is proposed in [36] to increase system flexibility, contribute to the supply of the network load in contingent events, and prevent the loss of the excessive power generated by the wind turbine. In [65], the network operator is in charge of operating the controllable energy resources to perform real-time grid management while maintaining the grid's stability after the P2P and day-ahead scheduling.

Additionally, DR mechanisms [37] (via incentives, mainly via time-of-use (TOU) tariffs [25,58,70]) are mostly included to provide flexible EH operation. However, these studies do not model the DR mechanisms from the market or network operation point of view.

The DR is addressed in the literature as follows: electrical and thermal demand response programs for shiftable loads are included in [19], the day-ahead and real-time demand-based prices during peak periods are considered in [103], shifting loads from high-price times to low-price times to reduce operational costs are modeled in [88], shiftable and transferrable loads are considered during peak demand to the off-peak periods in [71], deferrable and critical loads are also included in [26], electrical V2G flexibility from PHEVs is provided in the wholesale market in [54], fast ramping capabilities (fast ramp up- and downtime) are provided by PEM electrolyzer in [34], and the available electrical flexibility of DERs is modeled in [55]. TOU tariffs to end-users and white certificates (WC), derived from the Italian incentive scheme, are included in [93] for CHPs oriented to primary energy saving. Additionally, in [58], the TOU prices and the transmission and distribution costs are distributed among end-users. In [120], electricity-load shifting, and flexible electrical heating/cooling supply are included, and it demonstrates the effectiveness of the planning and design method to show the influence of nodal energy prices. Following another approach, Ref. [29] considers the EH participating in DR programs by changing its behavior in response to a TOU tariff. Reference [119] incorporates responsive thermal and electrical loads, and their effects on EH planning (optimal configuration and sizing) are investigated. Furthermore, Ref. [82] addresses TOU prices to end-users as an incentive scheme to reduce their operational costs by shifting load from high- to low-price times, while in [78], both electrical and thermal loads enhance the flexibility of demand-side load management based on users' comfort. In [27], emergency measures such as power curtailment and load shedding have to be taken for real-time adjustment. Other commodities may be considered as well, such as gas trading [19,23,25,27,28,30,31,33,35,36,39,40,44,49,54,55,57,58,70,72,74, 78,79,82–85,88,93,97,104,116,119,120,123,125]. However, the upstream gas network is only modeled by means of a gas price, not as an internal or external gas network.

In contrast, the downstream electricity and gas sub-networks inside or between the EHs are presented in [23,36,49,96,97,116,123], in view of the growing influence of RES in power systems and the mutual effects of the gas and electricity networks. For example, several EHs connected to a 33-bus radial distribution network and a 14-node gas network are considered in [36], where the internal system operator defines upper transaction limits for each EH in its day-ahead operation. Afterward, each EH optimizes its own operation in the wholesale markets. P2G is also studied in [49,116], and incorporating hydrogen into the gas network is studied in [23]. In [44], thermal needs are exchanged between different local energy communities through a common thermal network, while hot water exchange and gas tariff schemes are included in [70].

Table 19 summarizes the literature and the interaction considered with the external entities.

**Table 19.** Interaction of EHs with multiple markets in the literature and information available.

| Reference | DAM | IDM | Ancillary (Grid) Service | Gas/Fuel Trading | Demand Response | Other Markets or Resources |
|---|---|---|---|---|---|---|
| [63,69,86,94,114] | Price-based | | | Thermal needs | | Biogas [94] Hydrogen [69] Solar thermal [63] |
| [32,34,71,103] | Price-based | | | Thermal needs | DR, TOU (all) Fast ramp [34] | Hydrogen [32,34] |
| [28,31,33,35,39,40,44,50,72, 74,79,82–85,88,104,125] | Price-based | Price-based [35,50,84] | | Gas price-based | | Water [88] Biomass [39] |
| [19,26,27,29,54,55,57,58,70, 78,93,119,120] | Price-based | Price-based [27] | | Gas price-based | DR, TOU (all) V2G [55] Shedding [27] | |
| [25] | Network | | | DHN network | DR, TOU | |
| [30,49,96,97] | Network | | | Gas network | | |
| [23] | Network | | | Gas network | Shedding | Hydrogen |
| [36] | Network | | Contingency | Gas network | | |
| [116,123] | Network | Price-based | Reserve | Gas network | | |
| [66] | Price-based | Price-based | Grid fee | Gas price-based | | |
| [65] | Price-based | Price-based | System operator Control Reserve | | | |
| [131] | Price-based | | | | | |
| [129,130] | Price-based | Price-based | Reserve | Network and gas market | - | Green hydrogen and carbon markets |

*11.2. P2P Markets*

In the literature, three main P2P architectures [48,53,99] are proposed in the context of EHs according to how the decisions of the energy trading process are taken and their communication characteristics.

In a centralized P2P architecture, a central coordinator dictates the energy transactions and their prices among the peers in the community. The decision-making process is generally based on an optimization algorithm aimed at maximizing the overall benefit of the community. Peers only communicate with the coordinator, as seen in Figure 5. Once the transactions have been made effective, the central coordinator is in charge of distributing the obtained revenues among the participants [36,37]. The coordinator also acts as an intermediary between the community and the rest of the system. For instance, Ref. [52] proposes a novel centralized P2P energy trading model named operator-oriented P2P trading to lower the barriers to entering into the transaction while accommodating various customers. The operator decides the marginal/trading price and trading schedules that are defined with the objective of maximizing the social welfare given by the sum of the profits of all participants. In [65], another centralized P2P architecture is proposed for the real-time operation of the distribution network where the network operator has direct control over the DERs of crowdsources. The objective is to minimize the generator's cost function in addition to the thermal losses and crowdsources' disutility function designed to compensate for the inconvenience caused by rescheduling shapeable load. Reference [90] defines a centralized system that searches for optimization of the electricity and gas costs of a neighborhood, including technologies such as CHPs, HPs, heating, ventilation and air-conditioning (HVACs), EVs, and RES.

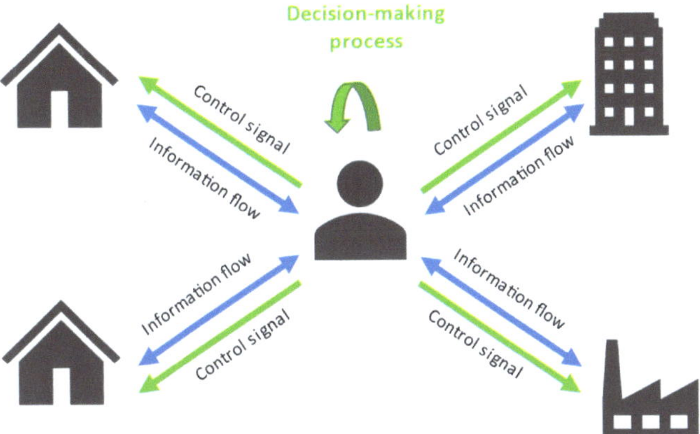



**Figure 5.** Centralized P2P architecture.

In a decentralized P2P architecture, peers can directly communicate with each other and negotiate energy transactions among themselves without the involvement of a central coordinator. The decision-making process for defining the amount of energy exchanged and the price is taken by each peer, generally based on an optimization algorithm to minimize its individual electricity costs or maximize its revenues, as shown in Figure 6. In addition to the centralized architecture described above, Reference [65] also proposes a solution based on a decentralized architecture for the day-ahead operation in which direct energy trading between crowdsources is carried out. These energy transactions are included within the optimization algorithm of the network operator as constraints. Reference [51] defines a decentralized architecture where customers can trade and exchange energy with other customers. All customers can communicate with each other using a two-way communication system. The optimization is based on two stages. The first stage allows a decision on whether a customer should participate or not in the P2P energy-sharing trading system based on the maximization of social welfare. The second stage allows for maximizing the payoff of the involved customers. Reference [91] considers a fully decentralized P2P system where smart energy buildings can share energy among themselves without the need to have a coordinator. A two-stage P2P energy-sharing strategy for a building cluster is developed. In the first stage, the optimal energy-sharing profiles of the buildings aim to minimize the total social energy cost. In the second stage, the clearing prices are calculated based on mutual energy sharing via a non-cooperative game.

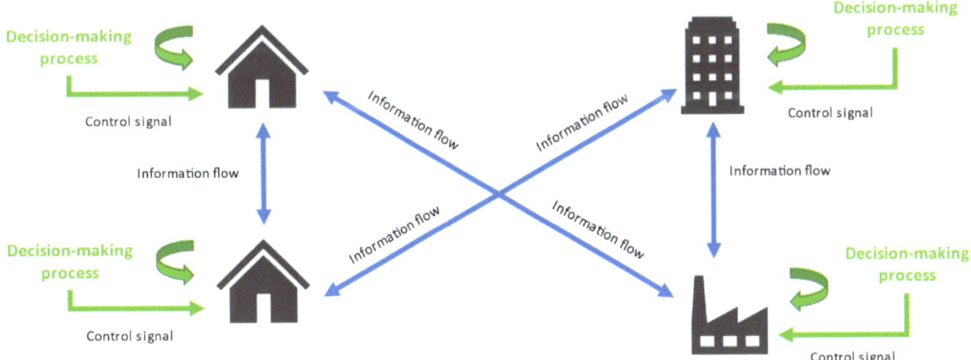



**Figure 6.** Decentralized P2P architecture.

In a distributed P2P architecture, a coordinator communicates with each peer and manages a local energy trading market set for the transactions among them. However, in contrast to a centralized P2P architecture, the coordinator does not have control over the amount of energy exchanged. It just tries to influence participating peers by sending suitable price signals. Each peer carries out the optimization process according to the received internal P2P market signal sent by the coordinator to minimize its individual electricity costs or maximize its individual revenues within the P2P energy-sharing system. It represents a hybrid approach between centralized and decentralized P2P architectures, as shown in Figure 7. For example, Ref. [81] proposes a distributed architecture where the operator only provides a local market platform with necessary functions, in which all the prosumers trade or share energy with each other in order to maximize their own benefits individually. Hence, prosumers have full authority over their distributed energy resources (DERs), eliminating the need for any supplementary incentives to encourage prosumer participation. The article proposes three distinct pricing approaches for determining the P2P market's internal trading price: the supply and demand ratio (SDR), mid-market rate (MMR), and bill sharing (BS).

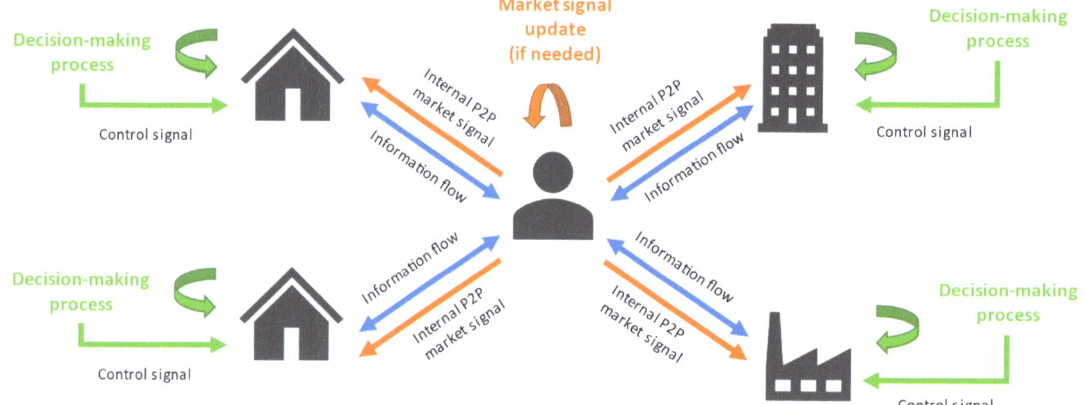


**Figure 7.** Distributed P2P architecture.

In [92], a P2P architecture is proposed, consisting of two types of agents: (i) P2P PV prosumers who trade energy among themselves, and (ii) an energy-sharing provider (ESP) responsible for coordinating the sharing activities. The ESP interacts with the utility grid by purchasing the required electricity or selling excess power production of the energy-sharing zone using a dynamic internal pricing model within the EH. In [56], a P2P distributed architecture is defined, utilizing blockchain technology to implement transactions. The transaction process involves four stages: (i) customer demand forecasting, (ii) initiation of the transaction, (iii) security check, and (iv) trading execution. In [57], a distributed P2P architecture is considered, where EHs participate in a local market managed by the LMO. EHs also have access to district markets (electricity utility, gas utility, district heating, district cooling) where they can trade various forms of energy. However, EHs prioritize exchanging the maximum energy in local markets due to more beneficial market clearing prices for both buyers and sellers.

Some other studies focused on the P2P electricity market and its pricing structure in which peers are able to sell their excess production and trade for a better electricity price than the one offered by the retailer, as seen in [41–44,54,58,62,93]. Table 20 shows the key factors among the three P2P architectures.

**Table 20.** Comparison of P2P architectures in the EH context.

| Centralized Architecture | Decentralized Architecture | Distributed Architecture |
| --- | --- | --- |
| Provides direct setpoints to which the consumption must adjust to; Allows for a more straightforward network operation; Most intrusive method; Allows for a coordinated response; More complexity in the optimization algorithm. It can imply a very high computational burden. | No supervisory figure in the energy exchange; Minimal exchange of information; Not able to perform coordinated actions for external actors. | Allows a certain degree of influence on consumer patterns; Cannot establish a specific setpoint; Less demand for communication infrastructure; It requires the definition of a suitable pricing mechanism by the coordinator to manage the internal energy trading market; Slow convergence of the P2P market algorithm to reach an agreement about the energy transactions may occur. |

A centralized P2P architecture offers several advantages, such as maximizing the total welfare of the community and providing better support for grid operator services. However, it may have higher management complexity and computational burden, particularly when many peers are involved. Additionally, confidentiality concerns may arise as prosumers need to share all information with the central coordinator. On the other hand, a decentralized P2P architecture enables prosumers to have complete control over their DERs and make decisions at the prosumer level. This architecture is also highly scalable. In contrast, as there is no coordinator, the overall welfare of the community is less optimized. Distributed P2P architecture shares the advantages of the previous two architectures. There exists a coordinator that, through sending suitable price signals, influences the participation of the prosumers while the decision-making process is still maintained at the prosumer level. The amount of information to be exchanged with the coordinator is less than in a centralized architecture, so prosumers can better maintain their privacy. It is very scalable and more compatible with the existing regulatory framework [48,53,99]. This architecture requires the definition of a suitable pricing mechanism by the coordinator to manage the internal energy trading market. Different approaches can be applied for the calculation of internal prices, such as (i) SDR, where the internal trading price is defined as a piecewise function [81,92]; (ii) MMR, where the internal trading price is set as the average of retail and export prices [81]; (iii) BS, where a pro-rata cost-sharing mechanism is defined in which the income and cost of each participant are proportional to its electricity production and consumption throughout the time horizon [81]; and (iv) auction-based pricing strategy, where EHs independently decide on the amount and price of energy to be traded in the local markets and express their interest in local trading through submitting their offers/bids to the coordinator [57,99].

*11.3. Key Challenges for the Interaction of EHs with Multiple MARKETS and Networks*

Several limitations and challenges are identified regarding the interaction of EHs with multiple markets. Firstly, although there are studies in the literature that consider electrical and gas networks, their approach does not elaborate and implement extensive market rules and constraints. In most cases, electrical and gas networks are referred to as internal EH or between EHs networks, without considering the upstream utility grids and associated market model. Therefore, there is a lack of studies that integrate grid distribution models and complex market mechanisms between several EHs.

Secondly, although most models consider electricity markets to purchase or sell electricity, authors use historical day-ahead electricity prices to model the electricity market in its simplest version. Additionally, few published articles consider electricity markets closer to operation, such as IDM or real-time markets, needed to adjust EH supply and demand under intrinsic uncertainty. In conclusion, a clear market framework is not presented (i.e., the framework of the European single market), which should include the main features of market mechanisms such as market bids, closure times, market clearing process, real-time power setpoint, or grid needs. There is a lack of sensible and realistic interaction with energy markets or upward networks. A further necessary step would be the specific-country analysis for a more realistic cost-benefit scenario.

Thirdly, several key challenges have been identified regarding thermal and gas networks. Most research papers usually have common characteristics as follows: (i) well-known natural gas price signal; (ii) natural gas or other fuel is only bought to satisfy the local thermal needs; and (iii) heating and cooling networks (if modeled) are only considered inside the EHs, and a central control is implemented to purchase and sell electricity in the utility grid. Regarding (i), in fact, only natural gas price and thermal load are considered in the optimized operation of the EHs in most papers, but without any gas market structure, gas transactions, gas scarcity, or shortage. According to (ii), gas or thermal energy is purchased according to internal EH needs. Moreover, hydrogen as an energy carrier is not commonly included in the EHs. Nowadays, a bidirectional electricity carrier is most common to purchase renewable generation, whereas bidirectional flows should be enabled for all energy carriers. Regarding (iii), thermal needs are mostly satisfied with internal heating and cooling networks fed by natural gas or other fossil fuel, but existing thermal or gas infrastructure downstream of the EH is not modeled.

Fourthly, DR is widely addressed, although it should be oriented more toward the upstream grid needs. In contrast, ancillary grid services are hardly considered in the literature under the scope of EHs to support the upstream utility network. After the consulted literature, the reserve market established by the market operator rules is conducted in [131] and to support contingent events in [36]. Few other research papers solve grid problems inside the EH without any market design, i.e., reserve availability [116,123] provided by a conventional generator to manage uncertainties inside EHs, but not to provide valuable services to the upstream grid. In conclusion, further research is required on how EHs may interact with multiple energy markets, assuming knowledge of grid network constraints, considering the market design, and evaluating mutual coupling with other energy vectors and markets.

Based on the reviewed literature on P2P, it is concluded that the establishment of a P2P energy-sharing mechanism benefits both prosumers, which can reduce their electricity bill and allow them to benefit from cleaner energy, and the system, which can obtain new sources of flexibility contributing to reducing peak load and improving system security through the participation in ancillary service markets [53]. However, there are several challenges that should be tackled to implement a P2P system efficiently.

The adoption of feasible and effective optimization algorithms and pricing schemes within the P2P system to automatize energy transactions is challenging. In the centralized architecture, there might be scalability problems with the optimization algorithm if many peers are involved because it could require a high computational effort to reach the optimal solution. In the distributed one, the proper P2P pricing mechanism should be adopted to incentivize prosumers' participation and help to maximize the overall welfare of the community. Several pricing schemes proposed in the literature can show convergence problems, such as SDR, MMR, and BS. These are based on iterative bidding processes in which the bids provided by the prosumers change in response to the dynamic internal price. It is possible that the energy bids and internal price values do not converge to a fixed point after a finite number of iterations. Therefore, it is necessary to implement mechanisms that ensure the convergence of these algorithms in the required time. It is essential that the implemented optimization algorithms include models that accurately simulate the behavior of the flexible resources and, therefore, can identify the available flexibility under different control actions in an accurate way. This represents a challenge since the flexibility depends on many factors, several of them subject to a high degree of uncertainty (e.g., end-users' behavior, weather conditions, etc.).

## 12. Business Models of EHs

From the reviewed works, it emerges that business models are mostly based on the different objectives of the systems under study. Many works focus on a single objective, usually consisting of the minimization of the cost of the energy carriers involved or of the total investment or operating cost of the systems. Among others, Ref. [19] aims to reduce the total cost of an EH by deciding on the electrical and thermal load dispatch, for electricity, heat generation, and storage units, to meet a specific electricity and heat demand at a minimum cost. The EH is allowed for such a purpose as to purchase/sell electricity from/to the upper grid. In [56], the business model aims to minimize the total energy expenditure of all individual customers in the microgrid. Similarly, in [65], the main objective is to minimize electricity costs by employing local resources from the community energy system in a way that does not violate any network constraints. In [75], the model employs a transactive energy scheme as a business model. Each building tends to charge more for the energy supplies to other buildings and pay less for the energy demands from other buildings; therefore, the clearing prices should be equilibrium prices such that each building will pay/charge for the energy demands/supplies. The objective is to minimize the daily operation cost, considered to be composed of three parts: net electricity purchasing cost, gas purchasing cost, and carbon emission cost. In [105,108], the business model is focused on the reduction of the total operating cost, while in [111], it is based on the reduction of the expected time-ahead energy costs.

Many works deal with two or more objectives, considering economic objectives associated with the minimization of the cost of emissions, cost of power loss, and $CO_2$ emissions. For instance, in [70], the business model is focused on the minimization of energy cost with the minimization of environmental impact, which is assessed in terms of $CO_2$ emissions, while in [97,109,110], the business model is focused on reducing the total cost of generation and environmental emissions.

Relating to the literature on energy markets and P2P schemes, in [48], three business models are discussed: C2C (consumer-to-consumer), on which a distributed P2P scheme would be based; B2C (business-to-consumer), which could portray a traditional scheme in which the consumer buys electricity from the retailer; and B2B (business-to-business), which would imply an exchange between different commercial entities. Regarding P2P energy trading, in [52], the business model is focused on the maximization of profits (P2P energy trading). In [57], the EHs have access to district markets, in which they can trade various forms of energy. Additionally, they have access to a local market, which allows the EHs to trade energy among themselves in a P2P market. EHs first try to exchange the maximum energy in the local markets because market clearing prices are assumed to be more beneficial for both seller and buyer EHs than the prices of district markets. As noted in [99], business models for energy sharing via local electricity markets are still very rarely put into commercial practice. This work presents the existing barriers in current regulatory frameworks focusing on the Portuguese energy market after analyzing the P2P energy-sharing business model of the project Community S (S stands for sharing, solar, storage, sustainable, and smart).

Table 21 summarizes the business models for all analyzed works.

**Table 21.** Business model per reference.

| Reference | Business Model | Objective |
|---|---|---|
| [19] | B2C, C2C | Reduce the total cost of the EH |
| [46,105,108] | B2C | Minimize operational costs |
| [48,57,87] | B2C, C2C | P2P exchange |
| [52] | B2C | Maximize profits (P2P energy trading) |
| [56] | B2C, C2C | Minimize the total energy expenditure |
| [60] | C2C | P2P exchange |
| [62] | B2C | Minimize costs and emissions |
| [65] | B2C, C2C | Minimize electricity costs |
| [67] | B2C, C2C | P2P exchange, minimize energy costs |
| [70] | B2C, C2C | Minimize energy costs and emissions |
| [75] | C2C | Minimize the daily operation cost |
| [90] | B2B, B2C | Trade between EH, minimize costs and emissions |
| [93] | B2C | Maximize the operator's profit and reduce the $CO_2$ emissions |
| [97,109,110] | B2C | Minimize costs and emissions |
| [99] | B2C, C2C | P2P energy sharing |
| [106] | B2C, C2C | Reduction of the fuel cost and emission |
| [111] | B2C | Reduction of the expected time-ahead energy costs |
| [124] | B2C | Minimize costs and emissions (literature review) |
| [125] | B2C | Maximize RES output |

Based on the reviewed papers, the business models of EHs have as a basis the reduction of the energy cost and/or of the total investment and operating costs of the EHs. In some papers, the reduction of the carbon emission cost is taken into consideration as well. In most cases, business models are limited to the electricity carrier, some include gas or heat, and they rarely focus on multi-carrier technical aspects such as the inclusion of additional energy carriers.

Relating to the literature on energy markets and P2P schemes, it emerges that business models are mostly limited to a single energy carrier, which is electricity, or do not go in-depth into the specificities of other energy markets. Therefore, the main challenge is to provide innovative business models that could reflect the different energy carriers present in the EHs and the synergies among them. This means fully exploiting the potential to participate in different markets, considering their specificities.

## 13. Other Collateral Concerns

### 13.1. Temporal Scope for the Operation Optimization of EHs

This section discusses the time horizons and time resolution that the analyzed works in the literature consider for the operation optimization of the EHs. The most commonly used time horizon among the papers considered in this study is either one year or one day, depending on the scope of the optimization. Many documents use one year [31,38,41,44,47–49,58,62,63,68–72,85,86,89,93,95,104,113,115,125] and focus on long-term energy system planning. However, this is not always performed in the same way. Indeed, a complete year can be too computationally challenging, and some models reduce the complexity by representing a year with representative days. The number of representative days varies depending on the model; references [41,44,71] simplify the annual scope by the use of one representative day per season, while [62] uses one week per month to study the whole year. Reference [68] uses three days for each month in a year corresponding to weekday, weekend, and peak, while [70,85,101] use only three typical days in total. Models with a stronger operational focus use more detailed descriptions and may choose only to use a horizon of one day, such as [19,23–25,28–30,33,34,40,42,43,46,48,50,54,55,57,61,74,75,80,83,84,90,92,94,96,98,102,105,114,116,123].

Other models need to account for the future trends in price and demands, e.g., using a longer horizon, five years, in the case of [64,66]. In addition, Ref. [66] compares the use of 4 day-types (summer weekday, summer weekend, winter weekday, and winter weekend) and 39 day-types (weekdays, Saturday, and Sunday for 13 periods for every 4 weeks). Finally, some documents use particular horizons, for example, seven weeks, as in [60].

In addition to the time horizon, the literature adequately describes the time resolution used in the various cases. The most common temporal resolution is one hour, as in [19–21,23,25,26,28–30,36,38,40,42,43,46,47,49,50,54,55,57,60,61,63,67,68,72,74,75,77,79,80,83,90,92,94,98,102,104,105,113,115,123,125]. An hourly resolution allows the capture of some of the variability in the demand and renewable generation and also fits well with the structure of the day-ahead markets. Nevertheless, Refs. [35,66,103] use a half-hourly resolution, as it is the time resolution in the UK's DAM. Even finer resolutions can also be rarely found. For example, Refs. [112,116,122] use 15 min, Refs. [65,91] use a time resolution between 5 and 15 min, Ref. [35] uses a time resolution of 1 min, and Ref. [87] uses a time resolution of half a minute. The greater the granularity, the deeper the level of detail that increases the reliability and accuracy of the model, although this causes higher computational efforts. In fact, some models use a lower resolution as a way to reduce the computational burden; for example, Ref. [101] uses 2 h, Ref. [86] 4 h, and Ref. [85] divides each day into 6 periods which do not have to be the same length.

Table 22 summarizes the different temporal resolutions used in the analyzed works in the literature dealing with the operation optimization of EHs.

**Table 22.** Temporal resolutions used in the analyzed works in the literature dealing with the operation optimization of EHs.

| Temporal Resolution | References |
| --- | --- |
| Half min | [87] |
| One min | [35] |
| Five min | [65,91] |
| Fifteen min | [112,116,122] |
| Half hour | [103] |
| One hour | [19–21,23,25,26,28–30,36,38,40,42,43,46,47,49, 50,54,55,57,60,61,63,67,68,72,74,75,77,79,80,83, 90,92,94,98,102,104,105,113,115,123,125] |
| Two hours | [101] |
| Four hours | [86] |
| Six periods per day | [85] |

*13.2. Spatial Scope for the Design and Operation Optimization of EHs*

The spatial scope varies greatly between models due to different goals, locations, technologies, and sectors (e.g., residential, industrial, commercial sectors) of the EH. In addition, it is not always reported precisely in the analyzed works and can focus on different types of information: for instance, some documents report a geographical location while others provide a physical description of the local system, and others provide no information at all.

Many models focus solely on an individual energy system, be it a single building ([66,125]) or a single area (neighborhood) [72,74,75,97,104,105]. The number of buildings can vary greatly, e.g., 4 in [101], 6 in [85], 25 in [113], and 29 in [47]. A few models focus more on the interaction between different energy systems inside a larger one. For example, in [46], 4 smart EHs are included in the IEEE 33-bus network. Similarly, the model presented in [36] is applied in a 33-bus radial distribution network and a 14-node gas network, 3 smart EHs, and a P2G system. The IEEE 30-bus network is also used in [96] with 5 generators, 4 tap-transformers, and 2 shunt capacitors.

A different approach Is taken in [86,90], where authors consider a grid of interconnected nodes. The microgrid considered in the latter has 21 nodes, 6 of which are EHs, each comprising a natural gas boiler, a CHP unit, an HP, and a heat storage unit. In other nodes of the microgrid, there are three wind generators, two batteries, and electrical loads.

Table 23 summarizes the different spatial scopes used in the reviewed papers found in the literature.

**Table 23.** Spatial scopes used in the analyzed works in the literature dealing with the design and operation optimization of EHs.

| Spatial Scope | References |
| --- | --- |
| Single building | [66,125] |
| Single area | [72,74,75,97,104,105] |
| Interaction between different energy systems inside a larger one | [36,46,47,85,96,101,113] |
| Grid of interconnected nodes | [30,86] |

## 14. Discussion and Conclusions

*14.1. Summary of Challenges and Limitations*

This review paper presents the optimal design and operation of EHs and their interaction with external networks and multiple markets, tackling all aspects of this multi-faceted theme.

Based on the reviewed literature, there are some important limitations and challenges related to the EH configuration that need to be addressed in order to optimize the planning and operation of such systems in the future, especially considering the RePowerEU plan objectives. Past research papers focus mainly on the electricity carrier as a backbone while limited other carriers and conversion technologies are considered in single studies. The optimal placement of the different energy technologies in the EH is an important aspect of the design optimization of EHs, especially in the case of thermal storage, whose location in a DHN can impact thermal losses and, therefore, energy consumption and related costs and emissions. However, in many cases, it is not addressed.

Also, based on the literature findings, the MILP approach is the most widely used to optimize the design and operation of multi-carrier energy systems. Nevertheless, some of the different variables that may be present in EHs respond to a nonlinear behavior, e.g., power flow variables, and to non-convexities, e.g., commitment status of energy conversion and storage units. The accurate representation of the physical phenomena taking place in the energy conversion and storage units of a multi-carrier energy system and of their design and operational decisions may require the use of MINLP formulations. Of course, the solution to MINLP problems is still today a challenging and computationally hard task, especially when the configuration is complex, as said. Usually, MINLP problems are relaxed and reformulated as MIQCP, MISOCP, or MILP problems and solved using proper decomposition techniques. However, very detailed models of energy technologies can limit the effectiveness of the optimization models and methods used to solve the problem, and a good compromise should be found between model fidelity and the complexity of the optimization process. As for the use of multi-objective approaches aimed at finding good compromise solutions that satisfy the needs of the different stakeholders taking part in the decision-making process of EHs, an ideal approach should be used where the optimization is performed before the decision-making. In this way, the final decision to select the trade-off between conflicting objectives on the Pareto front is left to the decision-maker.

Regarding the EHs' interaction with the markets, the most important limitation found in the literature is that ancillary grid services are hardly considered under the scope of EHs to support the upstream utility network. The day-ahead wholesale market is the most usual, but only a few papers consider intraday or real-time scheduling. Therefore, there is a lack of sensible and realistic interaction with markets or upward utility networks beyond EHs. A second limitation found in the studied works is the lack of a real electricity market setting in the EHs' interaction. Generally, when selling/purchasing electricity in the day-ahead electricity spot market, the EH or a proxy acting on EH's behalf, e.g., a market agent, must submit selling/buying bids before the market gate closure (usually the day before the electricity delivery/withdrawal). In between the gate closure time of the day-ahead electricity spot market and the time of delivery of the energy cleared in such a market, several intraday electricity markets are celebrated where the market agent representing the EH can submit selling or buying bids to modify the generation schedule of the day-ahead electricity spot market or the previous intraday market. This setting is mainly disregarded by the literature so far, whereas it has been partly addressed by the latest [130] where day-ahead and real-time optimization strategies have been combined for an aggregator representing multi-energy assets. Of course, more actors under different realistic business models shall be investigated along with different market settings, e.g., balancing, P2P, etc. The third limitation found in the revised works is analogous to the second one but refers to the natural gas market setting. Natural gas supply and consumption are also negotiated in day-ahead and intraday markets or sessions. The time horizon and temporal resolution of the day-ahead and intraday gas market sessions are not the same as those of the day-ahead and intraday electricity markets. Gas imbalances are also computed in the gas market. The gas sold/purchased in each market or market session in which the agent has participated is settled at the market clearing price (usually marginal), whereas the excess/deficit gas imbalances are cleared at the imbalance price (usually different for excess and deficit imbalances). Again, reference [130] considers not only electricity markets, but also the day-ahead gas market through multi-energy biddings. Still, the gas intraday is not considered, and thus none of the revised papers have considered the entire process so far. All papers—where the gas price is considered—use a single gas price, presumably corresponding to the day-ahead gas market price. The uncertainty of gas prices is not considered in any of the revised papers.

Finally, regarding the P2P markets, there are fruitful studies in the literature on the maximization of social welfare and increase in individual profits, which allow users to negotiate the excess electricity that is being produced as well as to choose who they are going to buy it from. Still, the interaction between users with other outside actors is either reduced to the minimum that is needed or nonexistent.

### 14.2. Research Pathways

Overall, EH is a promising concept that could serve the energy transition under the cross-sectoral view that is needed. Based on the findings, there is a need for studying realistic approaches, e.g., proper configuration, market interaction, actors' interaction, etc., so that EHs potential can be unfolded. In detail:

- Different types of EH configurations with a wide range of conversion technologies that compile general solutions and can be replicable and scalable should be considered. Energy storage and its flexibility potential are of high importance, and EHs' future configuration models need to consider the optimization of their sizing and placement, including potential alternative means of storage such as EVs and hydrogen. On top of that, as the interaction of different energy carriers affects the nonlinearity and nonconvexity of the problem, a more complex configuration shall lead to higher sensitivity of initial conditions that affect the problem formulation and solution. This challenge should be faced as the integrated grid approach needs the high interconnection of many different carriers;

- The optimization of the design and operation of an EH is a complicated task that has been considered in the past under different prisms as already analyzed. It is also well established that in order to address the needs of the most related actors in an EH, the multi-objective approach is the way forward. What is the most challenging so far is the accurate representation of physical phenomena that would add complexity to an already multi-faceted problem. Therefore, it seems that the optimization of design and operation should be tackled in layers and in a distributed way with loops of interaction that would allow the different layers to be in accordance. The layers could address physical carriers and/or hierarchical layers of governance where the complexity is built based on the pursuit of the actors involved. As an example, distributed optimization could be dealt with within the EH on the prosumers' side while being in good collaboration with the central optimization at the central level of the ILEC. An example of this innovative approach is tested under the approach proposed in the eNeuron H2020 project (November 2020–October 2024, ID: 957779), which has the main goal to develop an innovative toolbox for the optimal design and operation of local energy communities, integrating DERs and multiple energy carriers at different scales;

- Last but not least, regarding the EHs' interaction with the markets, a further necessary step would be the specific-country cost–benefit analysis feeding sustainable business models. Moreover, further research is required on how EHs may interact with multiple energy markets, assuming knowledge of grid network constraints, considering the market design, and evaluating mutual coupling with other energy carriers and markets, including gas markets. Especially for P2P markets, when considering users' interaction, local but centralized resources, as well as different energy carriers, should be involved. A good baseline for further investigation under the EH scope could be the consideration of the electricity–carbon integrated P2P market as presented in [132].

**Author Contributions:** Conceptualization, C.P., M.D.S. and G.G.; methodology, C.P.; validation, C.P., M.D.S. and G.G.; formal analysis, C.P., M.D.S., C.C., M.C., V.P., A.F.C.B., A.G.-G., N.R. and G.G.; writing—original draft preparation, C.P., M.D.S., C.C., M.C., V.P., A.F.C.B., A.G.-G., N.R. and G.G.; writing—review and editing, C.P., M.D.S., C.C., M.C., V.P., A.F.C.B., A.G.-G., N.R. and G.G.;

supervision, C.P., M.D.S. and G.G.; funding acquisition, C.P., M.D.S. and G.G. All authors have read and agreed to the published version of the manuscript.

**Funding:** This work has been carried out within the eNeuron project that has received funding from the European Union's Horizon 2020 Research and Innovation Programme under Grant Agreement No. 957779.

**Data Availability Statement:** Not applicable.

**Conflicts of Interest:** The authors declare no conflict of interest.

## Abbreviation

| Acronym | Meaning | Acronym | Meaning |
| --- | --- | --- | --- |
| AC | Air conditioners | LMO | Local market operator |
| ADMM | Alternating direction method of multipliers | LP | Linear programming |
| B2B | Business-to-business | MAGA | Multi-agent genetic algorithm |
| B2C | Business-to-consumer | MBFO | Modified bacterial foraging optimization |
| BS | Bill sharing | MECS | Multi-energy coupled systems |
| C2C | Consumer-to-consumer | MES | Multi-energy system |
| CAES | Compressed air energy storage | MGEM | Microgrid energy management |
| CD | Crowding distance | MILP | Mixed-integer linear programming |
| CES | Community energy storage | MINLP | Mixed-integer nonlinear programming |
| CHP | Combined heat and power | MIQCP | Mixed-integer quadratically constrained |
| DAM | Day-ahead electricity market | MISOCP | Mixed-integer second-order cone programming |
| DDRO | Data-driven distributionally robust optimization | MMR | Mid-market rate |
| DE | Differential evolution | MOEA | Multi-objective evolutionary algorithms |
| DER | Distributed energy resource | MO-MFEA-II | Multi-objective multifactorial evolutionary algorithm II |
| DHN | District heating network | MPC | Model predictive control |
| DRO | Distributionally robust optimization | MSGA-II | Multi-strategy gravitational search algorithm |
| DRP | Demand response programs | MTLBO | Modified teaching–learning-based optimization algorithm |
| EA | Evolutionary algorithms | NR | Non-dominated rank |
| EH | Energy hub | NSGA | Non-dominated sorting genetic algorithm |
| EHGHS | Electricity–hydrogen–gas heat integrated energy system | nZED | Net-zero energy districts |
| ESP | Energy-sharing provider | O&M | Operation and maintenance |
| ESS | Energy storage system | OEF | Optimal energy flow |
| ETIP SNET | European Technology & Innovation Platforms Smart Networks for Energy Transition | P2G | Power-to-gas |
| EU | European Union | P2P | Peer-to-peer |
| EV | Electric vehicle | PCC | Point Of common coupling |
| GA | Genetic algorithms | PCM | Phase change materials |
| GfG | Gas-fired generation | PHEV | Plug-in hybrid electric vehicle |
| GHG | Greenhouse gas | PRIMES | Price-induced narket equilibrium system |

| Acronym | Meaning | Acronym | Meaning |
|---------|---------|---------|---------|
| GIL | Grid integration level | PSO | Particle swarm optimization |
| HER | Heat-to-electricity ratio | PV | Photovoltaic |
| HES | Hybrid energy system | QPSO | Quantum particle swarm optimization |
| HP | Heat pump | RCGA | Real coded genetic algorithm |
| HVAC | Heating, ventilation, and air-conditioning | RES | Renewable energy sources |
| ICC | Integer cut constraints | RMILP | Robust MILP |
| IDM | Intraday market | SDR | Supply and demand ratio |
|  |  | TOU | Time of use |
| IGDT | Information gap decision theory | V2G | Vehicle-to-grid |
| ILECs | Integrated local energy communities | VIKOR | Vlsekriterijumska optimizacija i kompromisno resenj multi-criteria decision-making method |
| KPI | Key performance indicator | VRE | Variable renewable energy |
| LAES | Liquid air energy storage | WC | White certificates |
| LEC | Local energy community | ZCMES | Zero-carbon multi-energy system |


## References

1. Available online: https://ec.europa.eu/clima/policies/strategies/2030_en (accessed on 10 April 2022).
2. Di Somma, M.; Graditi, G. Challenges and opportunities of the energy transition and the added value of energy systems integration. In *Technologies for Integrated Energy Systems and Networks*; Wiley: Hoboken, NJ, USA, 2022; pp. 1–14. [CrossRef]
3. Rinaldi, R.; Losa, I.; De Nigris, M.; Prata, R.; Albu, M.; Kulmala, A.; Samovich, N.; Iliceto, A.; Amann, G.; Pastor, R.; et al. ETIP SNET VISION 2050 Integrating Smart Networks for the Energy Transition: Serving Society and Protecting the Environment. In Proceedings of the CIRED 2019, Madrid, Spain, 3–6 June 2019; p. 175. [CrossRef]
4. Di Somma, M.; Graditi, G. Integrated Energy Systems: The Engine for Energy Transition. In *Technologies for Integrated Energy Systems and Networks*; Wiley: Hoboken, NJ, USA, 2022; pp. 15–40. [CrossRef]
5. Available online: https://ec.europa.eu/clima/policies/strategies/analysis/models_en (accessed on 6 February 2023).
6. Jägemann, C.; Fürsch, M.; Hagspiel, S.; Nagl, S. Decarbonizing Europe's power sector by 2050—Analyzing the economic implications of alternative decarbonization pathways. *Energy Econ.* **2013**, *40*, 622–636. [CrossRef]
7. Available online: https://ec.europa.eu/energy/en/news/commission-proposes-new-rules-consumer-centred-cleanenergy-transition (accessed on 4 May 2023).
8. Available online: https://ec.europa.eu/info/strategy/priorities-2019-2024/european-green-deal/repowereu-affordable-secure-and-sustainable-energy-europe_en#documents (accessed on 14 June 2022).
9. Geidl, M.; Koeppel, G.; Favre-Perod, P.; Klockl, B.; Andersson, G.; Frohlich, K. Energy hubs for the future. *IEEE Power Energy Mag.* **2007**, *5*, 24–30. [CrossRef]
10. Comodi, G.; Spinaci, G.; Di Somma, M.; Graditi, G. Transition Potential of Local Energy Communities. In *Technologies for Integrated Energy Systems and Networks*; Wiley: Hoboken, NJ, USA, 2022; pp. 275–304. [CrossRef]
11. Mohammadi, M.; Noorollahi, Y.; Mohammadi-Ivatloo, B.; Yousefi, H. Energy hub: From a model to a concept—A review. *Renew. Sustain. Energy Rev.* **2017**, *80*, 1512–1527. [CrossRef]
12. Guelpa, E.; Bischi, A.; Verda, V.; Chertkov, M.; Lund, H. Towards future infrastructures for sustainable multi-energy systems: A review. *Energy* **2019**, *184*, 2–21. [CrossRef]
13. Mancarella, P. MES (multi-energy systems): An overview of concepts and evaluation models. *Energy* **2014**, *65*, 1–17. [CrossRef]
14. Mohammadi-Ivatloo, B.; Jabari, F. *Jabari, Operation, Planning, and Analysis of Energy Storage Systems in Smart Energy Hubs*; Springer International Publishing: Manhattan, NY, USA, 2018.
15. Papadimitriou, C.; Anastasiadis, A.; Psomopoulos, C.; Vokas, G. Demand response schemes in energy hubs: A comparison study. *Energy Procedia* **2019**, *157*, 939–944. [CrossRef]
16. Lasemi, M.A.; Arabkoohsar, A.; Hajizadeh, A.; Mohammadi-Ivatloo, B. A comprehensive review on optimization challenges of smart energy hubs under uncertainty factors. *Renew. Sustain. Energy Rev.* **2022**, *160*, 112320. [CrossRef]
17. Mohammadi, M.; Noorollahi, Y.; Mohammadi-Ivatloo, B.; Hosseinzadeh, M.; Yousefi, H.; Khorasani, S.T. Optimal management of energy hubs and smart energy hubs—A review. *Renew. Sustain. Energy Rev.* **2018**, *89*, 33–50. [CrossRef]
18. Akhtari, M.R.; Baneshi, M. Techno-economic assessment and optimization of a hybrid renewable co-supply of electricity, heat and hydrogen system to enhance performance by recovering excess electricity for a large energy consumer. *Energy Convers. Manag.* **2019**, *188*, 131–141. [CrossRef]
19. Lu, X.; Liu, Z.; Ma, L.; Wang, L.; Zhou, K.; Feng, N. A robust optimization approach for optimal load dispatch of community energy hub. *Appl. Energy* **2020**, *259*, 114195. [CrossRef]

20. Petkov, I.; Gabrielli, P.; Spokaite, M. The impact of urban district composition on storage technology reliance: Trade-offs between thermal storage, batteries, and power-to-hydrogen. *Energy* **2021**, *224*, 120102. [CrossRef]

21. Cannata, N.; Cellura, M.; Longo, S.; Montana, F.; Sanseverino, E.R.; Luu, Q.L.; Nguyen, N.Q. Multi-objective optimization of urban microgrid energy supply according to economic and environmental criteria. In Proceedings of the 2019 IEEE Milan PowerTech, Milan, Italy, 23–27 June 2019; pp. 1–6. [CrossRef]

22. Zhang, M.; Wu, Q.; Wen, J.; Lin, Z.; Fang, F.; Chen, Q. Optimal operation of integrated electricity and heat system: A review of modeling and solution methods. *Renew. Sustain. Energy Rev.* **2021**, *135*, 110098. [CrossRef]

23. Ghanbari, A.; Karimi, H.; Jadid, S. Optimal planning and operation of multi-carrier networked microgrids considering multi-energy hubs in distribution networks. *Energy* **2020**, *204*, 117936. [CrossRef]

24. Bidgoli, M.M.; Karimi, H.; Jadid, S.; Anvari-Moghaddam, A. Stochastic electrical and thermal energy management of energy hubs integrated with demand response programs and renewable energy: A prioritized multi-objective framework. *Electr. Power Syst. Res.* **2021**, *196*, 107183. [CrossRef]

25. Chen, H.; Liu, M.; Liu, Y.; Lin, S.; Yang, Z. Partial surrogate cuts method for network-constrained optimal scheduling of multi-carrier energy systems with demand response. *Energy* **2020**, *196*, 117119. [CrossRef]

26. Liu, T.; Zhang, D.; Wang, S.; Wu, T. Standardized modelling and economic optimization of multi-carrier energy systems considering energy storage and demand response. *Energy Convers. Manag.* **2019**, *182*, 126–142. [CrossRef]

27. Yong, W.; Wang, J.; Lu, Z.; Yang, F.; Zhang, Z.; Wei, J.; Wang, J. Day-ahead dispatch of multi-energy system considering operating conditions of multi-energy coupling equipment. *Energy Rep.* **2021**, *7*, 100–110. [CrossRef]

28. Wang, Z.; Hu, J.; Liu, B. Stochastic optimal dispatching strategy of electricity-hydrogen-gas-heat integrated energy system based on improved spectral clustering method. *Int. J. Electr. Power Energy Syst.* **2021**, *126*, 106495. [CrossRef]

29. Mansouri, S.A.; Ahmarinejad, A.; Javadi, M.S.; Catalão, J.P. Two-stage stochastic framework for energy hubs planning considering demand response programs. *Energy* **2020**, *206*, 118124. [CrossRef]

30. Shams, M.H.; Shahabi, M.; MansourLakouraj, M.; Shafie-Khah, M.; Catalão, J.P. Adjustable robust optimization approach for two-stage operation of energy hub-based microgrids. *Energy* **2021**, *222*, 119894. [CrossRef]

31. Shahrabi, E.; Hakimi, S.M.; Hasankhani, A.; Derakhshan, G.; Abdi, B. Developing optimal energy management of energy hub in the presence of stochastic renewable energy resources. *Sustain. Energy Grids Netw.* **2021**, *26*, 100428. [CrossRef]

32. Nosratabadi, S.M.; Hemmati, R.; Gharaei, P.K. Optimal planning of multi-energy microgrid with different energy storages and demand responsive loads utilizing a technical-economic-environmental programming. *Int. J. Energy Res.* **2021**, *45*, 6985–7017. [CrossRef]

33. Guo, X.; Bao, Z.; Yan, W. Stochastic model predictive control based scheduling optimization of multi-energy system considering hybrid CHPs and EVs. *Appl. Sci.* **2019**, *9*, 356. [CrossRef]

34. Preston, N.; Maroufmashat, A.; Riaz, H.; Barbouti, S.; Mukherjee, U.; Tang, P.; Wang, J.; Haghi, E.; Elkamel, A.; Fowler, M. How can the integration of renewable energy and power-to-gas benefit industrial facilities? From techno-economic, policy, and environmental assessment. *Int. J. Hydrog. Energy* **2020**, *45*, 26559–26573. [CrossRef]

35. Javadi, M.S.; Lotfi, M.; Nezhad, A.E.; Anvari-Moghaddam, A.; Guerrero, J.M.; Catalao, J.P.S. Optimal Operation of Energy Hubs Considering Uncertainties and Different Time Resolutions. *IEEE Trans. Ind. Appl.* **2020**, *56*, 5543–5552. [CrossRef]

36. Hosseini, S.; Ahmarinejad, A. Stochastic framework for day-ahead scheduling of coordinated electricity and natural gas networks considering multiple downward energy hubs. *J. Energy Storage* **2021**, *33*, 102066. [CrossRef]

37. Vahid-Ghavidel, M.; Javadi, M.S.; Gough, M.; Santos, S.F.; Shafie-Khah, M.; Catalão, J.P. Demand response programs in multi-energy systems: A review. *Energies* **2020**, *13*, 4332. [CrossRef]

38. Ye, Y.; Qiu, D.; Wu, X.; Strbac, G.; Ward, J. Model-Free Real-Time Autonomous Control for a Residential Multi-Energy System Using Deep Reinforcement Learning. *IEEE Trans. Smart Grid* **2020**, *11*, 3068–3082. [CrossRef]

39. Di Somma, M.; Yan, B.; Bianco, N.; Graditi, G.; Luh, P.; Mongibello, L.; Naso, V. Operation optimization of a distributed energy system considering energy costs and exergy efficiency. *Energy Convers. Manag.* **2015**, *103*, 739–751. [CrossRef]

40. Di Somma, M.; Graditi, G.; Heydarian-Forushani, E.; Shafie-Khah, M.; Siano, P. Stochastic optimal scheduling of distributed energy resources with renewables considering economic and environmental aspects. *Renew. Energy* **2018**, *116*, 272–287. [CrossRef]

41. Di Somma, M.; Yan, B.; Bianco, N.; Graditi, G.; Luh, P.; Mongibello, L.; Naso, V. Multi-objective design optimization of distributed energy systems through cost and exergy assessments. *Appl. Energy* **2017**, *204*, 1299–1316. [CrossRef]

42. Di Somma, M.; Yan, B.; Bianco, N.; Luh, P.B.; Graditi, G.; Mongibello, L.; Naso, V. Multi-objective operation optimization of a Distributed Energy System for a large-scale utility customer. *Appl. Therm. Eng.* **2016**, *101*, 752–761. [CrossRef]

43. Di Somma, M.; Caliano, M.; Graditi, G.; Pinnarelli, A.; Menniti, D.; Sorrentino, N.; Barone, G. Designing of Cost-Effective and Low-Carbon Multi-Energy Nanogrids for Residential Applications. *Inventions* **2020**, *5*, 7. [CrossRef]

44. Foiadelli, F.; Nocerino, S.; Di Somma, M.; Graditi, G. Optimal Design of der for Economic/Environmental Sustainability of Local Energy Communities. In Proceedings of the 2018 IEEE International Conference on Environment and Electrical Engineering and 2018 IEEE Industrial and Commercial Power Systems Europe (EEEIC/I&CPS Europe), Palermo, Italy, 12–15 June 2018. [CrossRef]

45. Nasir, M.; Jordehi, A.R.; Matin, S.A.A.; Tabar, V.S.; Tostado-Véliz, M.; Mansouri, S.A. Optimal operation of energy hubs including parking lots for hydrogen vehicles and responsive demands. *J. Energy Storage* **2022**, *50*, 104630. [CrossRef]

46. Majidi, M.; Zare, K. Integration of smart energy hubs in distribution networks under uncertainties and demand response concept. *IEEE Trans. Power Syst.* **2019**, *34*, 566–574. [CrossRef]

47. Orehounig, K.; Evins, R.; Dorer, V. Integration of decentralized energy systems in neighbourhoods using the energy hub approach. *Appl. Energy* **2015**, *154*, 277–289. [CrossRef]

48. Sousa, T.; Soares, T.; Pinson, P.; Moret, F.; Baroche, T.; Sorin, E. Peer-to-peer and community-based markets: A comprehensive review. *Renew. Sustain. Energy Rev.* **2019**, *104*, 367–378. [CrossRef]

49. Khani, H.; Sawas, A.; Farag, H.E. An estimation–Based optimal scheduling model for settable renewable penetration level in energy hubs. *Electr. Power Syst. Res.* **2021**, *196*, 107230. [CrossRef]

50. Cao, J.; Yang, B.; Zhu, S.; Ning, C.; Guan, X. Day-ahead chance-constrained energy management of energy hub: A distributionally robust approach. *CSEE J. Power Energy Syst.* **2021**, *8*, 812–825. [CrossRef]

51. Jiang, A.; Yuan, H.; Li, D. A two-stage optimization approach on the decisions for prosumers and consumers within a community in the Peer-to-peer energy sharing trading. *Int. J. Electr. Power Energy Syst.* **2021**, *125*, 106527. [CrossRef]

52. Heo, K.; Kong, J.; Oh, S.; Jung, J. Development of operator-oriented peer-to-peer energy trading model for integration into the existing distribution system. *Int. J. Electr. Power Energy Syst.* **2021**, *125*, 106488. [CrossRef]

53. Tushar, W.; Yuen, C.; Saha, T.K.; Morstyn, T.; Chapman, A.C.; Alam, M.J.E.; Hanif, S.; Poor, H.V. Peer-to-peer energy systems for connected communities: A review of recent advances and emerging challenges. *Appl. Energy* **2021**, *282*, 116131. [CrossRef]

54. Di Somma, M.; Ciabattoni, L.; Comodi, G.; Graditi, G. Managing plug-in electric vehicles in eco-environmental operation optimization of local multi-energy systems. *Sustain. Energy Grids Netw.* **2020**, *23*, 100376. [CrossRef]

55. Di Somma, M.; Graditi, G.; Siano, P. Optimal Bidding Strategy for a DER Aggregator in the Day-Ahead Market in the Presence of Demand Flexibility. *IEEE Trans. Ind. Electron.* **2018**, *66*, 1509–1519. [CrossRef]

56. Huang, H.; Nie, S.; Lin, J.; Wang, Y.; Dong, J. Optimization of peer-to-peer power trading in a microgrid with distributed PV and battery energy storage systems. *Sustainability* **2020**, *12*, 923. [CrossRef]

57. Khorasany, M.; Najafi-Ghalelou, A.; Razzaghi, R.; Mohammadi-Ivatloo, B. Transactive energy framework for optimal energy management of multi-carrier energy hubs under local electrical, thermal, and cooling market constraints. *Int. J. Electr. Power Energy Syst.* **2021**, *129*, 106803. [CrossRef]

58. Maroufmashat, A.; Sattari, S.; Roshandel, R.; Fowler, M.; Elkamel, A. Multi-objective Optimization for Design and Operation of Distributed Energy Systems through the Multi-energy Hub Network Approach. *Ind. Eng. Chem. Res.* **2016**, *55*, 8950–8966. [CrossRef]

59. Petkov, I.; Gabrielli, P. Power-to-hydrogen as seasonal energy storage: An uncertainty analysis for optimal design of low-carbon multi-energy systems. *Appl. Energy* **2020**, *274*, 115197. [CrossRef]

60. Rayati, M.; Sheikhi, A.; Ranjbar, A.M. Optimising operational cost of a smart energy hub, the reinforcement learning approach. *Int. J. Parallel Emergent Distrib. Syst.* **2015**, *30*, 325–341. [CrossRef]

61. Sogabe, T.; Malla, B.; Hioki, T.; Takahashi, K. Multi-carrier energy hub management through deep deterministic policy gradient over continuous action space. In Proceedings of the 33rd Annual Conference of the Japanese Society for Artificial Intelligence, Niigata, Japan, 4–7 June 2019; pp. 3–6. [CrossRef]

62. Buoro, D.; Casisi, M.; De Nardi, A.; Pinamonti, P.; Reini, M. Multicriteria optimization of a distributed energy supply system for an industrial area. *Energy* **2013**, *58*, 128–137. [CrossRef]

63. Comodi, G.; Giantomassi, A.; Severini, M.; Squartini, S.; Ferracuti, F.; Fonti, A.; Cesarini, D.N.; Morodo, M.; Polonara, F. Multi-apartment residential microgrid with electrical and thermal storage devices: Experimental analysis and simulation of energy management strategies. *Appl. Energy* **2015**, *137*, 854–866. [CrossRef]

64. Zou, H.; Tao, J.; Elsayed, S.K.; Elattar, E.E.; Almalaq, A.; Mohamed, M.A. Stochastic multi-carrier energy management in the smart islands using reinforcement learning and unscented transform. *Int. J. Electr. Power Energy Syst.* **2021**, *130*, 106988. [CrossRef]

65. Wang, S.; Taha, A.F.; Wang, J.; Kvaternik, K.; Hahn, A. Energy Crowdsourcing and Peer-to-Peer Energy Trading in Blockchain-Enabled Smart Grids. *IEEE Trans. Syst. Man Cybern. Syst.* **2019**, *49*, 1612–1623. [CrossRef]

66. Alvarado, D.C.; Acha, S.; Shah, N.; Markides, C.N. A Technology Selection and Operation (TSO) optimisation model for distributed energy systems: Mathematical formulation and case study. *Appl. Energy* **2016**, *180*, 491–503. [CrossRef]

67. Murty, V.V.S.N.; Kumar, A. Multi-objective energy management in microgrids with hybrid energy sources and battery energy storage systems. *Prot. Control Mod. Power Syst.* **2020**, *5*, 2. [CrossRef]

68. Mashayekh, S.; Stadler, M.; Cardoso, G.; Heleno, M. A mixed integer linear programming approach for optimal DER portfolio, sizing, and placement in multi-energy microgrids. *Appl. Energy* **2017**, *187*, 154–168. [CrossRef]

69. Liu, J.; Xu, Z.; Wu, J.; Liu, K.; Guan, X. Optimal planning of distributed hydrogen-based multi-energy systems. *Appl. Energy* **2021**, *281*, 116107. [CrossRef]

70. Ren, H.; Zhou, W.; Nakagami, K.; Gao, W.; Wu, Q. Multi-objective optimization for the operation of distributed energy systems considering economic and environmental aspects. *Appl. Energy* **2010**, *87*, 3642–3651. [CrossRef]

71. Alabi, T.M.; Lu, L.; Yang, Z. A novel multi-objective stochastic risk co-optimization model of a zero-carbon multi-energy system (ZCMES) incorporating energy storage aging model and integrated demand response. *Energy* **2021**, *226*, 120258. [CrossRef]

72. Gabrielli, P.; Gazzani, M.; Martelli, E.; Mazzotti, M. Optimal design of multi-energy systems with seasonal storage. *Appl. Energy* **2018**, *219*, 408–424, Correction in *Appl. Energy* **2018**, *212*, 720. [CrossRef]

73. Heendeniya, C.B.; Sumper, A.; Eicker, U. The multi-energy system co-planning of nearly zero-energy districts – Status-quo and future research potential. *Appl. Energy* **2020**, *267*, 114953. [CrossRef]

74. Jordehi, A.R.; Javadi, M.S.; Shafie-Khah, M.; Catalão, J.P. Information gap decision theory (IGDT)-based robust scheduling of combined cooling, heat and power energy hubs. *Energy* **2021**, *231*, 120918. [CrossRef]
75. Ma, T.; Wu, J.; Hao, L. Energy flow modeling and optimal operation analysis of the micro energy grid based on energy hub. *Energy Convers. Manag.* **2017**, *133*, 292–306. [CrossRef]
76. Maroufmashat, A.; Taqvi, S.T.; Miragha, A.; Fowler, M.; Elkamel, A. Modeling and optimization of energy hubs: A comprehensive review. *Inventions* **2019**, *4*, 50. [CrossRef]
77. Roberts, J.J.; Cassula, A.M.; Silveira, J.L.; Bortoni, E.D.C.; Mendiburu, A.Z. Robust multi-objective optimization of a renewable based hybrid power system. *Appl. Energy* **2018**, *223*, 52–68. [CrossRef]
78. Wang, Y.; Ma, Y.; Song, F.; Ma, Y.; Qi, C.; Huang, F.; Xing, J.; Zhang, F. Economic and efficient multi-objective operation optimization of integrated energy system considering electro-thermal demand response. *Energy* **2020**, *205*, 118022. [CrossRef]
79. Yan, R.; Wang, J.; Lu, S.; Ma, Z.; Zhou, Y.; Zhang, L.; Cheng, Y. Multi-objective two-stage adaptive robust planning method for an integrated energy system considering load uncertainty. *Energy Build.* **2021**, *235*, 110741. [CrossRef]
80. Nazari-Heris, M.; Mohammadi-Ivatloo, B.; Asadi, S. Optimal operation of multi-carrier energy networks with gas, power, heating, and water energy sources considering different energy storage technologies. *J. Energy Storage* **2020**, *31*, 101574. [CrossRef]
81. Zhou, Y.; Wu, J.; Long, C. Evaluation of peer-to-peer energy sharing mechanisms based on a multiagent simulation framework. *Appl. Energy* **2018**, *222*, 993–1022. [CrossRef]
82. Salehi, J.; Namvar, A.; Gazijahani, F.S. Scenario-based Co-Optimization of neighboring multi carrier smart buildings under demand response exchange. *J. Clean. Prod.* **2019**, *235*, 1483–1498. [CrossRef]
83. Motevasel, M.; Seifi, A.R.; Niknam, T. Multi-objective energy management of CHP (combined heat and power)-based micro-grid. *Energy* **2013**, *51*, 123–136. [CrossRef]
84. Li, P.; Guo, T.; Abeysekera, M.; Wu, J.; Han, Z.; Wang, Z.; Yin, Y.; Zhou, F. Intraday multi-objective hierarchical coordinated operation of a multi-energy system. *Energy* **2021**, *228*, 120528. [CrossRef]
85. Karmellos, M.; Georgiou, P.; Mavrotas, G. A comparison of methods for the optimal design of Distributed Energy Systems under uncertainty. *Energy* **2019**, *178*, 318–333. [CrossRef]
86. Rieder, A.; Christidis, A.; Tsatsaronis, G. Multi criteria dynamic design optimization of a small scale distributed energy system. *Energy* **2014**, *74*, 230–239. [CrossRef]
87. Bollenbacher, J.; Rhein, B. Optimal configuration and control strategy in a multi-carrier-energy system using reinforcement learning methods. In Proceedings of the 2017 International Energy and Sustainability Conference, Farmingdale, NY, USA, 19–20 October 2017; pp. 1–6. [CrossRef]
88. Pakdel, M.J.V.; Sohrabi, F.; Mohammadi-Ivatloo, B. Multi-objective optimization of energy and water management in networked hubs considering transactive energy. *J. Clean. Prod.* **2020**, *266*, 121936. [CrossRef]
89. Bartolini, A.; Carducci, F.; Muñoz, C.B.; Comodi, G. Energy storage and multi energy systems in local energy communities with high renewable energy penetration. *Renew. Energy* **2020**, *159*, 595–609. [CrossRef]
90. Çiçek, A.; Şengör, I.; Erenoğlu, A.K.; Erdinç, O. Decision making mechanism for a smart neighborhood fed by multi-energy systems considering demand response. *Energy* **2020**, *208*, 118323. [CrossRef]
91. Cui, S.; Wang, Y.-W.; Xiao, J.-W. Peer-to-peer energy sharing among smart energy buildings by distributed transaction. *IEEE Trans. Smart Grid* **2019**, *10*, 6491–6501. [CrossRef]
92. Liu, N.; Yu, X.; Wang, C.; Li, C.; Ma, L.; Lei, J. Energy-Sharing Model with Price-Based Demand Response for Microgrids of Peer-to-Peer Prosumers. *IEEE Trans. Power Syst.* **2017**, *32*, 3569–3583. [CrossRef]
93. Roberto, R.; De Iulio, R.; Di Somma, M.; Graditi, G.; Guidi, G.; Noussan, M. A multi-objective optimization analysis to assess the potential economic and environmental benefits of distributed storage in district heating networks: A case study. *Int. J. Sustain. Energy Plan. Manag.* **2019**, *20*, 5–20. [CrossRef]
94. Wu, T.; Bu, S.; Wei, X.; Wang, G.; Zhou, B. Multitasking multi-objective operation optimization of integrated energy system considering biogas-solar-wind renewables. *Energy Convers. Manag.* **2021**, *229*, 113736. [CrossRef]
95. Zhang, C.; Zhou, L.; Chhabra, P.; Garud, S.S.; Aditya, K.; Romagnoli, A.; Comodi, G.; Magro, F.D.; Meneghetti, A.; Kraft, M. A novel methodology for the design of waste heat recovery network in eco-industrial park using techno-economic analysis and multi-objective optimization. *Appl. Energy* **2016**, *184*, 88–102. [CrossRef]
96. Shabanpour-Haghighi, A.; Seifi, A.R. Multi-objective operation management of a multi-carrier energy system. *Energy* **2015**, *88*, 430–442. [CrossRef]
97. Geidl, M.; Andersson, G. Operational and structural optimization of multi-carrier energy systems. *Eur. Trans. Electr. Power* **2006**, *16*, 463–477. [CrossRef]
98. Fonseca, M.N.; Pamplona, E.D.O.; de Queiroz, A.R.; Valerio, V.E.D.M.; Aquila, G.; Silva, S.R. Multi-objective optimization applied for designing hybrid power generation systems in isolated networks. *Sol. Energy* **2018**, *161*, 207–219. [CrossRef]
99. Bjarghov, S.; Loschenbrand, M.; Ibn Saif, A.U.N.; Pedrero, R.A.; Pfeiffer, C.; Khadem, S.K.; Rabelhofer, M.; Revheim, F.; Farahmand, H. Developments and Challenges in Local Electricity Markets: A Comprehensive Review. *IEEE Access* **2021**, *9*, 58910–58943. [CrossRef]
100. Sadeghi, H.; Rashidinejad, M.; Moeini-Aghtaie, M.; Abdollahi, A. The energy hub: An extensive survey on the state-of-the-art. *Appl. Therm. Eng.* **2019**, *161*, 114071. [CrossRef]

101. Li, L.; Mu, H.; Li, N.; Li, M. Economic and environmental optimization for distributed energy resource systems coupled with district energy networks. *Energy* **2016**, *109*, 947–960. [CrossRef]
102. Xu, D.; Wu, Q.; Zhou, B.; Li, C.; Bai, L.; Huang, S. Distributed Multi-Energy Operation of Coupled Electricity, Heating, and Natural Gas Networks. *IEEE Trans. Sustain. Energy* **2019**, *11*, 2457–2469. [CrossRef]
103. Dan, M.; Srinivasan, S.; Sundaram, S.; Easwaran, A.; Glielmo, L. A Scenario-Based Branch-and-Bound Approach for MES Scheduling in Urban Buildings. *IEEE Trans. Ind. Inform.* **2020**, *16*, 7510–7520. [CrossRef]
104. Fazlollahi, S.; Mandel, P.; Becker, G.; Maréchal, F. Methods for multi-objective investment and operating optimization of complex energy systems. *Energy* **2012**, *45*, 12–22. [CrossRef]
105. Najafi-Ghalelou, A.; Nojavan, S.; Zare, K.; Mohammadi-Ivatloo, B. Robust scheduling of thermal, cooling and electrical hub energy system under market price uncertainty. *Appl. Therm. Eng.* **2019**, *149*, 862–880. [CrossRef]
106. Gong, D.-W.; Zhang, Y.; Qi, C.-L. Environmental/economic power dispatch using a hybrid multi-objective optimization algorithm. *Int. J. Electr. Power Energy Syst.* **2010**, *32*, 607–614. [CrossRef]
107. Alarcon-Rodriguez, A.; Ault, G.; Galloway, S. Multi-objective planning of distributed energy resources: A review of the state-of-the-art. *Renew. Sustain. Energy Rev.* **2010**, *14*, 1353–1366. [CrossRef]
108. Shabanpour-Haghighi, A.; Seifi, A.R. Energy Flow Optimization in Multicarrier Systems. *IEEE Trans. Ind. Inform.* **2015**, *11*, 1067–1077. [CrossRef]
109. Shabanpour-Haghighi, A.; Seifi, A.R.; Niknam, T. A modified teaching–learning based optimization for multi-objective optimal power flow problem. *Energy Convers. Manag.* **2014**, *77*, 597–607. [CrossRef]
110. Moeini-Aghtaie, M.; Abbaspour, A.; Fotuhi-Firuzabad, M.; Hajipour, E. A decomposed solution to multiple-energy carriers optimal power flow. *IEEE Trans. Power Syst.* **2013**, *29*, 707–716. [CrossRef]
111. Cesena, E.A.M.; Mancarella, P. Energy Systems Integration in Smart Districts: Robust Optimisation of Multi-Energy Flows in Integrated Electricity, Heat and Gas Networks. *IEEE Trans. Smart Grid* **2018**, *10*, 1122–1131. [CrossRef]
112. Nicholson, M. Smart Grids. In *The Power Makers' Challenge*; Springer: Berlin/Heidelberg, Germany, 2012; pp. 119–121. [CrossRef]
113. Ghorab, M. Energy hubs optimization for smart energy network system to minimize economic and environmental impact at Canadian community. *Appl. Therm. Eng.* **2019**, *151*, 214–230. [CrossRef]
114. Cao, Y.; Wang, Q.; Cheng, W.; Nojavan, S.; Jermsittiparsert, K. Risk-constrained optimal operation of fuel cell/photovoltaic/battery/grid hybrid energy system using downside risk constraints method. *Int. J. Hydrogen Energy* **2020**, *45*, 14108–14118. [CrossRef]
115. Wang, Y.; Zhang, N.; Zhuo, Z.; Kang, C.; Kirschen, D. Mixed-integer linear programming-based optimal configuration planning for energy hub: Starting from scratch. *Appl. Energy* **2018**, *210*, 1141–1150. [CrossRef]
116. Turk, A.; Wu, Q.; Zhang, M.; Østergaard, J. Day-ahead stochastic scheduling of integrated multi-energy system for flexibility synergy and uncertainty balancing. *Energy* **2020**, *196*, 117130. [CrossRef]
117. Ma, T.; Wu, J.; Hao, L.; Lee, W.-J.; Yan, H.; Li, D. The optimal structure planning and energy management strategies of smart multi energy systems. *Energy* **2018**, *160*, 122–141. [CrossRef]
118. Huang, W.; Zhang, N.; Yang, J.; Wang, Y.; Kang, C. Optimal configuration planning of multi-energy systems considering distributed renewable energy. *IEEE Trans. Smart Grid* **2017**, *10*, 1452–1464. [CrossRef]
119. Qiu, Z.; Wang, B.; Huang, J.; Xie, Z. Optimal configuration and sizing of regional energy service company's energy hub with integrated demand response. *IEEJ Trans. Electr. Electron. Eng.* **2019**, *14*, 383–393. [CrossRef]
120. Pan, G.; Gu, W.; Wu, Z.; Lu, Y.; Lu, S. Optimal design and operation of multi-energy system with load aggregator considering nodal energy prices. *Appl. Energy* **2019**, *239*, 280–295. [CrossRef]
121. Comodi, G.; Bartolini, A.; Carducci, F.; Nagaranjan, B.; Romagnoli, A. Achieving low carbon local energy communities in hot climates by exploiting networks synergies in multi energy systems. *Appl. Energy* **2019**, *256*, 113901. [CrossRef]
122. Capone, M.; Guelpa, E.; Verda, V. Multi-objective optimization of district energy systems with demand response. *Energy* **2021**, *227*, 120472. [CrossRef]
123. Zhang, Y.; Liu, Y.; Shu, S.; Zheng, F.; Huang, Z. A data-driven distributionally robust optimization model for multi-energy coupled system considering the temporal-spatial correlation and distribution uncertainty of renewable energy sources. *Energy* **2020**, *216*, 119171. [CrossRef]
124. Grond, M.O.W.; Luong, N.H.; Morren, J.; Slootweg, J.G. Multi-objective optimization techniques and applications in electric power systems. In Proceedings of the 2012 47th International Universities Power Engineering Conference (UPEC), Uxbridge, UK, 4–7 September 2012; pp. 1–6. [CrossRef]
125. Ascione, F.; Bianco, N.; De Masi, R.F.; De Stasio, C.; Mauro, G.M.; Vanoli, G.P. Multi-objective optimization of the renewable energy mix for a building. *Appl. Therm. Eng.* **2016**, *101*, 612–621. [CrossRef]
126. Arteconi, A.; Ciarrocchi, E.; Pan, Q.; Carducci, F.; Comodi, G.; Polonara, F.; Wang, R. Thermal energy storage coupled with PV panels for demand side management of industrial building cooling loads. *Appl. Energy* **2017**, *185*, 1984–1993. [CrossRef]
127. Good, N.; Karangelos, E.; Navarro-Espinosa, A.; Mancarella, P. Optimization Under Uncertainty of Thermal Storage-Based Flexible Demand Response With Quantification of Residential Users' Discomfort. *IEEE Trans. Smart Grid* **2015**, *6*, 2333–2342. [CrossRef]
128. Cesena, E.A.M.; Loukarakis, E.; Good, N.; Mancarella, P. Integrated Electricity– Heat–Gas Systems: Techno–Economic Modeling, Optimization, and Application to Multienergy Districts. *Proc. IEEE* **2020**, *108*, 1392–1410. [CrossRef]

129. Coelho, A.; Iria, J.; Soares, F.; Lopes, J.P. Real-time management of distributed multi-energy resources in multi-energy networks. In *Sustainable Energy, Grids and Networks*; Elsevier: Amsterdam, The Netherlands, 2023; Volume 34, p. 101022. [CrossRef]
130. Coelho, A.; Iria, J.; Soares, F. Network-secure bidding optimization of aggregators of multi-energy systems in electricity, gas, and carbon markets. *Appl. Energy* **2021**, *301*, 117460. [CrossRef]
131. Mancarella, P.; Chicco, G.; Capuder, T. Arbitrage opportunities for distributed multi-energy systems in providing power system ancillary services. *Energy* **2018**, *161*, 381–395. [CrossRef]
132. Li, J.; Ge, S.; Xu, Z.; Liu, H.; Li, J.; Wang, C.; Cheng, X. A network-secure peer-to-peer trading framework for electricity-carbon integrated market among local prosumers. *Appl. Energy* **2023**, *335*, 120420. [CrossRef]

*Brief Report*

# Discussion on Incentive Compatibility of Multi-Period Temporal Locational Marginal Pricing

Farhan Hyder [1], Bing Yan [1,*], Peter Luh [2,†], Mikhail Bragin [2], Jinye Zhao [3], Feng Zhao [3], Dane Schiro [3] and Tongxin Zheng [3]

[1]   Department of Electrical and Microelectronic Engineering, Rochester Institute of Technology, Rochester, NY 14623, USA; fh6772@rit.edu
[2]   Department of Electrical and Computer Engineering, University of Connecticut, Storrs, CT 06269, USA; peter.luh@uconn.edu (P.L.); mikhail.bragin@uconn.edu (M.B.)
[3]   Advanced Technology and Solutions, ISO New England, Holyoke, MA 01040, USA
*    Correspondence: bxyeee@rit.edu
†    P. Luh, who was the co-supervisor of this project, tragically passed away in November 2022. He was a professor emeritus of the Department of Electrical and Computer Engineering, University of Connecticut, Storrs, CT 06269, USA, and with the Department of Electrical Engineering, National Taiwan University, Taipei 10617, Taiwan. As a tribute to our dear friend and mentor, the remaining coauthors dedicate this paper to commemorating Dr. Luh's contributions and legacy.

**Abstract:** In real-time electricity markets, locational marginal prices (LMPs) can be determined by solving multi-interval economic dispatch problems to manage inter-temporal constraints (i.e., ramp rates). Under the current practice, the LMPs for the immediate interval are binding, while the prices for the subsequent intervals are advisory signals. However, a generator may miss the opportunity for higher profits, and compensatory uplift payments are needed at the settlement. To address these issues, the "temporal locational marginal pricing (TLMP)" that augments LMP by incorporating multipliers associated with generators' reported ramp rates was developed. It was demonstrated that it would result in zero uplift payments, showing great potential as a good pricing scheme. Numerical examples also showed that "the generators had incentives to reveal their ramp rates truthfully". In this paper, the incentive compatibility of TLMP with respect to ramp-rate reporting is discussed. Our idea is to develop numerical examples to investigate whether reporting the true ramp rates is the best option for generators. The results indicate that TLMP is not incentive compatible, and there are market-clearing scenarios where not reporting true ramp rates may be beneficial.

**Keywords:** multi-interval economic dispatch; locational marginal pricing; incentive compatibility; ramp-rate constraints

**Citation:** Hyder, F.; Yan, B.; Luh, P.; Bragin, M.; Zhao, J.; Zhao, F.; Schiro, D.; Zheng, T. Discussion on Incentive Compatibility of Multi-Period Temporal Locational Marginal Pricing. *Energies* **2023**, *16*, 4977. https://doi.org/10.3390/en16134977

Academic Editors: François Vallée and Yuji Yamada

Received: 29 May 2023
Revised: 23 June 2023
Accepted: 24 June 2023
Published: 27 June 2023

## 1. Introduction

In real-time electricity markets, locational marginal prices (LMPs) can be determined by solving rolling-window multi-interval economic dispatch (ED) with reported generator parameters and bids to manage inter-temporal constraints, i.e., ramp rates [1–3]. Under the rolling-window framework, LMPs for the immediate interval are binding and used at the market settlement, while the prices for subsequent intervals are advisory signals. It has been shown that multi-interval dispatch improves operational flexibility and system reliability as compared with single-interval dispatch since it considers system needs in future intervals [4–8]. However, the major challenge with the rolling-window multi-interval dispatch is the disparity between the settlement prices and advisory prices, as the ED problem is solved repeatedly with updated information to account for operational uncertainty. As a result, a generator may miss the opportunity for higher profits when it is asked to hold back generation to provide ramping support or to generate more, but the settlement prices may not support such dispatch decisions. Thus, out-of-market discriminatory uplift payments, such as lost opportunity costs (LOCs) are needed to compensate for generators at settlement based on the solutions to the profit maximization problems of individual generators. Otherwise, this might create a dispatch-following issue, where a generator may have incentives to deviate from the ISO dispatch. Therefore, a good pricing scheme should guarantee zero LOC while being incentive compatible.

Several approaches have been reported to reduce uplift payments [5–9]. In [5,6], the past opportunity costs that are represented by the dual variables of the past interval's optimization problem are added to the current interval's optimization objective. In this way, the past opportunity costs are reflected in the current interval's clearing price. A multi-settlement system is developed in [7,8] to coordinate between day-ahead (DA) and real-time (RT) markets in multi-interval pricing. Under this scheme, the DA schedule is financially binding, and the RT prices are used to settle the deviation from the DA market clearing. Market participants are only exposed to the RT price volatility by locking the DA clearing prices. In [9], a pricing model that minimizes uplift payments is developed, which uses prices as decision variables and coordinates between multi-period and single-period dispatches. However, none of the above-mentioned approaches [5–9] can guarantee zero LOC.

As reviewed in Section 2, the temporal locational marginal pricing (TLMP) was recently developed [10,11]. It augments LMPs by incorporating multipliers associated with generators' reported ramp-up and -down rates (which could be different), leading to individualized pricing, which is uncommon in power systems [5,6,8]. TLMP shows great potential as a good pricing scheme with zero LOC, regardless of rolling-window or one-shot (the prices for all the intervals are binding) dispatch, and of perfect or imperfect forecasts. With the same value for a generator's ramp-up and -down rates and linear generation costs, numerical testing shows that "the generators had incentives to reveal their ramp rates truthfully" [11]. However, rather than linear generation costs, piecewise linear or quadratic cost functions are usually used in most practical electricity markets.

The aim of this paper is to investigate the incentive compatibility of TLMP with respect to ramp-rate reporting through numerical examples. The incentive compatibility with respect to ramp-rate reporting is defined as a profit-maximizing generator that has no incentive to misreport its ramp rates. Following the testing examples in [10,11] as closely as possible, the incentive compatibility of TLMP is analyzed through numerical examples with different ramp-up and -down values and with piecewise linear and quadratic costs in Section 3. The incentive compatibility results with different costs are analyzed and discussed. Results show that a generator could be better off by not reporting its true ramp rates, leading to possible infeasibility in ED.

## 2. Temporal Locational Marginal Pricing

In this section, TLMP [10,11] is briefly reviewed. The ISO's one-shot ED problem is to minimize the total dispatch cost subject to the power balance as well as the ramp rate and generation capacity constraints of the bid-in generators but no transmission constraints for simplicity [10]. It is formulated (following Equation (3) in [10]) as

$$\underset{G=[g_{it}]}{\text{Min}}\ F(G),\text{with } F(G) \equiv \sum_{i=1}^{N}\sum_{t=1}^{T} f_{it}(g_{it}), \tag{1}$$

$$\text{s.t. } (\lambda_t)\ :\ \sum_{i=1}^{N} g_{it}\ =\ d_t,\quad \forall t \in 1,...,T, \tag{2}$$

$$\left(\mu_{it}^{D},\mu_{it}^{U}\right)\ :\ -\underline{r}_i \le g_{i(t+1)} - g_{it} \le \bar{r}_i,\quad \forall t \in 0,...,T-1, i \in 1,...,N, \tag{3}$$

$$\left(\rho_{it}^{Min},\rho_{it}^{Max}\right)\ :\ 0 \le g_{it} \le \bar{g}_i,\quad \forall t \in 1,...,T, i \in 1,...,N. \tag{4}$$

where $f_{it}$ is generator $i$'s bid-in cost at time $t$ (assumed convex and differentiable); $g_{it}$ is the generation level; $\overline{g_i}$ is the maximum generation limit (the minimum is assumed to be zero for simplicity); $\overline{r_i}$ and $\underline{r_i}$ are bid-in ramp-up/-down rates per time interval (could be different); and $d_t$ is the system demand. In the above, the dual variables are shown in front of the corresponding constraints.

The TLMP of generator $i$ at interval $t$ is defined as the marginal benefit of generator $i$ at $g_{it} = g_{it}^*$ (obtained by solving the above ED):

$$\pi_{it} = -\frac{\partial}{\partial g_{it}} F_{-it}(G^*), \tag{5}$$

where $F_{-it}(G) = F(G) - f_{it}(g_{it})$ is the partial cost that excludes generator $i$'s cost at $t$. With $g_{it}$ fixed at $g_{it}^*$, the modified ED is to minimize $F_{-it}(G)$. Based on the envelope theorem, TMLP is the sum of the multipliers associated with $g_{it}^*$ (Proposition 2 of [9]):

$$\pi_{it} = -\frac{\partial}{\partial g_{it}^*} F_{-it}(G^*) = \lambda_t^* + \Delta\mu_{it}^* - \Delta\mu_{i(t-1)}^* = \lambda_t^* + \Delta_{it}^*, \tag{6}$$

where $\Delta_{it}^* \equiv \Delta\mu_{it}^* - \Delta\mu_{i(t-1)}^*$ with $\Delta\mu_{it}^* \equiv \mu_{it}^{U*} - \mu_{it}^{D*}$ is the increment of the shadow prices associated with the ramp-rate constraints.

With optimal multipliers, the Lagrangian function in the dual space can be obtained as

$$\mathcal{L} = \sum_{i,t} \left( f_{it}(g_{it}) - (\lambda_t^* + \Delta_{it}^*)g_{it} + (\rho_{it}^{Max*} - \rho_{it}^{Min*})g_{it} \right) + \cdots \tag{7}$$

where the rest of the terms are independent of $g_{it}$. Now, Equation (7) clearly shows that under TLMP $\pi_{it} = \lambda_t^* + \Delta_{it}^*$, the multi-interval dispatch problem is decoupled into individual single-interval dispatch problems because the multipliers associated with the time-coupling ramp-rate constraints have been incorporated into TLMP.

To further understand TLMP, consider a special case when only the ramp-down constraint is binding at $t - 1$, i.e., LMP plus the marginal cost if the generator can ramp down more. TMLP is given as

$$\pi_{it} = \lambda_t^* + \mu_{i(t-1)}^{D*}. \tag{8}$$

Given TLMP, the profit maximization (PM) problem of generator $i$ is to maximize the total profit over all intervals without knowing other generators' costs. As described earlier in Equation (6), the multipliers associated with ramp rates are incorporated as a part of TLMP after solving ED. When solving PM, the multipliers associated with the ramp rates are zero according to the KKT conditions [10]. The multipliers with the capacity constraints at the minimum and maximum sides are the same as $\rho_{it}^{Min*}$ and $\rho_{it}^{Max*}$, respectively. Optimal generation in PM is thus identical to $g_{it}^*$ (Theorem 3 of [10]). Consequently, LOC is guaranteed to be zero, implying that TLMP satisfies market clearing and individual rationality conditions (Definition 2 of [10]). As the multipliers associated with the ramp-rate constraints in PM are all zero, the multi-interval dispatch is decoupled in time. LOC is thus zero regardless of rolling window or one shot, or perfect or imperfect forecasts (Theorems 3 and 4 of [10]).

The truthful reporting of ramp rates was discussed via numerical examples based on a three-generator system in [11]. For each generator, a linear marginal cost was considered, and the same value was used for its ramp-up and -down rates. Results showed that "under TLMP, profits of all generators grew as the revealed ramping limits grew to their true values" [11]. This implies that "the generators had incentives to reveal their ramp limits truthfully" [11]. However, linear costs may not be practical in current electricity markets. In addition, a generator's ramp-up and -down rates could be different.

### 3. Numerical Testing on Incentive Compatibility of TLMP

In this section, numerical examples are developed to investigate the incentive compatibility of TLMP with respect to ramp-rate reporting under piecewise linear and quadratic costs, following [10,11] as closely as possible.

#### 3.1. Data for Numerical Testing

Consider the three-generator system used in Section 5 (Performance) of [11]. The three generators are connected to a single bus, and their capacities, true ramp rates (same for up and down), and linear costs presented in [11] are shown in Table 1 below. It is shown in [11] that when the cost is linear, the profit of a generator grows when the revealed ramp rate grows to its true value. However, linear costs are not practical in the current electricity markets. Hence, in our study, for each generator, the piecewise linear cost is approximated from its linear cost and consists of two blocks (40 MW and 60 MW). Then, its quadratic cost function is approximated from the piecewise linear cost. The above two costs are also shown in Table 1. The system demand over 24 h to be shown later is approximated from the average demand curve presented in Figure 2 of [11] , which was generated from 300 scenarios of a CAISO load profile.

**Table 1.** Generator parameters.

| $G$ | Capacity (MW) | True Ramp Rate (MW/h) | Linear Costs ($/MW) | Piecewise Linear Costs ($/MW) | Quadratic Cost Functions ($) |
|---|---|---|---|---|---|
| $G_1$ | 100 | 25 | 28 | (28,29) | $0.008568g^2 + 28.7897g$ |
| $G_2$ | 100 | 60 | 30 | (30,31) | $0.007g^2 + 29.9626g$ |
| $G_3$ | 100 | 60 | 40 | (40,41) | $0.008568g^2 + 39.7897g$ |

Following [11], two generators report their true ramp rates (same for up and down), but the third generator might not report truthfully. In our study, it is assumed that the third generator reports its true ramp-up rate but it may not report its true ramp-down value. With the reported ramp rates, the ED problem is solved in a rolling-window manner with a window size of four intervals, where only the first interval is binding following [11]. Then, the PM problem is solved in a one-shot manner with the true ramp rates given TLMP for all intervals. The results with piecewise linear and quadratic costs are presented in Sections 3.2 and 3.3, respectively.

#### 3.2. Incentive Compatibility of TLMP with Piecewise Linear Costs

In this subsection, the piecewise linear costs presented in Table 1 are considered. It is assumed that generators $G_2$ and $G_3$ report their true ramp-up and -down rates (the same), and $G_1$ reports its true ramp-up rate (25 MW/h). It is also assumed that $G_1$ may report its ramp-down rate as 25 MW/h (the true value) or 5 MW/h (a low value). The ED problem is solved twice with the true and low values of the reported down-up rate of $G_1$. For each scenario, the $G_1$ PM problem is solved with its true ramp rate given the corresponding TLMP for all intervals. Then, the same process is repeated for scenarios where $G_2$ or $G_3$ may not report its true ramp-down value. The results are presented in Table 2 and Figure 1 below. It can be shown that each generator makes a higher profit by revealing a lower ramp-down rate. This implies that revealing ramping rates truthfully may not be in the best interest of the generators when their costs are piecewise linear.

**Table 2.** Profits with piecewise linear costs.

| $G$ | Report Truth | | Under-Report | |
| --- | --- | --- | --- | --- |
| | **RD (MW/h)** | **Profit ($)** | **RD (MW/h)** | **Profit ($)** |
| $G_1$ | 25 | 7460 | 5 | 8260 |
| $G_2$ | 60 | 3340 | 5 | 3420 |
| $G_3$ | 60 | 0 | 5 | 120 |

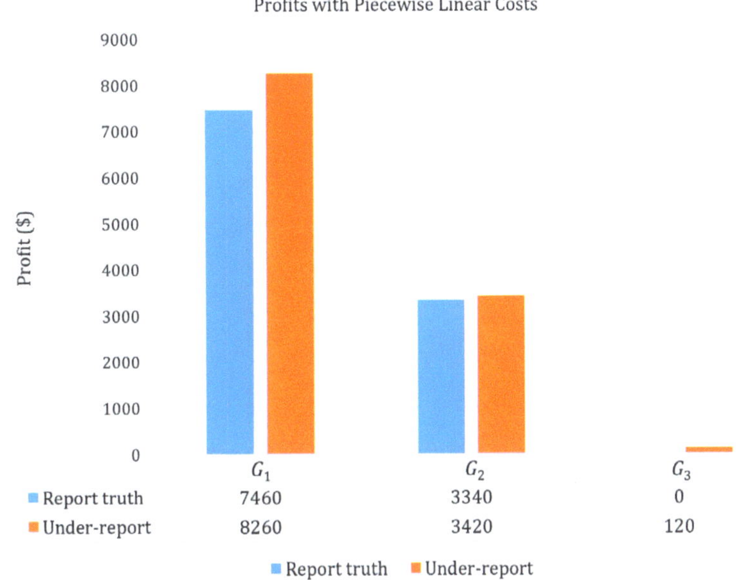


**Figure 1.** Generator profits with piecewise linear costs.

To further illustrate the results in Table 2, consider $G_2$ as an example. Figure 2 shows the TLMP values when $G_2$ reports a ramp-down rate of 60 MW/h (the true value) and when it reports 5 MW/h (a low value). With the low ramp-down rate, the TLMP values are higher during time intervals 4 to 8 and for intervals 18 and 21. This is because $G_2$ cannot ramp down fast enough when demand decreases for these intervals, resulting in binding ramping-down constraints and thus higher prices. This indicates that a generator can obtain higher prices by under-reporting its ramp-down rate.

Figure 3 shows the power output of $G_2$ when the reported ramp-down rate is 60 MW/h and when it is 5 MW/h. With the true ramp-down rate, it can be seen that the power output of $G_2$ becomes 0 MW when demand is lower than 100 MW (intervals 5 to 8) and $G_2$ does not get paid. However, with the low reported ramp-down rate, the power output of $G_2$ decreases slowly from 30 MW to 15 MW and does not reach 0 MW. As seen in Figure 2, the TLMP values for these intervals are higher than the marginal cost of $G_2$ ($30 for the first block). Therefore, $G_2$ is paid between intervals 5 and 8. The above shows that $G_2$ can get paid more by under-reporting its ramp-down rate.

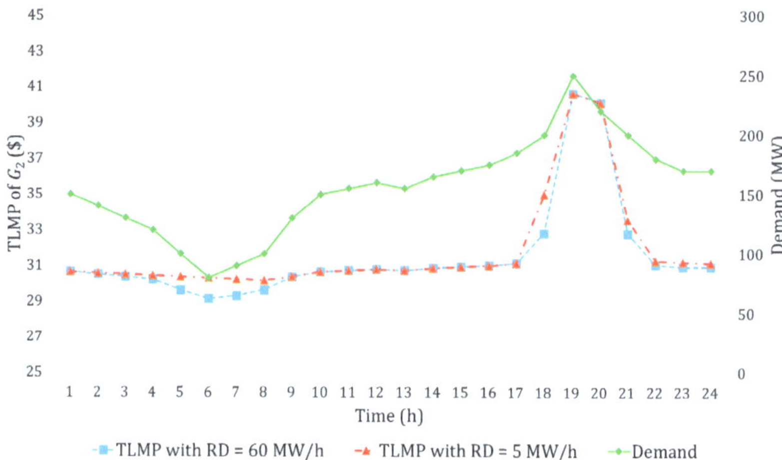

**Figure 2.** Demand and TLMP of $G_2$ under different reported ramp-down rates.

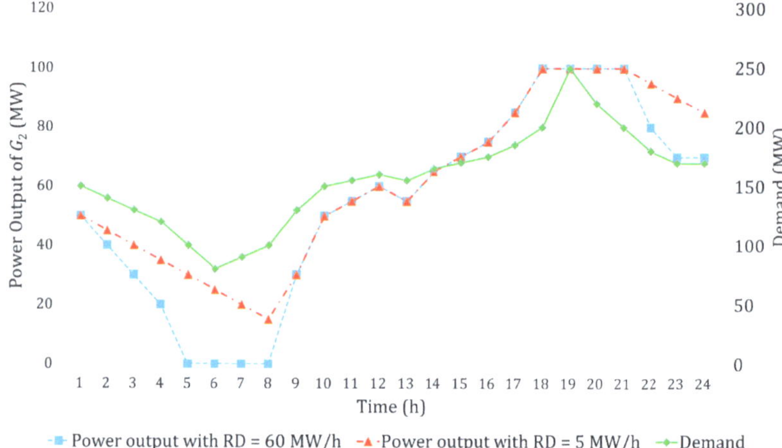

**Figure 3.** Demand and power output of $G_2$ under different reported ramp-down rates.

Figure 4 shows the profits of $G_2$ under different reported ramp-down rates. During intervals 4 to 8 and for intervals 18 and 21, the profits with the low reported ramp-down rate are higher than those with the true value. As mentioned early in Figure 2, when $G_2$ reports a low ramp-down rate, it obtains higher prices because it cannot ramp down fast enough when the demand decreases during these intervals. From Figure 3, for intervals 4 to 8, it is clear that the power output of $G_2$ when under-reporting its ramp-down rate is higher than that when reporting truthfully. During intervals 4 to 8, the combination of higher prices and higher power output results in higher profits for $G_2$ when it reports a low ramp-down rate. For intervals 18 and 21, high profits are caused by high prices. The results are similar for generators $G_1$ and $G_3$. This demonstrates that a generator can make higher profits by under-reporting its ramp-down rate when reporting its ramp-up rate truthfully under the TLMP pricing scheme.

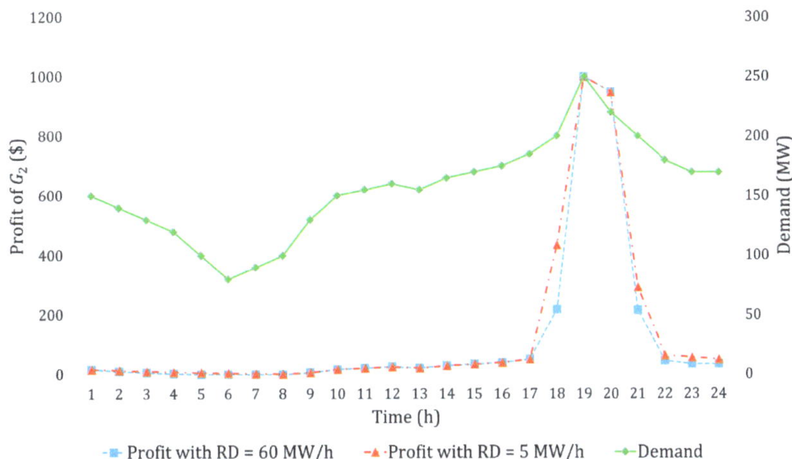

**Figure 4.** Demand and profits of $G_2$ under different reported ramp-down rates.

### 3.3. Incentive Compatibility of TLMP with Quadratic Costs

In this subsection, the quadratic cost functions presented in Table 1 are considered. Again, it is assumed that generators $G_2$ and $G_3$ report their true ramp-up and -down rates (the same), and $G_1$ reports their true ramp-up rate (25 MW/h). It is also assumed that $G_1$ may report its ramp-down rate as 25 MW/h (the true value) or 5 MW/h (a low value). The ED and PM problems are solved in the same way described above in Section 3.2. The ED problem is solved twice with the true and low values of the reported down-up rate of $G_1$. For each scenario, the $G_1$ PM problem is solved with its true ramp rate given the corresponding TLMP for all intervals. Then, the same process is repeated for scenarios where $G_2$ or $G_3$ may not report their true ramp-down value. The results are presented in Table 3 and Figure 5 below. Similar to what is presented in Section 3.2, each generator makes a higher profit by revealing a lower ramp-down rate. This implies that revealing ramping rates truthfully may not be in the best interest of the generators when their costs are quadratic under the TLMP pricing scheme.

In summary, when the generation costs are linear, TLMP is incentive compatible with respect to ramp-rate reporting, and the generator profits are not affected [11]. However, when the generation costs are piecewise linear or quadratic, TLMP is not incentive compatible. A generator might be able to make higher profits by under-reporting its ramp-down rate. This under-reporting of ramp-down rates could result in the possible infeasibility of ED. This may affect the reliability, stability, and overall performance of the grid, leading to operational difficulties within power systems.

**Table 3.** Profits with quadratic costs.

| $G$ | Report Truth | | Under-Report | |
|---|---|---|---|---|
| | **RD (MW/h)** | **Profit ($)** | **RD (MW/h)** | **Profit ($)** |
| $G_1$ | 25 | 6884 | 5 | 7075 |
| $G_2$ | 60 | 2774 | 5 | 3153 |
| $G_3$ | 60 | 25 | 5 | 76 |

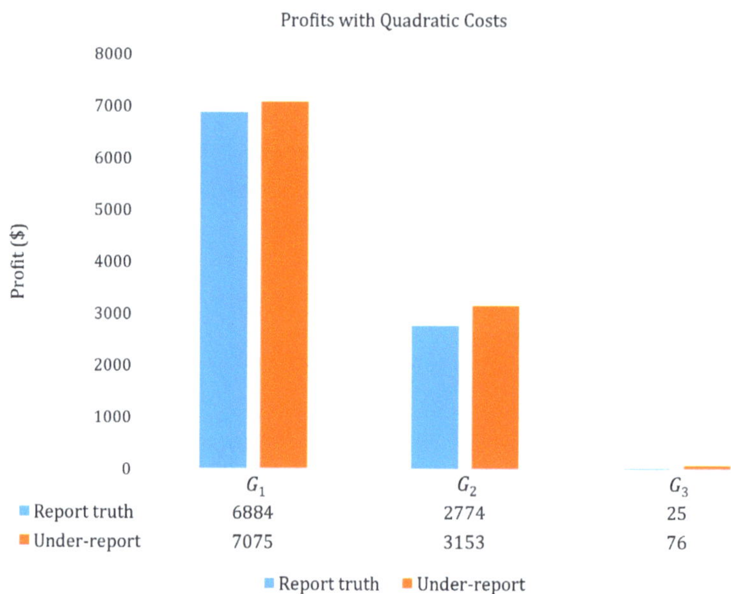


**Figure 5.** Generator profits with quadratic costs.

## 4. Conclusions

This paper discusses the incentive compatibility of TLMP with respect to ramp-rate reporting through numerical examples following [11], where it was shown that generators have the incentive to reveal their ramp rates truthfully when the marginal costs of generators are linear. As the linear costs used in [11] are not practical in the current electricity markets, piecewise linear and quadratic costs are considered. In addition, it is assumed that a generator may report different values for its ramp-up and -down rates. The results show that a generator can achieve higher profit by under-reporting its ramp-down rate while reporting its true ramp-up rate when costs are either piecewise linear or quadratic. It is implied that revealing the ramp rate truthfully may not be beneficial for a generator under the TLMP pricing scheme, resulting in the possible infeasibility of ED. This may affect the reliability, stability, and overall performance of the grid, leading to operational difficulties within power systems.

**Author Contributions:** Conceptualization, F.H., B.Y., P.L., J.Z., F.Z., D.S. and T.Z.; methodology, F.H., B.Y., P.L. and M.B.; software, F.H. and B.Y.; validation, F.H., B.Y., P.L. and M.B.; data curation, F.H.; supervision, T.Z., F.Z., P.L., M.B. and B.Y.; project administration, B.Y. and T.Z.; writing—original draft preparation, F.H., B.Y., P.L. and M.B.; writing—review and editing, F.H., B.Y., P.L., M.B., J.Z., F.Z., D.S. and T.Z.; visualization, F.H. All authors have read and agreed to the published version of the manuscript.

**Funding:** This work was supported in part by the National Science Foundation under Grants ECCS-1810108, and by a project funded by ISO New England. Any opinions, findings, conclusions or recommendations expressed in this paper are those of the authors and do not reflect the views of NSF or ISO New England.

**Data Availability Statement:** Data sharing not applicable.

**Conflicts of Interest:** The authors declare that they have no known competing financial interests or personal relationships that could have appeared to influence the work reported in this paper.

# References

1. New York Independent System Operator, Inc. Market Administration and Control Area Services Tariff. 2022. Available online: https://nyisoviewer.etariff.biz/ViewerDocLibrary/MasterTariffs/9FullTariffNYISOMST.pdf (accessed on 21 April 2023).
2. Corporation, C.I.S.O. California Independent System Operator Corporation Fifth Replacement. 2019. FERC electric tariff. Available online: http://www.caiso.com/Documents/Conformed-Tariff-asof-Sept28-2019.pdf (accessed on 21 April 2023).
3. Tong, J.; Ni, H. Look-ahead multi-time frame generator control and dispatch method in PJM real time operations. In Proceedings of the 2011 IEEE Power and Energy Society General Meeting, Detroit, MI, USA, 24–29 July 2011; p. 1.
4. Hogan, W.W. Electricity market design and zero-marginal cost generation. *Curr. Sustain./Renew. Energy Rep.* **2022**, *9*, 15–26. [CrossRef]
5. Hogan, W.W. Electricity market design: Optimization and market equilibrium. In Proceedings of the Workshop on Optimization and Equilibrium in Energy Economics, Institute for Pure and Applied Mathematics, Los Angeles, CA, USA, 13 January 2016.
6. Hua, B.; Schiro, D.A.; Zheng, T.; Baldick, R.; Litvinov, E. Pricing in multi-interval real-time markets. *IEEE Trans. Power Syst.* **2019**, *34*, 2696–2705. [CrossRef]
7. Schiro, D.A. Flexibility procurement and reimbursement: A multi-period pricing approach. In Proceedings of the FERC Technical Conference, Washington, DC, USA, 26 June 2017.
8. Zhao, J.; Zheng, T.; Litvinov, E. A multi-period market design for markets with intertemporal constraints. *IEEE Trans. Power Syst.* **2019**, *35*, 3015–3025. [CrossRef]
9. Yang, Z.; Wang, Y.; Yu, J.; Yang, Y. On the minimization of uplift payments for multi-period dispatch. *IEEE Trans. Power Syst.* **2020**, *35*, 2479–2482. [CrossRef]
10. Guo, Y.; Chen, C.; Tong, L. Pricing multi-interval dispatch under uncertainty part I: Dispatch-following incentives. *IEEE Trans. Power Syst.* **2021**, *36*, 3865–3877. [CrossRef]
11. Chen, C.; Guo, Y.; Tong, L. Pricing multi-interval dispatch under uncertainty part II: Generalization and performance. *IEEE Trans. Power Syst.* **2020**, *36*, 3878–3886. [CrossRef]

*Article*

# Use of Park's Vector Method for Monitoring the Rotor Condition of an Induction Motor as a Part of the Built-In Diagnostic System of Electric Drives of Transport

Oleg Gubarevych [1], Juraj Gerlici [2], Oleksandr Kravchenko [2,*], Inna Melkonova [3] and Olha Melnyk [4]

[1] Department of Electromechanics and Rolling Stock of Railways, Kyiv Institute of Railway Transport of State University of Infrastructure and Technologies, 04071 Kyiv, Ukraine; oleg.gbr@ukr.net

[2] Department of Transport and Handling Machines, Faculty of Mechanical Engineering, University of Zilina, 01026 Zilina, Slovakia; juraj.gerlici@fstroj.uniza.sk

[3] Department of Electrical Engineering, Volodymyr Dahl East Ukrainian National University, 91000 Kyiv, Ukraine; inna.mia.lg@gmail.com

[4] Department of Ship Power Units, Auxiliary Mechanisms of Ships and Their Operation, Kyiv Institute of Water Transport of State University of Infrastructure and Technologies, 04071 Kyiv, Ukraine; olga-melnik81@ukr.net

* Correspondence: oleksandr.kravchenko@fstroj.uniza.sk; Tel.: +421-41-513-2660

**Abstract:** The article is devoted to the use of Park's vector method for operational control of the rotor condition of induction motors of traction and auxiliary drives of railway rolling stock. In the course of the analysis, it was established that in order to increase the reliability and efficiency of the operation of vehicles, it is necessary to improve and implement diagnostic systems for monitoring the current state of the most damaged elements of induction electric motors built into the drive. This paper presents the development of a new approach to monitoring the state of a squirrel-cage rotor, which is based on the use of Park's vector approach. In the course of the research, the issue of taking into account the asymmetric power supply of the engine during the diagnostic period during industrial operation was solved, which affects the accuracy of determining the degree of damage to the rotor. On the basis of the conducted research, the algorithm of the module for diagnosing the state of the squirrel-cage rotor of an induction motor has been developed for practical use in the built-in on-board systems of vehicles, which allows us to determine the degree of damage and monitor the development of the rotor defect during operation, including in automated mode.

**Keywords:** induction motor; fault detection; Park's vector approach; rotor diagnostics; transport equipment monitoring; on-board diagnostic system

**Citation:** Gubarevych, O.; Gerlici, J.; Kravchenko, O.; Melkonova, I.; Melnyk, O. Use of Park's Vector Method for Monitoring the Rotor Condition of an Induction Motor as a Part of the Built-In Diagnostic System of Electric Drives of Transport. *Energies* **2023**, *16*, 5109. https://doi.org/10.3390/en16135109

Academic Editor: Silvio Simani

Received: 4 June 2023
Revised: 26 June 2023
Accepted: 30 June 2023
Published: 2 July 2023

## 1. Introduction

### 1.1. Motivation and Relevance

Increasing the operational efficiency and reliability of vehicles is an important modern problem, which depends not only on the fulfillment of logistical tasks but also on ensuring the safety of all types of transportation. The development of new, and the improvement of existing systems of current diagnosis of the condition of electrical equipment of vehicles is the main factor in solving this issue. Prompt detection of damage to elements of vehicle equipment provides an opportunity to take timely measures to eliminate or replace them during pre-planned repair or maintenance. A greater share of transport equipment failures, taking into account the difficult operating conditions, is due to the electric motors of drives of various mechanisms and devices, where the main element is the electric motor.

At present, induction motors with squirrel-cage rotors of various capacities are used mainly as part of the main and auxiliary equipment of modern vehicles.

They are used as traction engines in railway transport, for driving compressors, machine-fans used for cooling traction engines, and for other equipment on rolling stock of railways and water transport [1–4]. Despite the fairly high reliability of induction electric motors with a squirrel-cage rotor among other types of electric motors, they are variously vulnerable to damage that affects their performance and failure during operation [5–7]. Especially for the operation of the transport infrastructure, the reduction of emergency failures is an important influential factor in the implementation of logistics technologies [8]. In the locomotive industry, the assessment of the state of road safety and the reliability of transportation in most cases is carried out while taking into account the volume of work performed by the locomotive; that is, relative indicators of the volume of transportation are used [9]. The development and implementation of systems for the continuous monitoring of the state of the main elements of induction electric motors during their operation will increase the reliability and efficiency of the use of vehicles and decrease emergency failures.

*1.2. Literature Review*

The problems of controlling the current state during the operation of induction motors are the subject of research by many scientists. The authors of the research use various methods for monitoring the state of electrical machines, however, the need to monitor data on the state of the engine during operation, to predict the uptime remains relevant [10,11]. Thus, in [12], a method of constant monitoring of the condition of an induction motor using the analysis of electrical characteristics is proposed. The given technique is capable of predicting various types of failures, i.e., rotor failures, stator phase misalignments, and power cable failures in the early stages. The authors of [13] cite a study of defects that cause changes in the vibration signature over time. The methods of vibration monitoring considered in the work allow for diagnosing the condition of the electric machine from such damages as bearing defects and wear of the rotor and stator. However, research [14] indicates a number of limitations when using vibration diagnostic methods for industrial conditions. Ref. [15] gives an overview of the methods of early diagnosis of induction motor faults based on sounds and acoustic emission for four types of damage: bearings, rotor, stator, and element connections. Effective results of diagnostics of the main elements of an induction motor are obtained using current methods, which seem to be the most appropriate for practical implementation and use in diagnostic monitoring systems that meet modern operational requirements. The authors in [16] cite a study on engine diagnostics using the spectral characteristics of the engine stator current and engine speed. The diagnosis of the main elements of the engine–stator, rotor, and bearings–was performed using an automatically adjusted algorithm of reference vectors with arbitrary characteristics. Ref. [17] presents the results of research on the example of diagnosing damage to the rotor of an induction motor by five different techniques of signal processing using the stator current. The most effective results are obtained by Park's vector approach and the Hilbert transformation method. However, the use of the Hilbert transform method involves a quality power system used in stationary conditions, i.c., bench tests of equipment.

Works [18–20] show the effectiveness of using the current method of Park's vector approach to detect multiple malfunctions of induction motors. The main difference from the simple spectral analysis of current signals in the formation of Park's vector module spectra is that any characteristic frequency of the amplitude-modulated signal is taken into account in Park's vector spectrum only once. The harmonics in the current spectrum, which correspond to different types of faults, differ from each other. Thus, the detection of characteristic harmonics in the current spectrum reliably and unambiguously indicates the presence of specific defects in the electric machine. In [21], the authors presented the results of studies on the effective use of the Park vector method for diagnosing an interturn short circuit in the stator winding with the determination of the number of closed turns in a low-quality motor power supply system.

Thus, Park's vector method is the most effective and promising current method for detecting engine damage in the early stages for use in monitoring systems. However, the identification of malfunctions of various engine elements with the determination of the type and degree of their damage for various power and load conditions using Park's vector approach is complex and requires further research. The main disadvantages of using Park's method is the difficulty of detecting damage in the idling mode and its identification with an asymmetric power system. When using this method in the built-in monitoring systems of transport equipment, it is not expected to establish malfunctions in the idle mode. Conducting research on the development of control modules for the main elements of an induction motor for use in the built-in systems of on-board diagnostics of the induction electric machines of vehicles using Park's vector method is a relevant and promising modern task.

### 1.3. Contribution

This article is devoted to the development of an algorithm based on the proposed method of determining damage to the squirrel-cage rotor winding of an induction motor. Park's vector approach is used to detect core rotor damage. Damage to the rotor is determined by the thickness of the ring of the Park's pattern. The proposed algorithm takes into account the power supply of the machine from a non-sinusoidal and asymmetric voltage system, which is relevant for diagnostics in industrial conditions. Identification of the degree of damage to the rotor is determined by recalculating the stator currents to the orthogonal circular basis and calculating the difference between the maximum value of the Park's vector for the outer circle and the minimum value for the inner circle. The use of Park's method makes it possible to obtain information about the current state of the engine rotor from the moment of damage and to control the degree of development of the defect during operation. This will allow early diagnosis of damage to the rotor winding as part of the diagnostic system of transport equipment during the operation of induction motors to predict the period of trouble-free operation. The use of Park's vector approach has also shown effective results for diagnosing the stator winding and mechanical damage of an induction motor and can be considered one of the most reliable and convenient diagnostic methods.

### 1.4. Organization of the Paper

The work has the following structure: in Section 2, an analysis of the damage of the main elements of the induction motor is carried out with a more detailed consideration of rotor damage; Section 3 presents the theoretical principles of Park's vector approach, constructs the Park's vector diagram by means of mathematical modeling, and develops a methodology and algorithm for determining the degree of damage to the squirrel-cage rotor winding for use in built-in diagnostic systems of transport; Section 4 provides a discussion of the obtained results and the direction of further scientific research; conclusions are presented in Section 5.

## 2. The Main Elements of Monitoring the State of an Induction Electric Motor

The main elements of the engine that require state control during engine operation, which are among the most necessary to be included in the diagnostic monitoring system according to modern review studies, are stator winding, squirrel-cage rotor winding, and bearings [22,23].

According to the operational statistics [12,16,22], a quantitative analysis of the damage of asynchronous motors with a squirrel-cage rotor was carried out, the averaged results of which are combined into groups taking into account the most frequent failures and are presented in the Figure 1.

The largest share of failures of an induction motor with a squirrel-cage rotor (47%) is a consequence of damage to the stator winding. The main cause of failures is the occurrence of a turn-to-turn short circuit in the winding phase [20,21,24]. Turn-to-turn short-circuits refer to defects that are difficult to diagnose in the early stages, in which the induction motor continues to work with the deterioration of performance characteristics and energy indicators. At the same time, this defect can develop over time, and when a certain number of short-circuited turns is reached, insulation breakdown occurs as a result of which the electric machine fails. From this, it follows that the presence of a turn-to-turn short-circuit control module in the stator windings in the diagnostic system built into the drive is a necessary component. To determine the turn-to-turn shorting of the stator winding, a number of studies using current methods are presented [6,16,17]. The most effective results were obtained using the Park's vector method [19–21]. In [21], the authors presented research on the use of the Park's vector method to determine the number of closed turns in the stator phase winding of an induction motor using stator currents. In continuation of these studies, in [23], an algorithm and a functional scheme were developed for the practical implementation of the stator winding state control module to the diagnostic embedded system based on Park's vector approach.

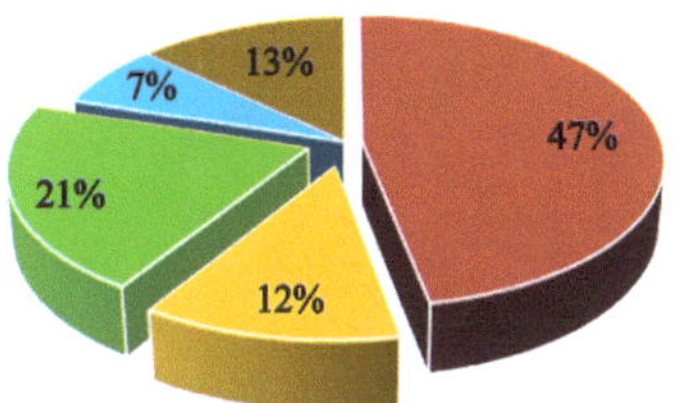


**Figure 1.** Distribution of induction motor failures.

The proportion of bearing damage is 21% of all electric machine failures (Figure 1). Given the small size of the air gap, bearing defects eventually lead to the engagement of the rotor with the stator, which leads to significant consequences in the condition of the engine. In addition, the appearance of vibration due to damage to the bearing has a destructive effect on all elements of the engine structure. Various methods of vibration diagnostics are widely used to control the vibration state of an electric machine.

Another important structural element of an induction motor, which affects the reliability of the machine and is subject to monitoring, is rotor winding. During operation, the rotor is exposed to centrifugal forces, thermal expansion, shock current loads, and electrodynamic forces, and in some cases, the "squirrel cage" of the rotor loses its structural integrity. Damage to the rotor winding, according to the given general statistics (see Figure 1), makes up 12% of engine failures, but for powerful traction engines, rotor failures have significantly greater values. Damage to the rotor winding manifests in the form of the breakage of some rods or the destruction of contact in the rods of the "squirrel cage". Structurally, the rods of the "squirrel cage" at the exit from the rotor grooves are connected to short-circuited rings located at some distance from the rotor core on both sides. When the rotor rotates, significant mechanical forces occur, which contribute to the creation of rod breaks at the exit from the rotor core or near short-circuited rings, especially during the start-up or limit load periods of rotors with copper or brass rods. For rotors with a winding made of aluminum, rod breaks occur more often in the grooves. Figure 2 shows images of rotor damage for various "squirrel cage" designs.

When operating machines with a broken "squirrel cage" structure, pulsation of currents occurs in the stator with a slip frequency, which creates vibration and affects the torque of the machine. At the same time, the frequency of rotation of the rotor fluctuates even with changes in small loads.

The occurrence of a contact failure in individual rods leads to an increase in the load on other rods remaining in the structure of the "squirrel cage" and increases their heating. Increased overheating during start-ups and significant engine loads leads to further destruction of the rods of the rotor winding structure, leading to emergency engine failure. In addition, the separated rods, under the action of centrifugal forces, are prone to displacement and exit from the grooves in the direction of the air gap (see Figure 2a,b), which leads to their engagement with the winding or the stator core and additional financial costs during engine restoration. The necessary current monitoring of the state of the rotor winding during operation as part of the built-in diagnostic system will reduce the probability of emergency failure and contribute to the timely identification of the appearance of a defect before the development and failure of the equipment. To solve the problem of early detection of the type and degree of damage to the rotor winding with a poor-quality power supply system in industrial conditions, the most effective is application of current diagnostic methods [17].

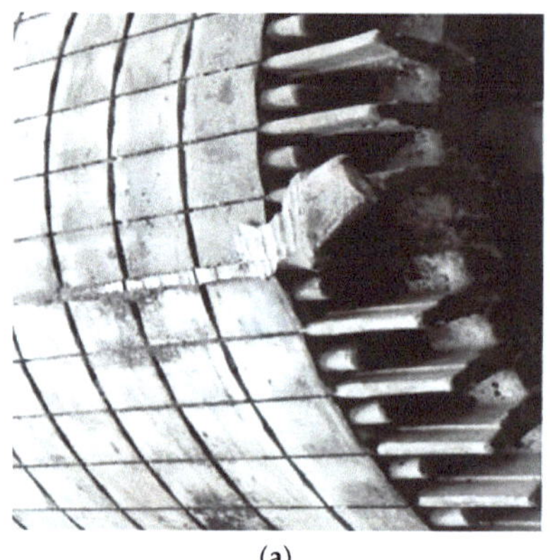

(a)      (b)

**Figure 2.** Damage to the rotor winding: (**a**) cast winding; (**b**) welded winding.

## 3. Development of an Algorithm for Determining the State of the Squirrel-Cage Rotor Winding of an Induction Motor

*3.1. Using the Park's Vector Method for Rotor Diagnostics*

A number of works [21,25,26] of modern researchers are devoted to the use of Park's vector approach, which belongs to current diagnostic methods for diagnosing elements of electric machines and drives.

The basis of Park's vector approach is the transformation of the three-phase coordinate system of the stator currents into a two-dimensional moving system (dq-coordinates). The trajectory described by the end of the created vector on the coordinate plane with the corresponding axes of the stator current $I_{sd}$ and $I_{sq}$ when the power supply frequency changes has diagnostic signs of both electrical and mechanical engine defects.

Park's vector currents $I_{sd}$ and $I_{sq}$ are determined from ratios based on the previously measured stator phase currents $I_A$, $I_B$, and $I_C$ [17]:

$$\text{Isd} = \sqrt{\frac{2}{3}}\text{IA} - \sqrt{\frac{1}{6}}\text{IB} - \sqrt{\frac{1}{6}}\text{IC} \tag{1}$$

$$\text{Isq} = \sqrt{\frac{1}{2}}\text{IB} - \sqrt{\frac{1}{2}}\text{IC} \tag{2}$$

Then, in the dq-coordinate system, the Park's vector for the engine describes a figure centered at the origin according to the equation:

$$\bar{\text{Is}} = \text{Isd} + \text{J} \cdot \text{Isq} \tag{3}$$

In the presence of a working electric machine, all currents are balanced and have no deviations from the normal mode. Then, under the condition of power supply from a strictly symmetrical voltage system, the Park's vector pattern has a regular circle centered at the origin of the dq system coordinates [18–20].

Depending on the degree of damage to the stator winding, rotor winding, bearings, or a violation of the symmetry of the supply voltage system, three-phase stator currents result in a changed form of the Park's vector pattern [27]. Despite the simplicity of damage detection using Park's vector approach, the identification of different types of damage from a graphical representation of the presence of off-state modes of operation is too difficult. The vector pattern is affected by the quality of the power supply, the engine's operating mode, the type of load, and many other factors. The most effective results were obtained for diagnosing the stator with the determination of the number of closed turns in case of a turn-to-turn short circuit in the phase of the stator winding, taking into account a poor-quality power supply system [21]. In these studies, the authors developed an algorithm for the practical use of the module for establishing the state and degree of the stator winding as part of the built-in diagnostic system [23]. The construction of a Park's vector ring is provided using a mathematical model of an induction motor with the addition of a Park's vector hodograph block.

The considered principle of using the Park's vector method also makes it possible to determine the defects of the rotor winding of an induction motor with high accuracy. In [26], the authors presented a study of the influence of this type of malfunction on the phase current of the machine using the Park's vector transformation method with analysis using complex wavelets. Determining the defect by the considered method is possible for stationary and bench conditions without the influence of external interference, especially if the induction motor operates at low slip values, when the characteristic frequency components of the rotor fault are very close to the main frequency component. The use of this method does not allow determination of the state of the rotor during operation with poor engine power. The development of this study is given in [18,26,28], where the analysis of the graphic drawing of the Park's vector in case of rotor damage is carried out using different approaches. In the studies cited in [26] regarding the diagnosis of rotor rod breakage in induction motors, an increase in the width of the Park's vector ring is used using more complex signal processing methods. In this work, it is shown that the result of defect detection depends on the magnetic poles and the number of rotor grooves, for which it is necessary to introduce a new approach with signal filtering, which greatly complicates the operational determination of the type of damage, and it is difficult to implement control in an automated mode. The authors of [28] use the deformation of the Park's vector pattern as an indicator for predicting the condition using different magnetic saturation in a working and faulty induction motor with a squirrel-cage rotor for rotor diagnostics. This approach does not allow us to accurately identify the degree of damage to the rotor and does not take into account the changes in parameters associated with a poor power supply system. This issue is partially resolved in [29], where the conducted studies established that the degree of failure depends on the width of the ring of the Park's vector pattern and gradually increases as the severity of the rotor malfunction increases. By monitoring the relative width of Park's vector patterns, it is possible to identify rotor defects and control their development in an automated control mode. The authors propose to measure the width of the ring by monitoring the amplitude of a specific frequency in the frequency spectrum of the Park's vector module, where rotor faults create spectral components in the left and right parts of the main frequency spectrum of linear currents. Control of these sidebands in the current spectra leads to an increase in the width of the Park's vector trajectory. However, the work does not take into account the effect of an asymmetric system of voltages on the change of spectral components, which are informative for assessing the degree of damage to the engine rotor.

Figure 3 shows the difference between the Park's vector patterns for an undamaged AIR132 M4 engine with a power of 11 kW (Figure 3a) and with simulated damage to the rotor rods (Figure 3b) with a symmetrical power supply system. A simulation model of an induction motor with a squirrel-cage rotor, made in the MATLab 2018b software environment and presented in [21,30] with the established adequacy of the simulation results [31], was used to construct the hodograph of the Park's vector. In order to obtain reliable research results, a stable mode of operation of the engine without the influence of transient processes was considered. 2018b

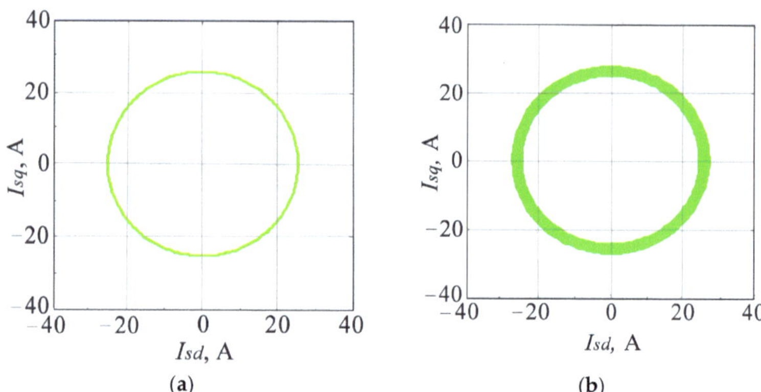

**Figure 3.** Hodographs of the Park's vector with a symmetrical induction power supply system of an electric machine: (**a**) for an intact rotor; (**b**) with one damaged rod rotor.

### 3.2. Development of a Technique for Determining Rotor Damage by the Park's Vector Hodograph Method

As can be seen from Figure 3a, with a symmetrical system of the power supply voltages of an induction motor, the hodograph of the Park's vector describes a circle, and the thickening of the Park's vector line means that there is damage in the rotor. It is possible to determine the presence of rotor damage by calculating the thickness of the Park's vector line. According to the conducted simulation, when the supply voltage system is not symmetrical, the trajectory of the Park's vector describes an ellipse even with symmetrical stator windings and an intact rotor. The effect of rotor damage for an unbalanced power system on the Park's vector pattern is also observed. If the rotor is not damaged, the ellipse will be described by a thin line, and if the rotor is damaged, the line of the ellipse will thicken. In addition to the creation of an ellipse, when the symmetry of the supply voltage is violated, the slope of the ellipse is created with an angle that depends on the degree of asymmetry of the supply voltages, which also affects the accuracy of the calculation of the thickness of the Park's vector pattern. To determine the thickness of the ellipse line—which depends on the accuracy of quantifying the degree of damage—in the presence of asymmetry in the supply voltage system, the trajectory of the Park's vector should be transferred from an orthogonal elliptical basis (Figure 4b) to an orthogonal circular one (Figure 4a).

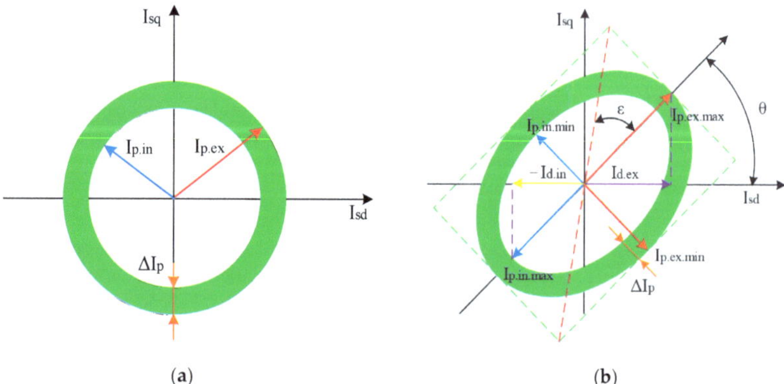

**Figure 4.** The trajectory of the Park's vector with a damaged rotor: (**a**) with a symmetrical power system; (**b**) with an asymmetric power system.

The following designations are accepted in Figure 4:

$I_{p.ex}$—Park's vector of the external circle (Figure 4a);

$I_{p.in}$—Park's vector of the inner circle (Figure 4a);

$I_{p.ex.max}$—the maximum value of the Park's vector for the external circle (Figure 4b);

$I_{p.ex.min}$—the minimum value of the Park's vector for the external circle (Figure 4b);

$I_{p.in.max}$—the maximum value of the Park's vector for the inner circle (Figure 4b);

$I_{p.in.min}$—the minimum value of the Park's vector for the inner circle (Figure 4b);

$I_{d.ex}$—projection of the Park's vector of the external circle onto the d-axis (Figure 4b);

$I_{d.in}$—projection of the Park's vector of the inner circle onto the d-axis (Figure 4b);

$\varepsilon$—angle of ellipticity;

$\theta$—angle of inclination of the ellipse.

To determine the degree of damage to the rotor with a symmetrical power system (see Figure 4a), the thickness of the circle line of the Park's vector is calculated according to the formula:

$$\Delta I_p = I_{p.ex} - I_{p.in}, \tag{4}$$

where:

$I_{p.ex}$—Park's vector of the external circle;

$I_{p.in}$—Park's vector of the inner circle.

Since, with the symmetry of the supply voltage system, the modulus of the Park's vector is the instantaneous value of the phase current of phase A, then expression (4) takes the following form:

$$\Delta I_p = I_{sAmax} - I_{sAmin}, \tag{5}$$

where:

$I_{sAmax}$—the maximum instantaneous value of the stator phase current of phase A;

$I_{sAmin}$—the minimum instantaneous value of the stator phase current of phase A.

A value of $\Delta I_p = 0$ will mean that the rotor is intact. A value of $\Delta i_p > 0$ will indicate the presence of damage in the rotor, and the larger the value of $\Delta i_p$, the more damaged the rotor.

When determining the thickness of the ellipse line in the presence of asymmetry in the supply voltage system, the calculation is performed in the following sequence to convert the Park's vector pattern from an orthogonal elliptical basis (Figure 4b) to an orthogonal-circular one (Figure 4a).

Transition from three-phase to two-phase dq coordinate system according to formulae:

$$\begin{cases} I_{sd} = I_{sA} \cdot \cos(\omega \cdot t + \varphi) - \frac{1}{\sqrt{3}} \cdot I_{sB} \cdot \sin(\omega \cdot t + \varphi) - \frac{1}{\sqrt{3}} \cdot I_{sC} \cdot \sin(\omega \cdot t + \varphi); \\ I_{sq} = I_{sA} \cdot \sin(\omega \cdot t + \varphi) + \frac{1}{\sqrt{3}} \cdot I_{sB} \cdot \cos(\omega \cdot t + \varphi) - \frac{1}{\sqrt{3}} \cdot I_{sC} \cdot \cos(\omega \cdot t + \varphi), \end{cases} \tag{6}$$

where:

$I_{sA}$—the instantaneous value of the stator phase current of phase A;

$I_{sB}$—the instantaneous value of the stator phase current of phase B;

$I_{sC}$—the instantaneous value of the stator phase current of phase C;

$\omega$—the angular frequency of the supply voltage;

$\varphi$—the phase shift between phase voltages and currents.

Formula (6) determines the maximum and minimum values of the Park's vector $I_{pmax}$ and $I_{pmin}$, respectively. The value with the max index corresponds to the Park's vector of the outer circle, and the value with the min index corresponds to the Park's vector of the inner circle.

When projecting the obtained maximum value of the outer circle of the Park's vector $I_{pmax}$ onto the d-axis, $I_{d.ex}$ is determined. Based on the values of the Park's vector for the outer circle $I_{pmax}$ and its projection on the d-axis $I_{d.ex}$, the angle of inclination of the ellipse of the drawing of the Park's vector is determined (see Figure 4b):

$$\theta = \arccos \frac{i_{d.ex}}{i_{pmax}}. \tag{7}$$

Since the ellipses of the external and inner circles have the same angles of inclination (Figure 4b), it is not necessary to determine the angle of inclination for the inner circle.

Then, the transition from an orthogonal elliptical basis (see Figure 4b) to an orthogonal circular one (see Figure 4a) can be carried out using the formulae:

$$i'_{d0ex} = (\cos \varepsilon_0 \cdot \cos \theta_0 - j \cdot \sin \varepsilon_0 \cdot \sin \theta_0) \cdot I_{pmax} \cdot \cos \theta + j \cdot (\cos \varepsilon_0 \cdot \sin \theta_0 - j \cdot \sin \varepsilon_0 \cdot \cos \theta_0) \cdot I_{pmax} \cdot \sin \theta, \tag{8}$$

$$\begin{aligned} i'_{q0ex} = \big(\cos(-\varepsilon_0) \cdot \cos(\theta_0 + \tfrac{\pi}{2}) - j \cdot \sin(-\varepsilon_0) \cdot (\theta_0 + \tfrac{\pi}{2})\big) \cdot I_{pmax} \cdot \cos \theta + \\ + j \cdot \big(\cos(-\varepsilon_0) \cdot \sin(\theta_0 + \tfrac{\pi}{2}) - j \cdot \sin(-\varepsilon_0) \cdot \cos(\theta_0 + \tfrac{\pi}{2})\big) \cdot I_{pmax} \cdot \sin \theta, \end{aligned} \tag{9}$$

where $\varepsilon_0$—the angle of ellipticity of the basic single vector along the d-axis in the new basis (along the q-axis, the value of this angle is $\varepsilon 0$. For an orthogonal-circular basis $\varepsilon_0 = \pi/4$); $\theta_0$—the angle of inclination of the basic single vector ellipse along the d-axis in the new basis (along the q-axis, the value of this angle is equal to $\theta_0 + \pi/2$). For an orthogonal-circular basis $\theta_0 = 0$.

The maximum value of the Park's vector for the inner circle is determined by the expression (Figure 4b):

$$I_{p.in.max} = \frac{I_{d.in}}{\cos \theta} = \frac{I_{pmin}}{\cos \theta \cdot \sin \theta}. \tag{10}$$

Substitution of expression (10) in (8) and (9) gives the following results:

$$i'_{d0in} = (\cos \varepsilon_0 \cdot \cos \theta_0 - j \cdot \sin \varepsilon_0 \cdot \sin \theta_0) \cdot \frac{i_{pmin}}{\sin \theta} \cdot \cos \theta + j \cdot (\cos \varepsilon_0 \cdot \sin \theta_0 - j \cdot \sin \varepsilon_0 \cdot \cos \theta_0) \cdot i_{pmax} \cdot \frac{i_{pmin}}{\cos \theta}, \tag{11}$$

$$\begin{aligned} i'_{q0in} = \big(\cos(-\varepsilon_0) \cdot \cos(\theta_0 + \tfrac{\pi}{2}) - j \cdot \sin(-\varepsilon_0) \cdot (\theta_0 + \tfrac{\pi}{2})\big) \cdot \tfrac{i_{pmin}}{\sin \theta} + \\ + j \cdot \big(\cos(-\varepsilon_0) \cdot \sin(\theta_0 + \tfrac{\pi}{2}) - j \cdot \sin(-\varepsilon_0) \cdot \cos(\theta_0 + \tfrac{\pi}{2})\big) \cdot \tfrac{i_{pmin}}{\cos \theta}. \end{aligned} \tag{12}$$

Then, the projections of the Park's vector of the external circle in the new basis can be defined as:

$$\begin{cases} i'_{d0ex} = \sqrt{\big(\mathrm{Re}(i'_{d0ex})\big)^2 + \big(\mathrm{Im}(i'_{d0ex})\big)^2}; \\ i'_{q0ex} = \sqrt{\big(\mathrm{Re}(i'_{d0ex})\big)^2 + \big(\mathrm{Im}(i'_{d0ex})\big)^2}. \end{cases} \tag{13}$$

Projections of the Park's vector of the inner circle in the new basis are defined as:

$$\begin{cases} i'_{d0in} = \sqrt{\big(\mathrm{Re}(i'_{d0in})\big)^2 + \big(\mathrm{Im}(i'_{d0in})\big)^2}; \\ i'_{q0in} = \sqrt{\big(\mathrm{Re}(i'_{d0in})\big)^2 + \big(\mathrm{Im}(i'_{d0in})\big)^2}. \end{cases} \tag{14}$$

If $I'_{d0ex} = I'_{q0ex}$ and $I'_{d0in} = I'_{q0in}$ are equal, and if $I'_{d0ex} \neq I'_{d0in}$ and $I'_{q0ex} \neq I'_{q0in}$, there will be damage in the rotor of the induction motor. When $I'_{d0ex} \neq I'_{q0ex}$ and $I'_{d0in} \neq I'_{q0in}$ and I $I'_{d0ex} \neq I'_{d0in}$ and $I'_{q0ex} \neq I'_{q0in}$, damage will occur in both the stator and the rotor. If $I'_{d0ex} \neq I'_{q0ex}$ and I' $I'_{d0in} \neq I'_{q0in}$ and $I'_{d0ex} = I'_{d0in}$ and $I'_{q0ex} = I'_{q0in}$, damage will occur only in the stator [21].

Since it was found that when only the rotor of the induction motor is damaged, $I'_{d0ex} = I'_{q0ex}$ and $I'_{d0in} = I'_{q0in}$, the thickness of the Park's vector line can be calculated using the expression:

$$\Delta i_p = i'_{d0ex} - i'_{d0in}, \tag{15}$$

or by expression:

$$\Delta i_p = i'_{q0ex} - i'_{q0in}. \tag{16}$$

Calculation of the figure thickness of the Park's vector according to expressions (15) and (16) according to the proposed method allows us to obtain the value of $\Delta I_p$, which can be used to quantitatively assess the degree of damage to the rotor winding of an induction electric machine with a sufficiently high reliability, regardless of the quality of the power supply system.

### 3.3. Algorithm for Using Park's Method for Rotor Diagnostics in Built-In Diagnostic Systems

In accordance with the proposed method of calculating the thickness of the Park's vector figure, an algorithm for diagnosing damage to the rotor of an induction motor in the built-in vehicle diagnostics system has been developed. When compiling the algorithm, we considered the option of sharing the same sensors for the stator interturn diagnostic module and determining the state of the rotor using the Park's vector method. The algorithm for diagnosing damage to the rotor of an induction motor for practical implementation is shown in Figure 5.

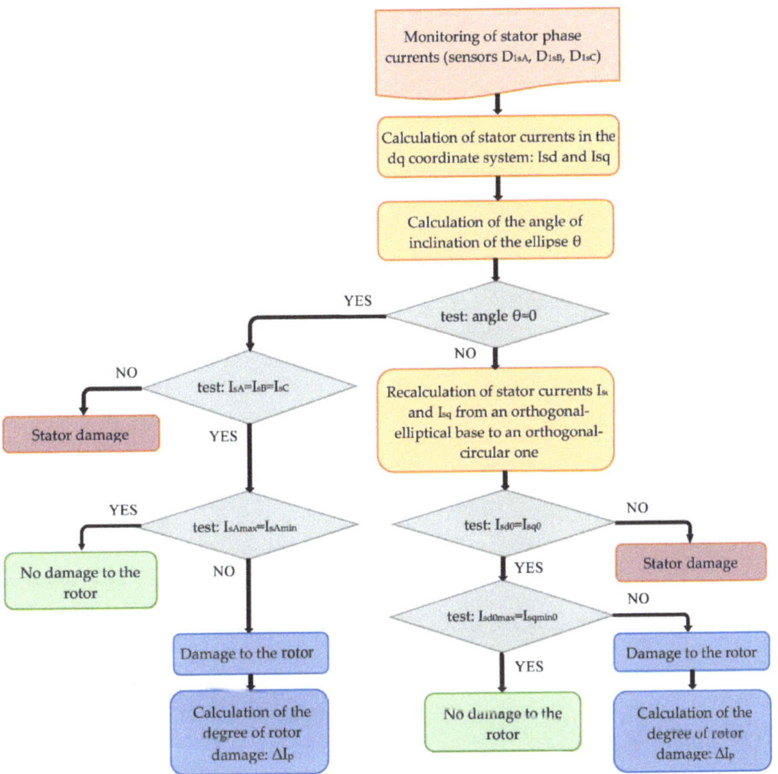


**Figure 5.** Algorithm for diagnosing damage to the rotor of an induction motor.

To obtain the value of the phase currents, three current sensors—$D_{IsA}$, $D_{IsB}$, and $D_{IsC}$—are used. It is typical for transport systems that an induction electric motor receives power from an autonomous voltage inverter, so the supply voltage system can be non-sinusoidal in nature. Therefore, to determine the amplitude and phase of the first (fundamental) harmonic of the stator phase currents ($I_{sA1}$, $I_{sB1}$, $I_{sC1}$, $\varphi_{IsA1}$, $\varphi_{IsB1}$, $\varphi_I$), a fast Fourier transform block is used to correctly obtain the stator current diagrams.

The amplitudes of the first harmonics of the stator phase currents ($I_{sA1}$, $I_{sB1}$, $I_{sC1}$) are sent to the block where they are converted from the three-phase coordinate system to the two-phase moving dq coordinate system with the receipt of $I_{sd}$ and $I_{sq}$ currents in accordance with (5). After determining $I_{pmax}$ and its projection on the d-axis ($I_{d.ex}$), the angle of inclination of the ellipse of the drawing of the Park's vector from (6) is determined. If the angle is θ = 0, then the induction motor has a symmetrical quality supply, and the stator currents are used to determine the damage element of the motor. In case of damage to the rotor winding, the difference between the maximum instantaneous phase current of the stator $I_{sAmax}$ and the minimum instantaneous phase current $I_{sAmin}$ gives the value of the parameter $\Delta I_p$ (5), which determines the degree of damage to the rotor.

At the set value of the angle of inclination of the shape of the trajectory of the Park's vector (Figure 4b), the electric machine receives low-quality power; therefore, in order to accurately determine the parameter $\Delta I_p$, it is necessary to recalculate the current values $I_{sd}$ and $I_{sq}$ for the transition from the orthogonal elliptical basis to the orthogonal circular basis according to the ratios (7)–(11).

After determining the values of the projections of the stator currents of the Park's vector figure on the d-axis ($I_{sd0}$) and on the q-axis ($I_{sq0}$), the damaged machine element is set. If the values of current projections are uneven, the stator winding of the machine is damaged. In the case of the $I_{sd0}=I_{sq0}$ equation, the maximum instantaneous value of the phase current projection $I_{sd0max}$ and the minimum instantaneous value of the stator phase current projection $I_{sd0min}$ are compared. When the values of the current projections differ, the parameter $\Delta I_p$ is calculated, which can be called the "rotor damage criterion" in the future. An increase in the $\Delta I_p$ parameter indicates an increase in the level of damage to the squirrel-cage rotor winding, the development of which can be monitored during the operation of the electric machine.

## 4. Discussion

The Park's vector hodograph method is the most promising method for monitoring the state of the main elements of an induction motor during operation using the automation of the diagnostic process. The advantage of the method is the ability to obtain reliable diagnostic parameters despite the quality of the power supply system being disturbed at the earliest stages of their appearance. In the proposed rotor condition control algorithm, it is possible to establish the current technical condition of the rotor by comparing the width of the ring of the Park's vector circle with the reference value for this type of engine and to control the development of the degree of damage to the rotor by increasing the width of the ring of the Park's vector figure. To control the development of rotor defects during the operation of the electric machine, the damage criterion $\Delta I_p$ is used. When powering the machine through an inverter with a non-sinusoidal signal, the algorithm provides for the use of a fast Fourier transform block to determine the amplitude and phase of the first (main) harmonic of the phase currents to obtain correct diagnostic results. The main disadvantage of using Park's method in diagnostics is the difficulty of detecting damage in idling modes or with a slight engine load. However, for the use of the method in the built-in diagnostic systems of vehicles, this is not decisive. In order to identify the degree of severity of damage to the squirrel-cage rotor winding and to determine the number of damaged rods based on the value of the parameter $\Delta I_p$, it is necessary to conduct additional research by means of mathematical modeling. At the same time, already at this stage, it is possible to predict the period of trouble-free operation of the engine based on the deviation of the value of $\Delta I_p$. In addition, when using Park's method, some current sensors and a part of the calculation blocks are used to determine the state of the stator winding to ensure the diagnostic processes of both important engine elements.

The use of Park's vector method is also possible for the detection of mechanical damage, which demonstrates the universality and perspective of the method for use in the built-in diagnostic monitoring systems of all important elements of an induction motor.

The next work will be devoted to research on determining the number of damaged rotor winding rods via the proposed method. This is necessary for making a prognosis of trouble-free operation during the period of operation of the electric machines of vehicles and planning the recovery period in the event of damage.

## 5. Conclusions

The work provides statistics of damage to the main important elements of induction electric machines that are subject to control during the period of operation to ensure the efficiency and safety of transportation. Research on the development of a methodology and algorithm for the practical application of Park's vector approach for diagnosing rotor damage during engine operation is also presented.

In the course of research, a procedure was developed for recalculating the values of the amplitudes of the phase currents of the stator—obtained with a poor-quality power supply system—to the actual values of the currents of the symmetrical system of the supply voltages of the induction motor in order to correctly determine the degree of damage to the rotor.

A means of determining the thickness of the circle of the trajectory of the Park's vector is proposed for determining the degree of rotor damage by the difference between the maximum and minimum values of the phase currents of the Park's vector pattern, which correspond to the outer and inner size of the circle of the vector trajectory.

The algorithm of the module for monitoring the presence of damage in the squirrel-cage rotor winding has been developed for practical implementation in diagnostic systems with possible use together with the module for monitoring the stator from common sensors and part of the blocks.

The proposed control approach allows us to determine the degree of damage to the squirrel-cage rotor winding and to monitor the development of the defect during operation using an automated mode, regardless of the quality of the power supply system.

**Author Contributions:** Conceptualization, O.G. and O.K.; methodology, O.G., O.K. and J.G.; software, O.G. and I.M.; validation, O.G. and I.M.; formal analysis, O.K., I.M. and O.M.; investigation, O.G., O.K. and I.M.; resources, O.G., I.M. and O.M.; writing—original draft preparation, O.G. and O.K.; writing—review and editing, J.G. and O.K.; visualization, O.G., O.K. and I.M.; supervision, O.G. and J.G. All authors have read and agreed to the published version of the manuscript.

**Funding:** This publication was issued thanks to support from the Cultural and Educational Grant Agency of the Ministry of Education of the Slovak Republic in the projects, "Implementation of modern methods of computer and experimental analysis of properties of vehicle components in the education of future vehicle designers" (Project No. KEGA 036ŽU-4/2021). This research was also supported by the Slovak Research and Development Agency of the Ministry of Education, Science, Research and Sport of the Slovak Republic in Educational Grant Agency of the Ministry of Education of the Slovak Republic in the project and VEGA 1/0513/22 "Investigation of the properties of railway brake components in simulated operating conditions on a flywheel brake stand" and project "Scholarships for Excellent Researchers Threatened by the Military Conflict in Ukraine", code 09103-03-V01-00129.

**Data Availability Statement:** Not applicable.

**Conflicts of Interest:** The authors declare no conflict of interest.

## References

1. Tuychieva, M. Control of electric locomotives with asynchronous electric motors under asymmetric operating conditions in Uzbekistan. *IOP Conf. Ser. Earth Environ. Sci.* **2020**, *614*, 012060. [CrossRef]
2. Lovska, A.; Fomin, O.; Píštěk, V.; Kučera, P. Dynamic load modelling within combined transport trains during transportation on a railway ferry. *Appl. Sci.* **2020**, *10*, 5710. [CrossRef]
3. Kuznetsov, V.; Kardas-Cinal, E.; Gołębiowski, P.; Liubarskyi, B.; Gasanov, M.; Riabov, I.I.; Kondratieva, L.; Opala, M. Method of Selecting Energy-Efficient Parameters of an Electric Asynchronous Traction Motor for Diesel Shunting Locomotives—Case Study on the Example of a Locomotive Series ChME3 (ЧМЭ3, ČME3, ČKD S200). *Energies* **2022**, *15*, 317. [CrossRef]
4. Goolak, S.; Tkachenko, V.; Šťastniak, P.; Sapronova, S.; Liubarskyi, B. Analysis of Control Methods for the Traction Drive of an Alternating Current Electric Locomotive. *Symmetry* **2022**, *14*, 150. [CrossRef]
5. Balakrishna, P.; Khan, U. An Autonomous Electrical Signature Analysis-Based Method for Faults Monitoring in Industrial Motors. *IEEE Trans. Instrum. Meas.* **2021**, *70*, 1–8. [CrossRef]
6. Wu, Y.; Jiang, B.; Wang, Y. Incipient winding fault detection and diagnosis for squirrel-cage induction motors equipped on CRH trains. *ISA Trans.* **2020**, *99*, 488–495. [CrossRef] [PubMed]
7. Kondratieva, L.; Bogdanovs, A.; Overianova, L.; Riabov, I.; Goolak, S. Determination of the Working Energy Capacity of the On-Board Energy Storage System of an Electric Locomotive for Quarry Railway Transport During Working with a Limitation of Consumed Power. *Arch. Transp.* **2023**, *65*, 119–135. [CrossRef]

8.  Turpak, S.M.; Taran, I.O.; Fomin, O.V.; Tretiak, O.O. Logistic technology to deliver raw material for metallurgical production. *Sci. Bull. Natl. Min. Univ.* **2018**, *1*, 162–169. [CrossRef]

9.  Bodnar, B.; Ochkasov, O.; Bodnar, E.; Hryshechkina, T.; Keršys, R. Safety performance analysis of the movement and operation of locomotives. In Proceedings of the Transport means 2018: 22nd International Scientific Conference, Trakai, Lithuania, 3–5 October 2018; Kaunas University of Technology: Kaunas, Lithuania, 2018. Pt. II. pp. 839–843. Available online: https://transportmeans.ktu.edu/wp-content/uploads/sites/307/2018/02/Transport-means-II-A4-2018-09-25.pdf (accessed on 25 March 2023).

10. Gubarevych, O.; Goolak, S.; Golubieva, S. Classification of Defects, Systematization and Selection of Methods for Diagnosing the Stator Windings Insulation of Asynchronous Motors. *Rev. Roum. Sci. Techn.–Électrotechn. et Énerg* **2022**, *67*, 445–450.

11. Choudhary, A.; Goyal, D.; Shimi, S.L.; Akula, A. Condition Monitoring and Fault Diagnosis of Induction Motors: A Review. *Arch. Comput. Methods Eng.* **2019**, *26*, 1221–1238. [CrossRef]

12. Rauf1, A.; Usman, M.; Butt, A.; Ping, Z. Health Monitoring of Induction Motor Using Electrical Signature Analysis. *J. Dong Hua Univ.* **2022**, *39*, 265–271. [CrossRef]

13. Gundewar, S.K.; Kane, P.V. Condition Monitoring and Fault Diagnosis of Induction Motor. *J. Vib. Eng. Technol.* **2021**, *9*, 643–674. [CrossRef]

14. Gubarevych, O.; Gerlici, J.; Gorobchenko, O.; Kravchenko, K.; Zaika, D. Analysis of the features of application of vibration diagnostic methods of induction motors of transportation infrastructure using mathematical modeling. *Diagnostyka* **2023**, *24*, 2023111. [CrossRef]

15. AlShorman, O.; Alkahatni, F.; Masadeh, M.; Irfan, M.; Glowacz, A.; Althobiani, F.; Kozik, J.; Glowacz, W. Sounds and acoustic emission-based early fault diagnosis of induction motor. A review study. *Adv. Mech. Eng.* **2021**, *13*, 1–19. [CrossRef]

16. Yatsugi, K.; Pandarakone, S.E.; Mizuno, Y.; Nakamura, H. Common Diagnosis Approach to Three-Class Induction Motor Faults Using Stator Current Feature and Support Vector Machine. *IEEE Access* **2023**, *11*, 24945–24952. [CrossRef]

17. Abdelhak, G.; Sid Ahmed, B.; Djekidel, R. Fault diagnosis of induction motors rotor using current signature with different signal processing techniques. *Diagnostyka* **2022**, *23*, 2022201. [CrossRef]

18. Wei, L.; Rong, X.; Wang, H.; Yu, S.; Zhang, Y. Method for Identifying Stator and Rotor Faults of Induction Motors Based on Machine Vision. *Math. Probl. Eng.* **2021**, *2021*, 6658648. [CrossRef]

19. Vilhekar, T.G.; Ballal, M.S.; Suryawanshi, H.M. Application of Multiple Parks Vector Approach for Detection of Multiple Faults in Induction Motors. *J. Power Electron.* **2017**, *17*, 972–982. [CrossRef]

20. Wei, S.; Zhang, X.; Xu, Y.; Fu, Y.; Ren, Z.; Li, F. Extended Park's vector method in early inter-turn short circuit fault detection for the stator windings of offshore wind doubly-fed induction generators. *IET Gener. Transm. Distrib.* **2020**, *14*, 3905–3912. [CrossRef]

21. Gerlici, J.; Goolak, S.; Gubarevych, O.; Kravchenko, K.; Kamchatna-Stepanova, K.; Toropov, A. Method for Determining the Degree of Damage to the Stator Windings of an Induction Electric Motor with an Asymmetric Power System. *Symmetry* **2022**, *14*, 1305. [CrossRef]

22. Sheikh, M.A.; Bakhsh, S.T.; Irfan, M.; bin Mohd Nor, N.; Nowakowski, G. A Review to Diagnose Faults Related to Three-Phase Industrial Induction Motors. *J. Fail. Anal. Prev.* **2022**, *22*, 1546–1557. [CrossRef]

23. Gubarevych, O.; Goolak, S.; Melkonova, I.; Yurchenko, M. Structural diagram of the built-in diagnostic system for electric drives of vehicles. *Diagnostyka* **2022**, *23*, 2022406. [CrossRef]

24. Sarkar, S.; Das, S.; Purkait, P. Wavelet and SFAM based classification of induction motor stator winding short circuit faults and incipient insulation failures. In Proceedings of the IEEE 1st International Conference on Condition Assessment Techniques in Electrical Systems (CATCON), Kolkata, India, 6–8 December 2013; pp. 237–242. Available online: http://toc.proceedings.com/21152webtoc.pdf (accessed on 25 March 2023).

25. Spyropoulos, D.V.; Mitronikas, E.D. Induction motor stator fault diagnosis technique using Park vector approach and complex wavelets. In Proceedings of the 2012 XXth International Conference on Electrical Machines, Marseille, France, 2–5 September 2012; pp. 1730–1734.

26. Gyftakis, K.N.; Cardoso, A.J.M.; Antonino-Daviu, J.A. Introducing the Filtered Park's and Filtered Extended Park's Vector Approach to detect broken rotor bars in induction motors independently from the rotor slots number. *Mech. Syst. Signal Process.* **2017**, *93*, 30–50. [CrossRef]

27. Goolak, S.; Riabov, I.; Tkachenko, V.; Kondratieva, L. Concerning the Diagnosis of Asymmetric Modes in the Traction Induction Drive of the Electric Locomotive. In Proceedings of the 2022 IEEE 3rd KhPI Week on Advanced Technology (KhPIWeek), Kharkiv, Ukraine, 3–7 October 2022; pp. 1–6.

28. Chaouch, A.; Chouitek, M.; Mohamed Reda, M.A.; Belaid, M. Current Park's Vector Pattern Technique for Diagnosis of Broken Rotor Bars Fault in Saturated Induction Motor. *J. Electr. Eng. Technol.* **2023**. [CrossRef]

29. Messaoudi, M.; Flah, A.; Alotaibi, A.A.; Althobaiti, A.; Sbita, L.; El-Bayeh, C.Z. Diagnosis and fault detection of rotor bars in squirrel cage induction motors using combined Park's vector and extended Park's vector approaches. *Electronics* **2022**, *11*, 380. [CrossRef]

30. Goolak, S.; Riabov, I.; Gorobchenko, O.; Yurchenko, V.; Nezlina, O. Improvement of the model of an asynchronous traction motor of an electric locomotive by taking into account power losses. *Prz. Elektrotechnicznythis* **2022**, *98*, 1–10. [CrossRef]
31. Gubarevych, O.; Golubieva, S.; Melkonova, I. Comparison of the results of simulation modeling of an asynchronous electric motor with the calculated electrodynamic and energy characteristics. *Przegląd Elektrotechniczny* **2022**, *98*, 61–66. [CrossRef]

Article

# Characterizing the Wake Effects on Wind Power Generator Operation by Data-Driven Techniques

**Davide Astolfi [1], Fabrizio De Caro [2] and Alfredo Vaccaro [2,*]**

[1] Department of Engineering, University of Perugia, Via G. Duranti 93, 06125 Perugia, Italy; davide.astolfi.green@gmail.com

[2] Department of Engineering, University of Sannio, Piazza Roma 21, 82100 Benevento, Italy; fdecaro@unisannio.it

* Correspondence: vaccaro@unisannio.it

**Abstract:** Wakes between neighboring wind turbines are a significant source of energy loss in wind farm operations. Extensive research has been conducted to analyze and understand wind turbine wakes, ranging from aerodynamic descriptions to advanced control strategies. However, there is a relatively overlooked research area focused on characterizing real-world wind farm operations under wake conditions using Supervisory Control And Data Acquisition (SCADA) parameters. This study aims to address this gap by presenting a detailed discussion based on SCADA data analysis from a real-world test case. The analysis focuses on two selected wind turbines within an onshore wind farm operating under wake conditions. Operation curves and data-driven methods are utilized to describe the turbines' performance. Particularly, the analysis of the operation curves reveals that a wind turbine operating within a wake experiences reduced power production not only due to the velocity deficit but also due to increased turbulence intensity caused by the wake. This effect is particularly prominent during partial load operation when the rotational speed saturates. The turbulence intensity, manifested in the variability of rotational speed and blade pitch, emerges as the crucial factor determining the extent of wake-induced power loss. The findings indicate that turbulence intensity is strongly correlated with the proximity of the wind direction to the center of the wake sector. However, it is important to consider that these two factors may convey slightly different information, possibly influenced by terrain effects. Therefore, both turbulence intensity and wind direction should be taken into account to accurately describe the behavior of wind turbines operating within wakes.

**Keywords:** wind energy; wind turbines; wakes; data analysis; SCADA; condition monitoring.

**Citation:** Astolfi, D.; De Caro, F.; Vaccaro, A. Characterizing the Wake Effects on Wind Power Generator Operation by Data-Driven Techniques. *Energies* **2023**, *16*, 5818. https://dx.doi.org/10.3390/en16155818

Academic Editors: Marialaura Di Somma, Jianxiao Wang and Bing Yan

Received: 12 July 2023
Revised: 31 July 2023
Accepted: 2 August 2023
Published: 5 August 2023

## 1. Introduction

Wake interactions between nearby wind turbines represent the most important cause of producible energy loss in an operating wind farm. It is well known that the wind intensity downstream of a rotor gets reduced and that, if nearby wind turbines are not sufficiently spaced, the velocity deficit does not completely recover, thus affecting power production. In particular, on the one hand for offshore installations, it is convenient to maintain the layout of a wind farm as sufficiently compact to reduce Operation and Maintenance (OandM) costs but, on the other hand, if the layout is too compact the wake losses might reach 10–20% of the Annual Energy Production (AEP) [1,2].

In this regard, the offshore Lillgrund wind farm has become a paradigmatic test case which has been extensively studied in the literature [3,4]. It is composed of 48 Siemens SWT-2.3-93 wind turbines, with 2.3 MW of rated power. The layout is approximately square and the lowest distances between nearby wind turbines are 3.3 and 4.3 rotor diameters. In [5], it is estimated that the wake effects account for a 28% AEP loss. Another paradigmatic test case is the Horns Rev wind farm [6–9]. In that wind farm, the turbine spacing is higher (7,

9.3, and 10.4 rotor diameters) and the particular interest of the test case is in the fact that the wind farm is very large (80 Vestas V80 wind turbines).

The behavior of wind turbines under wake is also a factor to be accounted for by Transmission System Operators (TSOs) because the increasing wind energy penetration requires wind power plants to provide ancillary services [10–12]. For example, in [13], a control algorithm is formulated to distribute the power contribution of each turbine to minimize the wake effects and thus maximize the power reserve. On the other way round, if a wind turbine is requested to provide frequency support services, its wake behavior is affected [14]. Taking into account the wake effects is also an improvement for short-term wind power forecasts [15,16], which have crucial importance in electrical grid management.

Given this premise, it is evident that the analysis of wind turbine wakes is a topic that has attracted an extremely vast amount of literature, dealing with several aspects. Some of the most important are wind tunnel analysis [17–19], wind farm control [20,21], numerical simulations [22–26], and so on.

Nevertheless, some aspects are overlooked in the literature, namely those dealing with the exploitation of wind turbine Supervisory Control And Data Acquisition (SCADA) data [27,28] for the characterization of wind turbine operation under wake. To the best of the authors' knowledge, there are only a few papers on the topic. In [29], the authors relate meteorological data from a long-range lidar measurement campaign to the key SCADA parameters of offshore wind turbines. The main result of the work is that there is a good correlation between the standard deviation of the active power in the units of the average power, and in the ambient turbulence intensity (TI). Similar considerations are formulated in [30]. In [31], data-driven power curve models are formulated for a cluster of wind turbines extracted from a larger wind farm. It is shown that a graph model, accounting for wake interactions between wind turbines, largely diminishes the mismatch between model estimates and measurements.

Based on the aforementioned considerations, this paper aims to contribute to the identification of crucial SCADA parameters for characterizing wind turbine behavior in wake conditions. This work distinguishes itself from state-of-the-art ones in the type of employed data and in the methodologies. In particular, in [29,30], the fluctuations of wind turbine operation under wake are put in relation to meteorological data collected by a LiDAR. Yet, the use of this kind of sensor for wind farm OandM is at present still discouraged by economic considerations and it is therefore valuable to identify what can be understood by using solely SCADA data, as in this present work. The work in [31] moves from a consideration similar to the starting point of this work, which is the mismatch between nominal and real-world power curves in different environmental conditions [32]. In [33], a meaningful example is reported, which is the power curve of two wind turbines of the same model placed in different environments (moderately vs. highly turbulent). Those two curves appear remarkably different and this occurs not only for performance issues, but also because the measurement of the nacelle wind speed is a critical point. Not only are the nacelle wind speed measurements taken behind the rotor span and must be renormalized to estimate the free stream wind speed [34], but these measurements are also influenced by environmental conditions, including the turbulence generated by wake interactions. In fact, for example, the works of [35,36] highlight the effects of turbulence intensity on wind turbine power curves. Therefore, if on the one hand it is reasonable to construct power curve models taking into account the wake interactions (as done in [31]), on the other hand, for a deeper comprehension of the behavior of the wind turbines, it is meaningful to also consider curves based only on operation variables, developing further the approach used in previous studies such as [37].

Particularly, this work presents a real-world test case discussion using one year of SCADA data from an onshore wind farm located in Italy. The wind farm consists of nine turbines, each with a rated power of 3.3 MW. The turbine behavior under wake conditions is analyzed separately from operation under free stream conditions, based on the computation of waked sectors defined by the International Electrotechnical Commission (IEC). Several meaningful test cases are then examined.

As an anticipation of the specific contributions collected in this study, it can be stated that, also using solely SCADA data, it is possible to highlight how turbulence intensity plays a significant role in determining notable differences in operational behavior under wake conditions. Recent state-of-the-art works confirm this view and analyze in detail the role of turbulence intensity by employing mainly numerical simulations [38,39]. It is therefore valuable to investigate what can be concluded in this regard through the analysis of a real-world industrial case. In this present work, turbulence intensity is shown to have a strong correlation with wind direction in the wake sector, meaning that the closer the wind direction aligns with the line connecting the involved turbines, the higher the turbulence intensity. However, this study demonstrates that these two variables, namely distance from the wake center and turbulence intensity, provide slightly different information. Therefore, a non-trivial insight achieved in this work is that both variables are required for accurate data-driven modeling of wind turbine power output in waked operations.

The structure of this paper is as follows: Section 2 provides a description of the employed methodology, whereas Section 3 analyzes the test case wind farm and the dataset. Section 4 discusses the obtained experimental results, and finally, Section 5 summarizes the main conclusions.

## 2. Methodology

### 2.1. The Data-Driven Approach

The methodology formulated in this work consists of several key steps, including the identification of the wake sector, characterization of wind turbine operation within the wake, and quantification of wake losses. Figure 1 provides an illustrative example of the target of the proposed methodology, showcasing a reference wind turbine generator (WTG) surrounded by two other WTGs. The red and green colored angular sectors represent the wake sectors of the reference WTG with respect to the neighboring WTGs (j-th and k-th). Particularly, these colored angular sectors are identified and analyzed to estimate the wake energy losses.

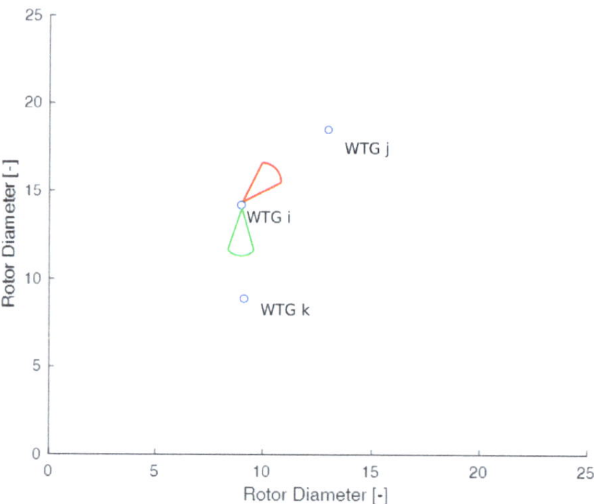


**Figure 1.** Visualization of wake sectors, where wind turbine i is affected by the operations of j (red) and k (blue).

The key steps of the proposed method go in sequence and proceed as follows:

- Identify the wake sectors by employing SCADA data (Section 2.2).

- Characterize the behavior of the wind turbines under wake through the analysis of appropriate operation curves (Section 2.3). This step is inspired by the IEC analysis of the power curve but generalizes it to further curves which are meaningful for the comprehension of the behavior under wake. Furthermore, this analysis allows for identifying limitations for the use of the power curve and therefore motivates the following step. Indeed, the nacelle wind speed measurements are affected by the amount of turbulence intensity and therefore it is unreliable to employ two average power curves measured under different conditions (e.g., in free stream vs. wake) to estimate the amount of power production loss caused by the wake.

- Formulate a reliable data-driven method for estimating the production loss caused by the wake interactions (Section 2.4.1). This is achieved by training a data-driven regression for the power of the target wind turbine as a function of the power of a reference wind turbine by using the free stream data. When simulating through the model the power of the target wind turbine when it is in wake, the quantification of the wake loss is substantially different from the model estimate (which is a simulation of the power that should have been produced if the wind turbine was not in wake) and the produced power.

- Employ the knowledge matured with the previous steps by formulating a method for characterizing, in general, the behavior under wake (Section 2.4.2). This step starts by employing only the measurements which are not biased by the presence of the wake. Namely, the nacelle wind speed measurements are excluded and only the operation variables and features elaborated from them are considered. A Sequential Features Selection is employed for identifying the key factors for describing accurately the power variability under wake.

*2.2. Identification of the Wake Sectors*

The center of a waked sector $\theta_{ij}^{teo}$ is the direction connecting straightly two wind turbines $i$ and $j$ and the amplitude in degrees of the sector [40] is defined in Equation (1):

$$\alpha = 1.3 \frac{180 \arctan\left(2.5\frac{D}{L} + 0.15\right)}{\pi} + 10, \tag{1}$$

where $D$ is the rotor diameter and $L$ is the distance between the wind turbines. For the sake of clarity, Figure 2 shows an example of $\alpha$ and $\theta_{ij}^{teo}$. Through Equation (1), by elaborating the nacelle wind direction measurements, it is possible to establish if a wind turbine is operating in free stream or subjected to a single or multiple wakes. Namely, the procedure goes as follows:

- Consider a target $i$-th wind turbine;
- Set a counter to 0;
- For each wind turbine $j = 1, \ldots, N$, where $N$ is the number of wind turbines in the farm and $j \neq i$, compute $\alpha_{ij}$ using Equation (1);
- For each wind turbine $j = 1, \ldots, N$, where $N$ is the number of wind turbines in the farm and $j \neq i$, compute the theoretical angle of the wake sector center as $\theta_{ij}^{teo} = \arctan\frac{y_i - y_j}{x_i - x_j}$;
- If the nacelle wind direction $\theta_i$ of the target $i$-th wind turbine is comprised in the interval $[\theta_{ij}^{teo} - \frac{1}{2}\alpha_{ij}, \theta_{ij}^{teo} + \frac{1}{2}\alpha_{ij}]$, increase the counter.

If, upon cycling $j$ from 1 to $N$, the counter is 0, the measurement corresponds to the operation in the free stream of the $i$-th wind turbine. If the counter is 1, the operation is under a single wake (i.e., only of one wind turbine), and so on. In this study, only sectors of free stream operation and single wake have been considered, and the corresponding datasets are indicated in general as $D_{free}$ and $D_{wake}$.

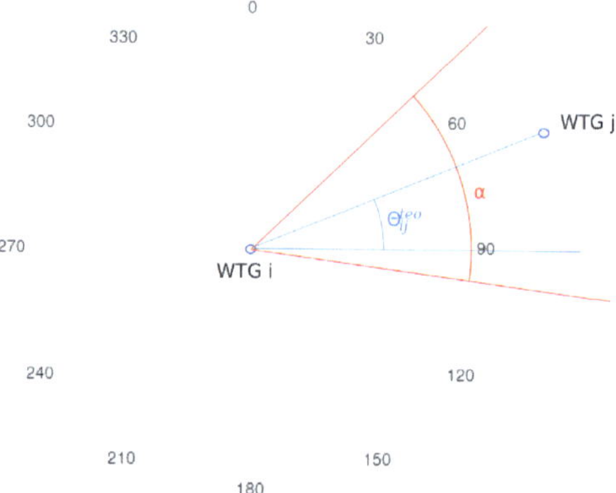


**Figure 2.** Visualization of $\alpha$ and $\theta_{ij}^{teo}$.

*2.3. Characterization of Wind Turbine Operation under Wake*

The common ground for a general comprehension of the wind turbine behavior under wake is the generalization of the binning method, which the IEC has codified for the analysis of the power curve [41].

In the case of the power curve, the point is simply averaging the power measurements per interval of the nacelle wind speed. The amplitude of the bin is typically selected as 0.5 or 1 m/s. The former is selected in this work. Therefore, given $N_j$ measurements occurring in the $j$-th bin of wind intensity, the average $P_j$ is simply given in Equation (2):

$$P_j = \frac{1}{N_j} \sum_{i=1}^{N_j} P_{i,j} \tag{2}$$

where $P_{i,j}$ is the $i$-th measurement occurring in the $j$-th bin. The average power curve is therefore given by the points $(v_j, P_j)$, where $v_j$ is the center of the $j$-th wind speed bin. The data are pre-filtered from cut-in ($v_{cut-in}$) to rated wind speed ($v_{rated}$). In principle, other pre-processing methods could be necessary in case of power curtailments [42], but this is not the case for the selected wind farm.

The above method can be generalized by considering whatever couple of quantities $(X, Y)$ whose relation is considered relevant. The principle is the same, i.e., averaging $Y$ per interval of $X$.

In order to characterize the waked sectors, some features can be computed from the raw SCADA measurements, as, for example, the turbulence intensity, which is defined in Equation (3):

$$I = \frac{v_\sigma}{v} \tag{3}$$

as the ratio between the standard deviation ($v_\sigma$) of the nacelle wind speed on a 10-minute time basis and the average wind speed ($v$) on the same time interval.

Another meaningful feature that can be computed from the raw data is the angular distance between the center of the wake sector and the measured wind direction $\theta$. This quantity is defined in Equation (4):

$$\theta_d = \theta^{teo} - \theta. \tag{4}$$

The curves considered of interest for the purposes of this present work are therefore summarized in Table 1, where the range of $X$ and the bin amplitudes are reported.

**Table 1.** Analyzed operation curves.

| Curve | X | Y | X Range | X bin |
|---|---|---|---|---|
| Power Curve | $v$ | $P$ | $[v_{cut-in}, v_{rated}]$ m/s | 0.5 m/s |
| Turbulence Intensity Curve | $v$ | $I$ | $[v_{cut-in}, v_{rated}]$ m/s | 0.5 m/s |
| Rotor-Power Curve | $\omega$ | $P$ | $[\omega_{min}, \omega_{max}]$ rpm | 0.5 rpm |
| Power-Blade Pitch Curve | $P$ | $\beta$ | $[0, P_{rated}]$ kW | 0.1 $P_{rated}$ |
| Angular Distance-Turbulence Intensity | $\theta_d$ | $I$ | $[\theta_c - 0.5\theta_a, \theta_c + 0.5\theta_a]$ | 5° |
| Angular Distance-Residuals | $\theta_d$ | $R$ | $[\theta_c - 0.5\theta_a, \theta_c + 0.5\theta_a]$ | 5° |

The curves indicated in Table 1 have an intuitive explanation, except for those involving the use of the angular distance with respect to the center of the wake. The necessary details about the use of such a curve are reported in Section 4.

*2.4. Characterization of the Waked Sectors and of the Wake Losses*

2.4.1. Estimation of the Wake Losses

The computation of the wake losses needs an estimate of how much a wind turbine would have produced if it were not in wake. In this regard, the idea proposed in this work is learning this from the data by exploiting the fact that the wind turbines are grouped in clusters. Namely, the procedure goes as follows:

- Filter the data for the selected wind turbine pair where both turbines are operating in free stream conditions. To achieve this, as discussed in Section 2.2, ensure that for each turbine pair $i$ and $j$, the wind directions $\theta_i$ and $\theta_j$ do not fall within any wake sector created by other turbines in the farm.
- Create a filtered dataset, denoted as $D_{free}$, containing the selected turbine pairs.
- Train a data-driven model with the power of the reference upstream wind turbine $P_{upstream}$ as the input and the power of the target downstream wind turbine $P_{downstream}$ as the output.
- Consider the dataset describing the downstream wind turbine affected by a single wake of an upstream one, referred to as $D_{wake}$ for brevity.
- Use the trained model to simulate the output, denoted as $y_{free}$, by inputting the power of the reference wind turbine from the $D_{wake}$ dataset.
- Compare the model estimates $y_{free}$ with the actual measurements $y$.

The selected type of model is a Support Vector Regression (SVR) with Gaussian Kernel. It has been selected because it has desirable characteristics for this kind of application [43–45]: robustness to outliers, relatively fast convergence, and nonlinearity.

The rationale of this approach to the estimate of the wake losses is that, by training the model with the $D_{free}$ dataset, the model learns the relation between the power of the reference and target wind turbines when both are in free stream. By feeding as input to the model the data $D_{wake}$, the model simulates how much the target wind turbine would produce if it were not in wake for a given power of the reference wind turbine (which is in a free stream in the $D_{wake}$ dataset). In other words, the estimate of the wake losses can be retrieved from the difference between measurements and model estimates. In particular, the loss relative to the Annual Energy Production is estimated in Equation (5):

$$E_{perc.loss.} = \frac{\sum_{D_{wake}} y - y_{free}}{\sum_{D_{tot}} y},\tag{5}$$

where $y$ is the power measurements of the target wind turbine and $y_{free}$ are the estimates of the model, which simulates the free stream data-driven relation. The numerator is the sum of the residuals in the $D_{wake}$ dataset, while the sum at the denominator should be intended on the whole yearly dataset considered in this study (indicated as $D_{tot}$).

### 2.4.2. Features Classification

This analysis is aimed at identifying the key SCADA parameters which are required for a thorough characterization of the operation under wake. In general, in the wind power sector, there are several problems that can be stated as the necessity of determining the most relevant features for predicting an output which in general has a multivariate dependence on several factors [46]. In particular, for the present application, due to the critical points related to the wind speed measurements which are affected by the level of turbulence intensity, we have decided to employ only operation variables which are not affected by such kinds of biases. Namely, the starting set of features is listed in Table 2 and constitute a matrix of $P$ features: $\mathcal{X} = \{x_1, \ldots, x_j, \ldots, x_P\}$. The amount of turbulence intensity is resembled in the variability of the operation variables, such as the rotational speed and blade pitch. Notice that this selection is quite standard, in that all the modern wind turbines collect the measurements needed to construct the features in Table 2. Therefore, the selection can be considered general and not linked to the specific test case.

**Table 2.** Starting set of features for the characterization of the operation under wake.

| |
|---|
| Rotor Speed (Average, Minimum, Maximum, Std Dev) $\omega$ (rpm) |
| Blade Pitch (Average, Minimum, Maximum, Std Dev) $\beta$ (°) |
| Direction Distance with respect to the wake center $\theta_d$ (°) |

In order to classify the above features, these preliminary steps are applied:

- Select a portion of a waked dataset; reasonably, the most populated;
- Divide it in a training and testing portion (two-thirds and one-third);
- Feed the input variables of Table 2 in the training dataset to a sequence of SVRs whose output is the power of the downstream wind turbine in waked operation.

Thereafter, the importance of the features is classified through a Sequential Features Selection (SFS) algorithm, whose objective is determining sequentially what features provide a decrease in a loss function and by how much. At each round of the algorithm, the most desirable input variable to the model (and thus the most important feature at that round) is the one which, if added to the model, leads to the highest decrease in a loss function. The loss function selected in this work is the Root Mean Square Error (RMSE), which is defined in Equation (6):

$$\text{RMSE} = \sqrt{\frac{\sum_i^N (R[i] - \bar{R})^2}{N}},\tag{6}$$

where $R[i]$ is the $i$-th difference between measurement $y$ and model estimates $\hat{y}$ and $\bar{R}$ are the average residual on the considered testing set. Namely, the precise steps of the SFS are the following:

- Initialize a matrix $X_{M=0}$ with null dimensions to store the most significant predictors from the set $\mathcal{X}_{M=0} = x_1, \ldots, x_j, \ldots, x_P$, where $P$ represents the number of input variables. Set a vector of output $y$, a counter variable $M$ set to 1, and $\text{RMSE}_{M=0}$ to $\infty$.
- Repeat until $\text{RMSE}_M > \text{RMSE}_{M-1}$:

    1. Iterate for each $j$-th variable $x_j$, where $j \in [1, |\mathcal{X}_{M-1}|]$:

        (a) Divide the training set $\mathcal{X}_{M-1}$ using $K$-fold cross-validation.

        (b) For each $k \in [1, \mathcal{K}]$, build a SVR-based model using the $k$-th training set, and merge the variables of $X_{M-1}$ with the $j$-th variable $x_j$ in the set $\mathcal{X}_{M-1}$. Test each $j$-th model on each one of the $\mathcal{K} - 1$ folds.

        (c) Compute the average loss function across the $\mathcal{K}$ sample-sets for each $j$-th model.

        (d) Sort the $|\mathcal{X}_{M-1}|$ models and select the variable from $\mathcal{X}_{M-1}$ associated with the lowest loss function.

2. Remove the selected variable from $\mathcal{X}_{M-1}$ and merge it with the variables in $\mathbf{X}_{M-1}$ to create $\mathbf{X}_M$.
3. Increase the counter $M$ of one unit.

- End the algorithm when the loss function stops decreasing.

The interest in the application of this algorithm is twofold:

- The most meaningful features for modeling the power of the test case wind turbine in wake are identified using a particular dataset;
- One could employ the above features on other waked sectors and inquire how much the selected set is appropriate.

The selected accuracy metrics are the RMSE (defined in Equation (6)) and the Mean Absolute Error (MAE), which is defined in Equation (7):

$$\text{MAE} = \frac{1}{N}\sum_{i}^{N}|R[i]|, \tag{7}$$

as simply the average of the absolute differences between model estimates and measurements.

## 3. Case Study

The test case wind farm features nine wind turbines with $D = 117$ meters of rotor diameter. The machines have variable rotational speeds which are controlled through hydraulic blade pitch actuation. The rated power of each wind turbine is 3.3 MW. The wind farm is sited onshore on a gentle terrain and the layout is reported in Figure 3, where the inter-turbine distance is indicated in units of rotor diameters.

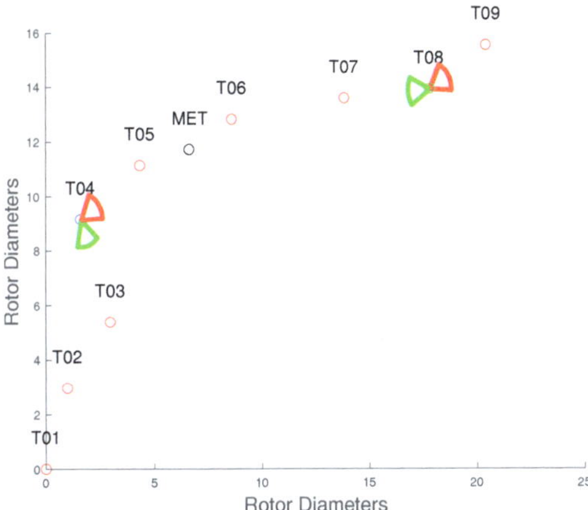


**Figure 3.** The layout of the selected wind farm. The target wind turbines are indicated, with the waked sectors with respect to the nearby ones.

The available SCADA-collected data have ten minutes of averaging time and go from 1 January to 31 December 2020. The measurement channels at disposal are listed in Table 3.

**Table 3.** SCADA-collected measurements at disposal for the study.

| |
| --- |
| Nacelle wind speed $v$ (Average, Minimum, Maximum, Std Dev) (m/s) |
| Nacelle wind direction $\theta$ (Average) ° |
| Rotor Speed $\omega$ (Average, Minimum, Maximum, Std Dev) (rpm) |
| Blade Pitch $\beta$ (Average, Minimum, Maximum, Std Dev) (°) |
| Active Power $P$ (Average, Minimum, Maximum, Std Dev) (kW) |

### 3.1. Selection of the Wake Sectors

The waked sectors have been computed for each wind turbine with each other turbine in the farm. Based on these methods, the sectors indicated in Table 4 have been selected (and plotted in Figure 3) for the analyses of this work. In those sectors, the downstream wind turbine is under the wake of only one wind turbine, namely the upstream one indicated in Table 4.

Evidently, a wind turbine layout is designed in order to minimize the occurrence of operation under wake and therefore the waked sectors in general occur rarely. These target wind turbines (T04 and T08) have been selected because they were more characterized by the occurrence of measurements describing their operation under a single wake of a nearby wind turbine. The wind roses measured by the nacelle anemometer of T04 and T08 are indeed reported, respectively, in Figures 4 and 5, and the population of the sectors is reported in Table 4.

**Table 4.** Selected single wake sectors.

| Sector Name | Upstream | Downstream | Center $\theta_c$ | Amplitude $\theta_a$ | N. Measurements |
|---|---|---|---|---|---|
| T04 North | T05 | T04 | 54° | 64.3° | 1897 |
| T04 South | T03 | T04 | 160° | 59° | 5529 |
| T08 North | T09 | T08 | 58.7° | 66.9° | 1626 |
| T08 West | T07 | T08 | 265° | 59.3° | 1064 |

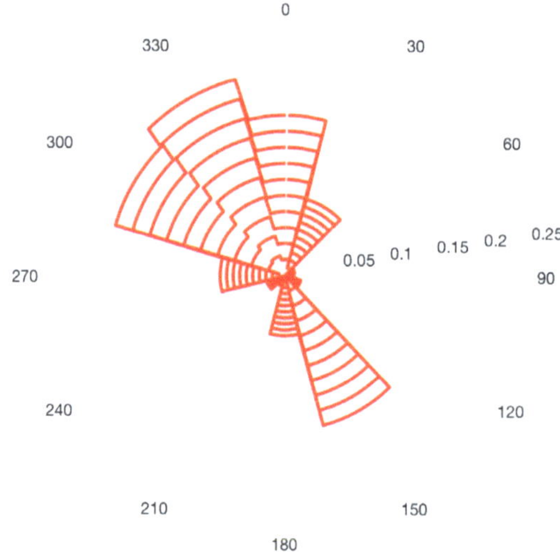


**Figure 4.** The wind rose measured by the T04 nacelle anemometer.

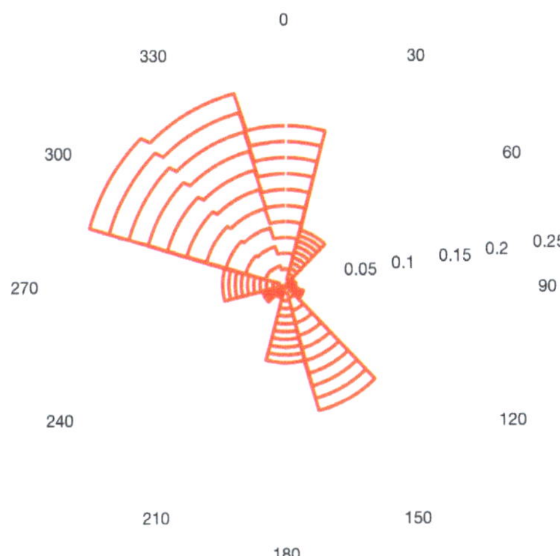


**Figure 5.** The wind rose measured by the T08 nacelle anemometer.

### 3.2. Operation Curves

In Table 5, the values of the $X$ variables listed in Table 1 are reported for the specific case study of this work.

**Table 5.** Range of the variables for the operation curve analysis

| Variable | Value |
| --- | --- |
| $v_{cut-in}$ | 4 m/s |
| $v_{rated}$ | 13 m/s |
| $\omega_{min}$ | 6 rpm |
| $\omega_{max}$ | 13 rpm |
| $P_{rated}$ | 2 MW |

### 3.3. Estimation of the Wake Losses

In Table 6, the input and output variables of the model for the wake losses estimation are reported for the case of interest, in relation to Table 4 and Figure 3.

**Table 6.** Input and output for the data-driven model for estimating the wake losses.

| Sector Name | Input | Output |
| --- | --- | --- |
| T04 North | $P$ T05 | $P$ T04 |
| T04 South | $P$ T03 | $P$ T04 |
| T08 North | $P$ T09 | $P$ T08 |
| T08 West | $P$ T07 | $P$ T08 |

### 3.4. Features Classification

For the selected case study, the features classification for the characterization of the operation under wake has been run using the T04 South case because it is the most populated (see Table 4). The crosscheck of the method has been pursued by computing accuracy metrics on two datasets: the testing part of the T04 South dataset and the T04 North dataset. This comparison is particularly interesting in this test case, in light of the different characteristics of these two waked sectors (which are discussed in detail in the following Section 4).

## 4. Experimental Results

### 4.1. Characterization of Wind Turbine Operation under Wake

Given the line of reasoning in [29], which is summarized in Section 1, the first analysis is inquiring if the selected waked sectors can be distinguished as regards the turbulence intensity. Therefore, in Figure 6, the average curve of the turbulence intensity as a function of the wind speed is reported for the four datasets of Table 4. Interestingly, it arises that for each target wind turbine, there is a sector with higher turbulence (T04 South and T08 North) and a sector with lower turbulence (T04 North and T08 West).

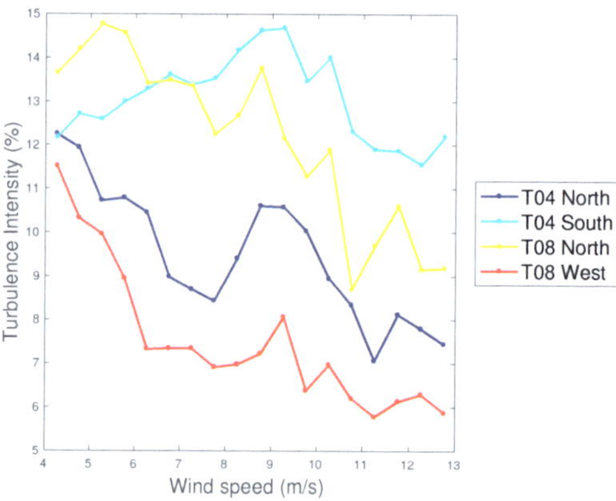


**Figure 6.** The average turbulence intensity for the two waked sectors for the two target wind turbines.

Therefore, in Figure 7 we investigate if there are differences related to the waked sectors in the most important operation curve, which of course is the power curve. For brevity, the curve for only T04 is reported, but the situation is similar also for T08. The interpretation of the power curve of Figure 7 is somehow counter-intuitive. From that curve, one would be led to argue that the power curve measured in wake is slightly better than in free stream and the effect is slightly higher for the waked sector with higher turbulence. Likely, as discussed for example in [33], the increased turbulent kinetic energy is not captured by the nacelle anemometer and this leads to an effect of under-estimation of the wind intensity. Therefore, the fact that the power curve in the waked sectors in Figure 7 looks slightly higher than in free stream is most likely due to wind speed measurement issues. This supports the fact that a consistent interpretation of wind turbine performance in complex conditions requires the analysis of further operation curves [37].

In Figures 8 and 9, two fundamental curves describing wind turbine operation are reported, which are the rotor speed-power and the power-blade pitch curve (see Table 1). From Figure 8, it arises that when there is higher turbulence (thus in wake), the extracted power for a given rotational speed is slightly higher. It looks as if, in the full aerodynamic load regime, higher turbulence for a given wind speed could even be slightly favorable for power extraction. From Figure 9, it arises that the average blade pitch does not change remarkably in wake or in free stream up to more or less 2 MW, which for this wind turbine model is the point at which the rotational speed saturates. When in partial aerodynamic load, if the turbulence is higher then the average blade pitch is also higher, which means that the aerodynamic efficiency is lower, and thus a higher incoming wind kinetic energy is required to extract a certain power output. In other words, the increased turbulence in waked operation is remarkably unfavorable in the partial aerodynamic load. An explanation of this behavior is that, when the wind turbine regulates the rotational speed and the blade pitch (full aerodynamic load), it can follow the rapid fluctuations induced by the wake in the form of increased turbulence. When the rotational speed is held fixed and only the blade pitch can vary (partial aerodynamic load), the wind turbine is not capable anymore to follow the variability of the highly turbulent wind, and the efficiency of the power conversion is therefore lower.

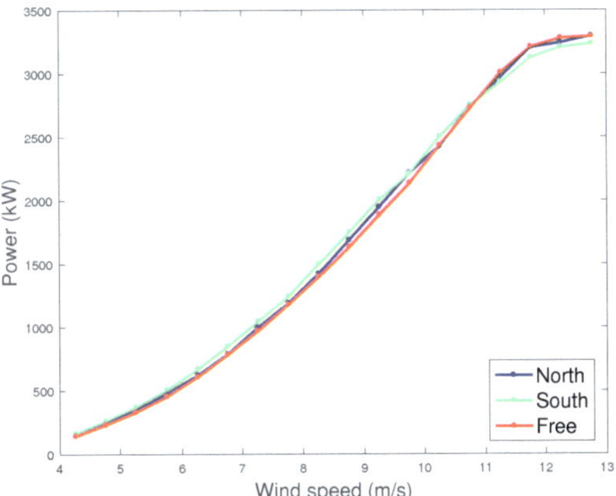

**Figure 7.** The average power curve for the target T04 wind turbine, for the two waked sectors, and for the free stream case.

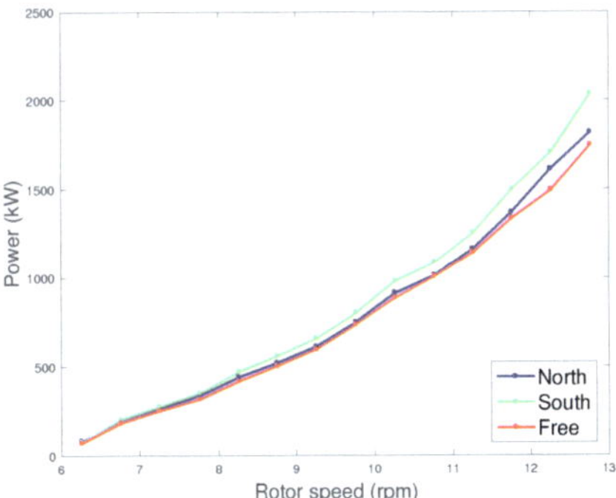

**Figure 8.** The average rotor speed-power curve for the target T04 wind turbine, for the two waked sectors and for the free stream case.

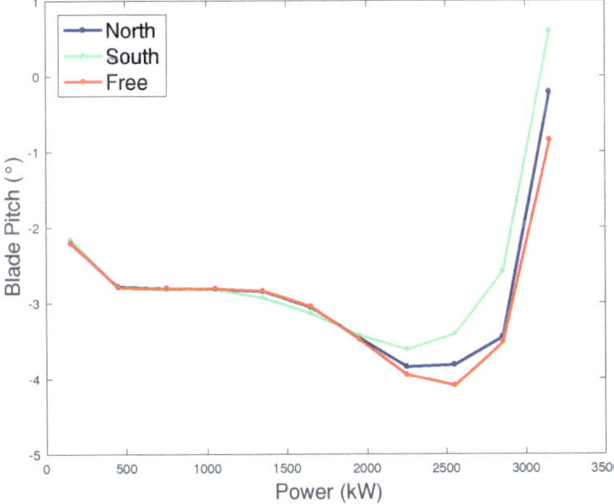

**Figure 9.** The average power-blade pitch curve for the target T04 wind turbine, for the two waked sectors, and for the free stream case.

*4.2. Characterization of the Waked Sectors and of the Wake Losses*

A meaningful quantity for characterizing the operation under wake is how much the wind direction deviates with respect to the center of the wake (Equation (4)), i.e., to the values reported in Table 4. In Figure 10, we report the distribution of such a quantity for the four considered cases. The four waked sectors have quite different features. The T04 North case is completely skewed negatively, while T04 South is similar but with a non-negligible occurrence of measurements along the center of the wake. The T08 North has a practically uniform distribution from $-30°$ to $+20°$, while the T08 West is completely positively skewed.

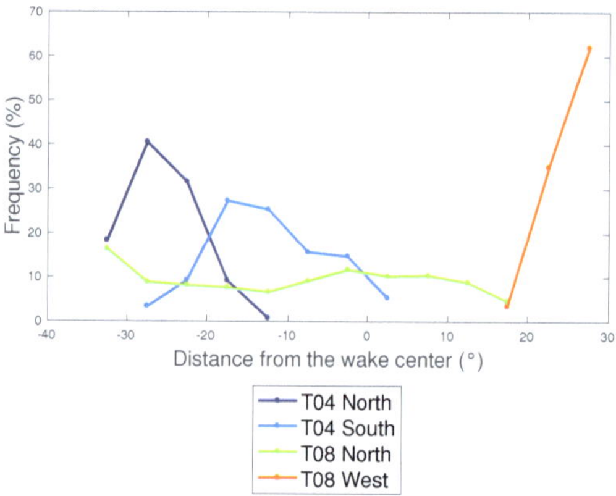


**Figure 10.** The distribution of distances from the center of the waked sectors for the four test cases.

Figure 11 reports the average turbulence intensity as a function of the distance from the wake center for the four considered waked sectors. The turbulence intensity is measured through the nacelle anemometer of the turbine downstream (T04 and T08, respectively). Figure 11 is interesting because for three cases out of four (T04 North, T04 South, and T08 West) the turbulence intensity is observed to increase when the wind direction approaches the center of the wake, which is a reasonable result. For the T08 North case, the turbulence intensity is quite constant as a function of the direct distance from the wake center. This is most probably due to the effect of the terrain. It is interesting to notice that this kind of effect, which is somehow expected in complex terrain [47], indeed, occurs also in cases such as the one selected in this work, where the terrain is quite gentle. This result provides a qualitative indication of the fact that it is very likely that turbulence intensity and direction distance from the wake center are well-correlated quantities, but this is not assured and in general, those two quantities might convey slightly different information.

Figure 12 reports the difference between the power measurements of the downstream wind turbine in waked operation and the corresponding simulation in free stream conditions as a function of the angular distance with respect to the wake center. As described in general in Section 2 and specified in Section 3 for this test case, the free stream simulation for the target T04 or T08 wind turbines is obtained by generating the output of a data-driven model taking as input the power of the upstream wind turbine (see Table 4) when both wind turbines are upstream. In Figure 12, such a set of residuals is averaged per intervals of $\theta_d$, thus leading to the Angular Distance–Residual curve indicated in Table 1. Figure 12 quite fairly agree with Figure 11 because of the higher the turbulence intensity and the higher the wake losses. The most relevant loss occurs for T04 in the South sector at the center of the wake (up to 250 kW on average). It is interesting to notice that there is practically no loss when the absolute value of the direction distance with respect to the wake center is higher than $20°$, for the T04 North, T04 South, and T08 West cases. Instead, for the T08 North case there is a relevant loss along all the wake sectors, which is likely related to the fact that the turbulence intensity does not decrease when the distance from the wake center increases. The case of T04 South is further analyzed in Figure 13, where the residuals between model estimates and measurements are averaged per power intervals of the upstream wind turbine T03. From Figure 13, it arises that the wake losses are higher for powers of T03 higher than 2 MW. This can be seen as a different way of characterizing the behavior reported in Figure 9: in this case, using the relative performance with respect to the upstream wind turbine.

The wake losses reported in Figure 12 can then be averaged and reported to the measured AEP (Equation (5)), thus obtaining the estimates of Table 7. Coherently with the above results, the wake sectors non-negligibly affecting the AEP are T04 South and T08 North, i.e., those characterized by higher average turbulence intensity.

Finally, the analysis of the factors influencing the behavior under wake is pursued by determining the input variables which are required for modeling the power with the lowest possible error. As described in Section 2 and specified in Section 3, a Sequential Features Selection is employed starting from the dataset T04 South, which is selected because it is the most populated. The selected input variables are reported in Table 8. The level of turbulence intensity is accounted for by the presence of minimum, maximum, and standard deviations on the 10-minutes time interval. It is worth noticing that the distance from the wake center is selected as the input variable. This means that this variable provides additional information which is not merely already contained in the variability over the time interval of the rotational speed and the blade pitch.

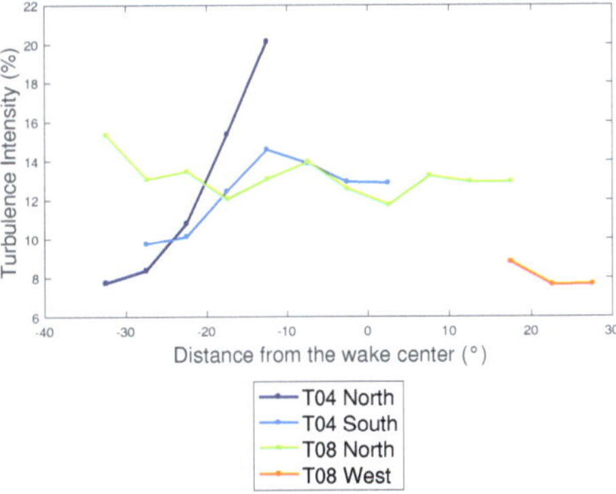


**Figure 11.** The distribution of turbulence intensity as a function of the distance from the center of the waked sectors for the four test cases.

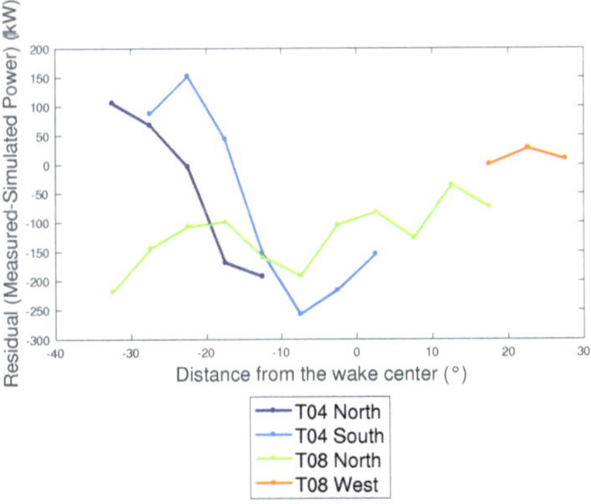


**Figure 12.** The distribution of the residuals between power measurements and simulations as a function of the distance from the center of the waked sectors for the four test cases.

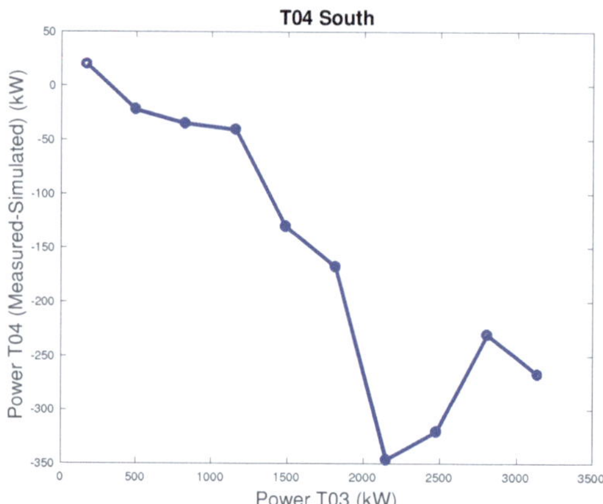


**Figure 13.** The residuals between model estimates and measurements for T04 South case, as a function of the power of the upstream wind turbine (T03).

**Table 7.** Wake losses in percentage of the AEP, as estimated from the data-driven model.

| Case | Production Loss |
| --- | --- |
| T04 North | +0.11% |
| T04 South | −1.13% |
| T08 North | −0.59% |
| T08 West | +0.04% |

**Table 8.** Selected input variables for modeling the power under wake operation.

| Rotor Speed (Average, Minimum) (rpm) |
| --- |
| Blade Pitch (Average, Minimum, Maximum, Std Dev) (°) |
| Direction Distance with respect to the wake center (°) |

Finally, the so-obtained model is tested on a portion of the T04 South dataset and on the T04 North dataset. The accuracy metrics for such testing are reported in Table 9. It arises that the metrics for the T04 North dataset are only in the order of 10% higher than for the testing subset of the T04 South dataset. Considering that a part of the T04 South dataset is the training dataset and that the behavior of the T04 North wake sector is peculiar as regards the distribution of turbulence intensity (Figure 11), this result is remarkable in the sense that it tells that the set of input variables indicated in Table 8 captures features of the wake behavior that can be considered quite general.

**Table 9.** The accuracy metrics of the model for the power of wind turbine T04. The model is trained on two-thirds of the data from T04 South and tested on the remaining third and on the T04 North wake sector.

| Metric | kW |
| --- | --- |
| MAE T04 South | 56.3 |
| MAE T04 North | 61.1 |
| RMSE T04 South | 97.4 |
| RMSE T04 North | 107.1 |

## 5. Conclusions

This work has dealt with the characterization of wind turbine operation under wake through SCADA data analysis and has been organized as a real-world test case discussion. Actually, two wind turbines from an onshore wind farm have been selected and their behavior in the waked sectors has been analyzed and compared to the free stream operation. The general motivation of this work is that the identification of the key SCADA parameters describing the behavior of wind turbines in wake is an overlooked topic, which otherwise would be important for advanced wind farm control applications and for managing the power variability in case the wind farms are requested to contribute to grid ancillary services.

The characterization of wind turbine operation in wake has been pursued through the analysis of appropriate operation curves and through data-driven methods. The turbulence intensity is individuated as the key factor determining the observed behavior. Actually, the operation curve analysis highlights that a wind turbine in wake not only loses production because the wind intensity decreases while passing through the upstream rotor but also because, for a certain wind intensity, the higher turbulence induced by the wake stresses the wind turbine control. In particular, in the full aerodynamic load operation (i.e., when the wind turbine regulates the blade pitch and the rotational speed) the increased turbulence does not cause appreciable losses, while it does when the aerodynamic load is partial (i.e., the rotational speed is rated and the wind turbine regulates the load through the blade pitch). Indeed, a deviation of approximately $1°$ of blade pitch in the partial aerodynamic load region is observed for the wake sectors characterized by higher turbulence intensity.

The above interpretation has been confirmed by a data-driven model for the wake losses estimate, taking as input the power of the nearby wind turbine. The model is trained to learn the data-driven relation between a couple of nearby wind turbines when both are in free stream and is employed to simulate the output when the downstream wind turbine is in the wake of the upstream one. The difference between model estimates and measurements therefore allows for estimating the losses. It results that the wake losses are negligible for the two sectors with average turbulence intensity less than 10%, while the impact is meaningful for the two sectors (T04 South and T08 North) for which the average turbulence intensity is in the order of 13%.

The waked sectors have been subsequently characterized by analyzing how the turbulence intensity distributes as a function of the wind direction, specifically focusing on the proximity of wind intensity to the center of the wake. It was observed that, in general, the closer the direction to the center of the wake, the higher the turbulence intensity. However, it is important to consider the distance of the wind direction from the wake center and the turbulence intensity as separate factors when explaining the observed power variability. Terrain effects can make the relationship between these two quantities non-trivial also in the absence of evident complexity, as observed in this work in one of the analyzed waked sectors.

In general, therefore, the recommendation arising from this work is that a robust comprehension of how the power of wind turbine generators varies in the presence of wakes requires non-trivial data analysis methods. It should be taken into account that in an offshore environment there could be further factors to take into account, for example the influence of waves [48,49], but the approach proposed in this study can be easily generalized by including further features affecting the extracted power. The results of this study can be beneficial for further advancements in the general field of wake active control and for short or ultra-short-term wind power forecasts. Actually, for those applications, it is important to determine through real-world test cases what are the factors determining the behavior of wind farms in highly variable and highly uncertain operation conditions.

**Author Contributions:** Methodology, D.A. and F.D.C.; Software, D.A.; Validation, D.A. and F.D.C.; Formal analysis, A.V.; Writing—original draft, A.V. All authors have read and agreed to the published version of the manuscript.

**Funding:** This research received no external funding.

**Data Availability Statement:** Data may be obtained upon a reasonable request made to the Authors, subject to the requisite permission being obtained from ENGIE.

**Acknowledgments:** The authors thank the ENGIE Italia company for providing the data and for the technical support.

**Conflicts of Interest:** The authors declare no conflict of interest.

## Abbreviations

The following abbreviations are used in this manuscript:

| | |
|---|---|
| AEP | Annual Energy Production |
| IEC | International Electrotechnical Commission |
| MAE | Mean Absolute Error |
| O& M | Operation & Maintenance |
| RMSE | Root Mean Square Error |
| SCADA | Supervisory Control And Data Acquisition |
| SFS | Sequential Features Selection |
| SVR | Support Vector Regression |
| TSO | Transmission System Operators |
| WTG | Wind Turbine Generator |

## References

1. Barthelmie, R.J.; Hansen, K.; Frandsen, S.T.; Rathmann, O.; Schepers, J.; Schlez, W.; Phillips, J.; Rados, K.; Zervos, A.; Politis, E.; et al. Modelling and measuring flow and wind turbine wakes in large wind farms offshore. *Wind Energy Int. J. Prog. Appl. Wind Power Convers. Technol.* **2009**, *12*, 431–444. [CrossRef]
2. Gaumond, M.; Réthoré, P.E.; Bechmann, A.; Ott, S.; Larsen, G.C.; Peña, A.; Hansen, K.S. Benchmarking of wind turbine wake models in large offshore wind farms. In Proceedings of the Science of Making Torque from Wind Conference, Oldenburg, Germany, 9–11 October 2012.
3. Nilsson, K.; Ivanell, S.; Hansen, K.S.; Mikkelsen, R.; Sørensen, J.N.; Breton, S.P.; Henningson, D. Large-eddy simulations of the Lillgrund wind farm. *Wind Energy* **2015**, *18*, 449–467. [CrossRef]
4. Walker, K.; Adams, N.; Gribben, B.; Gellatly, B.; Nygaard, N.G.; Henderson, A.; Marchante Jimémez, M.; Schmidt, S.R.; Rodriguez Ruiz, J.; Paredes, D.; et al. An evaluation of the predictive accuracy of wake effects models for offshore wind farms. *Wind Energy* **2016**, *19*, 979–996. [CrossRef]
5. Sebastiani, A.; Castellani, F.; Crasto, G.; Segalini, A. Data analysis and simulation of the Lillgrund wind farm. *Wind Energy* **2021**, *24*, 634–648. [CrossRef]
6. Wu, Y.T.; Porté-Agel, F. Modeling turbine wakes and power losses within a wind farm using LES: An application to the Horns Rev offshore wind farm. *Renew. Energy* **2015**, *75*, 945–955. [CrossRef]
7. Hansen, K.S.; Barthelmie, R.J.; Jensen, L.E.; Sommer, A. The impact of turbulence intensity and atmospheric stability on power deficits due to wind turbine wakes at Horns Rev wind farm. *Wind Energy* **2012**, *15*, 183–196. [CrossRef]
8. Gaumond, M.; Réthoré, P.E.; Ott, S.; Pena, A.; Bechmann, A.; Hansen, K.S. Evaluation of the wind direction uncertainty and its impact on wake modeling at the Horns Rev offshore wind farm. *Wind Energy* **2014**, *17*, 1169–1178. [CrossRef]
9. Hasager, C.B.; Rasmussen, L.; Peña, A.; Jensen, L.E.; Réthoré, P.E. Wind farm wake: The Horns Rev photo case. *Energies* **2013**, *6*, 696–716. [CrossRef]
10. Cole, M.; Campos-Gaona, D.; Stock, A.; Nedd, M. A critical review of current and future options for wind farm participation in ancillary service provision. *Energies* **2023**, *16*, 1324. [CrossRef]
11. Zhong, C.; Wang, H.; Jiang, Z.; Tian, D. A unified optimization control of wind farms considering wake effect for grid frequency support. *Wind Eng.* **2023**, *47*, 0309524X231163823. [CrossRef]
12. Bhyri, A.K.; Senroy, N.; Saha, T.K. Enhancing the grid support from DFIG-Based wind farms during voltage events. *IEEE Trans. Power Syst.* **2023**, 1–12. [CrossRef]
13. Siniscalchi-Minna, S.; Bianchi, F.D.; De-Prada-Gil, M.; Ocampo-Martinez, C. A wind farm control strategy for power reserve maximization. *Renew. Energy* **2019**, *131*, 37–44. [CrossRef]
14. Singh, N.; De Kooning, J.D.; Vandevelde, L. Dynamic wake analysis of a wind turbine providing frequency support services. *IET Renew. Power Gener.* **2022**, *16*, 1853–1865. [CrossRef]
15. Guo, N.Z.; Shi, K.Z.; Li, B.; Qi, L.W.; Wu, H.H.; Zhang, Z.L.; Xu, J.Z. A physics-inspired neural network model for short-term wind power prediction considering wake effects. *Energy* **2022**, *261*, 125208. [CrossRef]

16. Wang, Y.; Shen, R.; Ma, Y.; Ma, M.; Zhou, Q.; Lu, Q.; Zhang, J. Research on Ultra-short term Forecasting Technology of Wind Power Output Based on Wake Model. *J. Phys. Conf. Ser.* **2022**, *2166*, 012041. [CrossRef]

17. Iungo, G.V.; Viola, F.; Camarri, S.; Porté-Agel, F.; Gallaire, F. Linear stability analysis of wind turbine wakes performed on wind tunnel measurements. *J. Fluid Mech.* **2013**, *737*, 499–526. [CrossRef]

18. Chamorro, L.P.; Porté-Agel, F. A wind-tunnel investigation of wind-turbine wakes: Boundary-layer turbulence effects. *Bound. Layer Meteorol.* **2009**, *132*, 129–149. [CrossRef]

19. Chamorro, L.P.; Porté-Agel, F. Effects of thermal stability and incoming boundary-layer flow characteristics on wind-turbine wakes: A wind-tunnel study. *Bound. Layer Meteorol.* **2010**, *136*, 515–533. [CrossRef]

20. Houck, D.R. Review of wake management techniques for wind turbines. *Wind Energy* **2022**, *25*, 195–220. [CrossRef]

21. Nash, R.; Nouri, R.; Vasel-Be-Hagh, A. Wind turbine wake control strategies: A review and concept proposal. *Energy Convers. Manag.* **2021**, *245*, 114581. [CrossRef]

22. Makridis, A.; Chick, J. Validation of a CFD model of wind turbine wakes with terrain effects. *J. Wind Eng. Ind. Aerodyn.* **2013**, *123*, 12–29. [CrossRef]

23. Sanderse, B.; Van der Pijl, S.; Koren, B. Review of computational fluid dynamics for wind turbine wake aerodynamics. *Wind Energy* **2011**, *14*, 799–819. [CrossRef]

24. Castellani, F.; Vignaroli, A. An application of the actuator disc model for wind turbine wakes calculations. *Appl. Energy* **2013**, *101*, 432–440. [CrossRef]

25. Dar, A.S.; Porté-Agel, F. An analytical model for wind turbine wakes under pressure gradient. *Energies* **2022**, *15*, 5345. [CrossRef]

26. Vahidi, D.; Porté-Agel, F. A physics-based model for wind turbine wake expansion in the atmospheric boundary layer. *J. Fluid Mech.* **2022**, *943*, A49. [CrossRef]

27. Pandit, R.; Astolfi, D.; Hong, J.; Infield, D.; Santos, M. SCADA data for wind turbine data-driven condition/performance monitoring: A review on state-of-art, challenges and future trends. *Wind Eng.* **2023**, *47*, 422–441. [CrossRef]

28. Chesterman, X.; Verstraeten, T.; Daems, P.J.; Nowé, A.; Helsen, J. Overview of normal behavior modeling approaches for SCADA-based wind turbine condition monitoring demonstrated on data from operational wind farms. *Wind Energy Sci.* **2023**, *8*, 893–924. [CrossRef]

29. Mittelmeier, N.; Allin, J.; Blodau, T.; Trabucchi, D.; Steinfeld, G.; Rott, A.; Kühn, M. An analysis of offshore wind farm SCADA measurements to identify key parameters influencing the magnitude of wake effects. *Wind Energy Sci.* **2017**, *2*, 477–490. [CrossRef]

30. Gonzalez, E.; Valldecabres, L.; Seyr, H.; Melero, J. On the effects of environmental conditions on wind turbine performance: An offshore case study. *J. Phys. Conf. Ser.* **2019**, *1356*, 012043. [CrossRef]

31. Hammer, F.; Helbig, N.; Losinger, T.; Barber, S. Graph machine learning for predicting wake interaction losses based on SCADA data. *J. Phys. Conf. Ser.* **2023**, *2505*, 012047. [CrossRef]

32. Ciulla, G.; D'Amico, A.; Di Dio, V.; Brano, V.L. Modelling and analysis of real-world wind turbine power curves: Assessing deviations from nominal curve by neural networks. *Renew. Energy* **2019**, *140*, 477–492. [CrossRef]

33. Astolfi, D. Perspectives on SCADA data analysis methods for multivariate wind turbine power curve modeling. *Machines* **2021**, *9*, 100. [CrossRef]

34. Carullo, A.; Ciocia, A.; Malgaroli, G.; Spertino, F. An Innovative Correction Method of Wind Speed for Efficiency Evaluation of Wind Turbines. *Acta Imeko* **2021**, *10*, 46–53. [CrossRef]

35. Honrubia, A.; Vigueras-Rodríguez, A.; Gómez-Lázaro, E. The influence of turbulence and vertical wind profile in wind turbine power curve. In *Progress in Turbulence and Wind Energy IV*; Springer: Berlin/Heidelberg, Germany, 2012; pp. 251–254.

36. Hedevang, E. Wind turbine power curves incorporating turbulence intensity. *Wind Energy* **2014**, *17*, 173–195. [CrossRef]

37. Astolfi, D. Wind Turbine Operation Curves Modelling Techniques. *Electronics* **2021**, *10*, 269. [CrossRef]

38. Zhang, Y.; Li, Z.; Liu, X.; Sotiropoulos, F.; Yang, X. Turbulence in waked wind turbine wakes: Similarity and empirical formulae. *Renew. Energy* **2023**, *209*, 27–41. [CrossRef]

39. De Cillis, G.; Cherubini, S.; Semeraro, O.; Leonardi, S.; De Palma, P. The influence of incoming turbulence on the dynamic modes of an NREL-5MW wind turbine wake. *Renew. Energy* **2022**, *183*, 601–616. [CrossRef]

40. Gasch, R.; Twele, J. *Wind Power Plants: Fundamentals, Design, Construction and Operation*; Springer Science & Business Media: Berlin/Heidelberg, Germany, 2011.

41. IEC. *Power Performance Measurements of Electricity Producing Wind Turbines*; Technical Report 61400–12; International Electrotechnical Commission: Geneva, Switzerland, 2005.

42. De Caro, F.; Vaccaro, A.; Villacci, D. Adaptive wind generation modeling by fuzzy clustering of experimental data. *Electronics* **2018**, *7*, 47. [CrossRef]

43. Pandit, R.; Kolios, A. SCADA Data-Based Support Vector Machine Wind Turbine Power Curve Uncertainty Estimation and Its Comparative Studies. *Appl. Sci.* **2020**, *10*, 8685. [CrossRef]

44. Dhiman, H.S.; Deb, D.; Carroll, J.; Muresan, V.; Unguresan, M.L. Wind turbine gearbox condition monitoring based on class of support vector regression models and residual analysis. *Sensors* **2020**, *20*, 6742. [CrossRef]

45. Vidal, Y.; Pozo, F.; Tutivén, C. Wind turbine multi-fault detection and classification based on SCADA data. *Energies* **2018**, *11*, 3018. [CrossRef]

46. Astolfi, D.; De Caro, F.; Vaccaro, A. Condition Monitoring of Wind Turbine Systems by Explainable Artificial Intelligence Techniques. *Sensors* **2023**, *23*, 5376. [CrossRef] [PubMed]

47.  Elgendi, M.; AlMallahi, M.; Abdelkhalig, A.; Selim, M.Y. A review of wind turbines in complex terrain. *Int. J. Thermofluids* **2023**, *17*, 100289. [CrossRef]
48.  Xiao, S.; Yang, D. Large-eddy simulation-based study of effect of swell-induced pitch motion on wake-flow statistics and power extraction of offshore wind turbines. *Energies* **2019**, *12*, 1246. [CrossRef]
49.  Sacie, M.; Santos, M.; López, R.; Pandit, R. Use of state-of-art machine learning technologies for forecasting offshore wind speed, wave and misalignment to improve wind turbine performance. *J. Mar. Sci. Eng.* **2022**, *10*, 938. [CrossRef]

*Article*

# Optimal Scheduling of Virtual Power Plant with Flexibility Margin Considering Demand Response and Uncertainties

Yetuo Tan [1], Yongming Zhi [2], Zhengbin Luo [3], Honggang Fan [4,*], Jun Wan [5] and Tao Zhang [6]

[1] Jiangxi Port Group Co., Ltd., Nanchang 332000, China
[2] China Water Resources Pearl River Planning, Surveying and Designing Co., Ltd., Guangzhou 510610, China
[3] Jiangxi Transportation Institute Co., Ltd., Nanchang 330200, China
[4] State Key Laboratory of Hydroscience and Engineering, Department of Energy and Power Engineering, Tsinghua University, Beijing 100084, China
[5] Jiangxi Jiepai Navigation and Electricity Hub Management Office, Yingtan 335000, China
[6] Department of Electrical Engineering, Tsinghua University, Beijing 100084, China
* Correspondence: fanhonggang@tsinghua.org.cn

**Abstract:** The emission reduction of global greenhouse gases is one of the key steps towards sustainable development. Demand response utilizes the resources of the demand side as an alternative of power supply which is very important for the power network balance, and the virtual power plant (VPP) could overcome barriers to participate in the electricity market. In this paper, the optimal scheduling of a VPP with a flexibility margin considering demand response and uncertainties is proposed. Compared with a conventional power plant, the cost models of VPPs considering the impact of uncertainty and the operation constraints considering demand response and flexibility margin characteristics are constructed. The orderly charging and discharging strategy for electric vehicles considering user demands and interests is introduced in the demand response. The research results show that the method can reduce the charging cost for users participating in reverse power supply using a VPP. The optimizing strategy could prevent overload, complete load transfer, and realize peak shifting and valley filling, solving the problems of the new peak caused by disorderly power utilization.

**Keywords:** virtual power plant (VPP); flexibility margin; demand response; uncertainties; integrated energy system; renewable energy

**Citation:** Tan, Y.; Zhi, Y.; Luo, Z.; Fan, H.; Wan, J.; Zhang, T. Optimal Scheduling of Virtual Power Plant with Flexibility Margin Considering Demand Response and Uncertainties. *Energies* **2023**, *16*, 5833. https://doi.org/10.3390/en16155833

Academic Editor: Javier Muñoz Antón

Received: 5 May 2023
Revised: 13 July 2023
Accepted: 28 July 2023
Published: 7 August 2023

## 1. Introduction

With the increasing energy crisis and pollution problems, new technologies such as the smart grid, energy internet, energy hub, integrated energy system (IES), and virtual power plant (VPP) have been introduced to realize the multi-energy coordinated supply and cascade utilization of energy [1,2]. Meanwhile, a high proportion of wind power and photovoltaic power generation are connected to the power grid, resulting in a large increase in flexibility demands [3]. The traditional scheduling strategy relies on the improvement of a rotating reserve capacity to ensure the stable operation of a power system which is unable to cope with the rapidity of net load changes. Therefore, demand responses and flexibility loads have gradually become one of the research hotspots of current power system optimization scheduling. Moreover, the concept of a virtual power plant was proposed to integrate different energy resources such as distributed generations, energy storage systems, and flexibility loads to provide system support services [4,5].

A VPP benefits from the electricity market or dynamic pricing to shift energy demand [6–8]. A VPP always focuses on economic benefits and the optimization of VPP operation is closely related to it. Many researchers have conducted a lot of research on it and have also achieved many excellent results. The scheduling optimization of VPPs usually aims to minimize operating costs and maximize operating benefits. Moreover, a lot of papers focus on multiple objectives such as cost, benefits, and power grid stability through methods such as the fuzzy multi-objective method [9]. The currently used optimal algorithms include the linear optimization algorithm [10], mixed integer linear programming algorithm [11,12], hierarchical optimization algorithm [13], differential evolution algorithm [14], adaptive heuristic algorithm [15,16], and robust optimization algorithm [17]. Li et al. analyzed the feasibility of VPPs by means of local renewable energy plant construction and the updating of high-efficiency appliances located at electricity customers [18]. Some scholars use the data envelopment analysis method to consider the comprehensive efficiency of the candidate units for economy, environmental protection, stability, and reliability, and they select the units to build a VPP according to the results [19]. Sousa et al. proposed a simulated annealing approach to address energy resource management from the point of view of a VPP, and the results showed that a VPP can purchase additional energy from a set of external suppliers [20].

The VPPs aggregate a lot of equipment which include wind power, photovoltaic power (PV), electric boilers, air conditioners, electric vehicles, flexibility loads, and so on [21–23]. Moreover, the uncertainties of renewable energy output, energy demand, and market price bring a huge challenge to the optimal scheduling of VPPs [24,25]. The uncertainty of renewable energy output mainly includes wind and photovoltaic power. The uncertainty of wind power output is mainly due to the randomness of wind speed, and the uncertainty of photovoltaic power output is mainly due to solar radiation. Moreover, the weather can affect renewable energy output, especially on a rainy day. Energy demands are uncertainty in VPP optimization problems which derive from prediction and measurement errors. The uncertainties of market price include electricity price, natural gas price, and heating price which have very strong fluctuations. A lot of optimization approaches considering uncertainty have been studied by different scholars. These include the Monte Carlo simulation [26,27], robust optimization [28], rolling horizon, stochastic dominance [29], fuzzy chance constraint programming constraints [30], and point estimation methods [31]. Some scholars focused on the fluctuation problem of VPP output. Hooshmand et al. [32] introduced the user side of power stations in a virtual power plant and built a double-layer model to increase revenues and to provide backup service to the energy system.

Previous research has already studied the optimization of VPP operation and achieved a lot of results. However, some studies only considered the uncertainty of wind power and PV, and the method could only handle the constraint conditions without stochastic variables. In this paper, a flexibility margin considering demand response and uncertainties is analyzed with a stochastic chance constrained planning method. Moreover, the demand response of electric vehicles and traditional loads are optimized to guide customers' power consumption behavior.

This paper is structured as follows: Section 2 describes an overview of the VPP's structure, which includes the model formulation, constraint conditions, and objective function. Section 3 describes the flexibility margin considering demand response. Section 4 gives an example to analyze the VPP. Section 5 concludes this research study by describing challenges and future work.

## 2. VPP Structure

As shown in Figure 1, the VPP consists of a distributed photovoltaic system, combined heat and power system, gas-fired boiler, absorption refrigeration unit, refrigeration unit, electric boiler, electric vehicle, cooling storage, electric storage, thermal storage, electrical load, cooling load, heating load, electricity market, and so on. The VPP operator is obligated to satisfy the demands of consumers by purchasing energy from the electricity market. In the electricity market, the VPP operator allows consumers to participate in the market to alleviate supply pressure, inducing load reductions by incentivizing consumers. Moreover, the VPP serves as a backup that shifts loads from peak to off-peak periods.

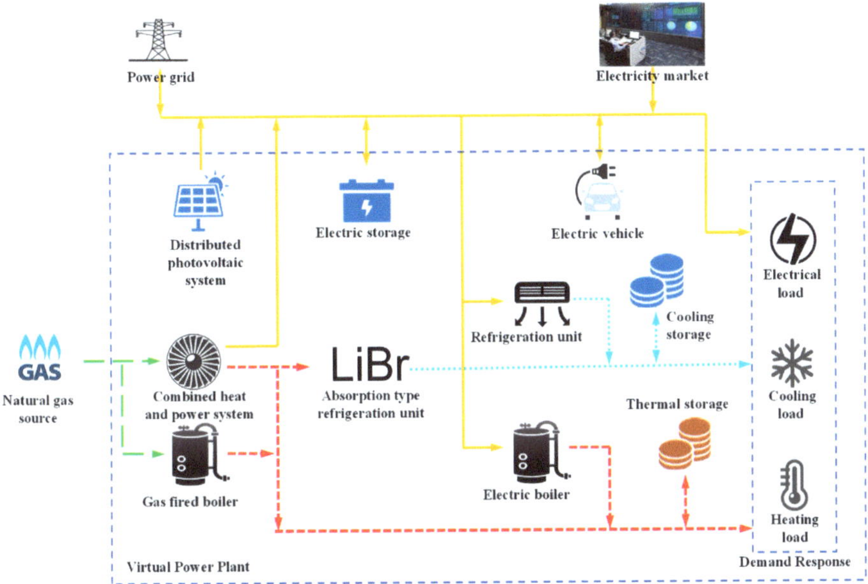


**Figure 1.** Basic structure of VPP.

### 2.1. Model Formulation of the VPP

### 2.1.1. Distributed Photovoltaic System

The power output of the distributed photovoltaic system is greatly affected by environmental factors. The power output is determined by light intensity and ambient temperature in an ideal situation which is shown as follows:

$$P_{PV} = f_{pv} P_{PVR} \frac{G}{G_S} [1 + \alpha_{PV}(T - T_S)] \tag{1}$$

where $P_{PV}$ represents the power output of the photovoltaic power system, MW. $f_{pv}$ and $P_{PVR}$ represent the reduction coefficient and rated power output in the standard state. $G$ and $G_S$ represent the illumination intensity of the current position and the standard state. $\alpha_{PV}$ is the power temperature reduction coefficient in the standard state. $T$ and $T_S$ represent the temperature on the surface of the photovoltaic panel and the temperature of the photovoltaic surface in the standard state.

### 2.1.2. Combined Heat and Power System

The combined heat and power (CHP) system generates electricity and heating energy by burning natural gas. Collecting the heating energy could improve the energy utilization rate of the gas turbine field in the CHP. Moreover, the output of heating and power energy are proportional to the consumption of natural gas. The calculation formulas are as follows:

$$P_{EGT} = \eta_E F_{GT} \tag{2}$$

$$P_{HGT} = \eta_H F_{GT} \tag{3}$$

$$\eta_E + \eta_H + \eta_{loss} = 1 \tag{4}$$

where $P_{EGT}$ and $P_{HGT}$ are the electric power output and thermal power output by the CHP, MW. $\eta_E$, $\eta_H$, $\eta_{loss}$ indicate the electric efficiency, thermal efficiency, and heat loss rate of the CHP, respectively. $F_{GT}$ represents the energy of gas combustion.

### 2.1.3. Gas-Fired Boiler

The gas-fired boiler consumes natural gas to produce thermal power which meets thermal balance. The thermal output of the gas boiler is proportional to the natural gas consumption, which is as follows:

$$P_{GB} = \eta_{GB} F_{GB} \tag{5}$$

where $P_{GB}$ is the thermal output power of the gas boiler. $\eta_{GB}$ indicates the gas utilization efficiency of the gas boiler. $F_{GB}$ is the consumption of natural gas.

### 2.1.4. Refrigeration Unit

The electric refrigeration unit could supply cooling to the consumer, and the output of the refrigerator is proportional to the input electric power which is as follows:

$$P_{CEC} = \eta_{EC} P_{EC} \tag{6}$$

where $P_{EC}$ is the cooling output of the electric refrigeration unit. $\eta_{EC}$ indicates the utilization efficiency of the electric refrigeration unit.

The absorption refrigeration unit utilizes the working medium to release cooling. The cooling output is directly proportional to the input thermal power and electric power, which are as follows:

$$P_{HRC} = \eta_{RC} P_{RC} \tag{7}$$

where $P_{HRC}$ and $P_{RC}$ are the cooling output and heating input of the absorption refrigeration unit. $\eta_{RC}$ is the refrigeration efficiency.

### 2.1.4.1. Energy Storage Unit

The VPP includes electric, heating, and cooling storage units, which meet the loads' demands. The energy storage unit has the function of balancing peaks and valleys which could improve the coefficient of energy utilization. The mathematical models are as follows:

$$E_{ES}(t) = (1 - \eta_{ES}) E_{ES}(t-1) + (P_{ESC} \eta_{ESC} - P_{ESD} / \eta_{ESD}) \Delta t \tag{8}$$

$$\alpha_{ESC} + \alpha_{ESD} \le 1 \tag{9}$$

$$0 \le P_{ESC} \le \alpha_{ESC} P_{ESC\ max} \tag{10}$$

$$0 \le P_{ESD} \le \alpha_{ESD} P_{ESD\ max} \tag{11}$$

$$E_{ES\ min} \le E_{ES}(t) \le E_{ES\ max} \tag{12}$$

where $E_{ES}(t)$ and $E_{ES}(t-1)$ represent the electric energy stored by the electric energy storage unit at time $t$ and time $t-1$. $P_{ESC}$ and $P_{ESD}$ are the charging power and discharge power. $\eta_{ES}$, $\eta_{ESC}$, $\eta_{ESD}$ are the self-discharge ratio, charging efficiency, and discharge efficiency, respectively. $P_{ESC\ max}$ and $P_{ESD\ max}$ are the rated charging power and the rated discharge power, respectively. $E_{ES\ min}$ and $E_{ES\ max}$ are the lower and upper climbing limits.

Moreover, Heating exchanges of thermal storage unit are as follows:

$$Q_{TS}(t) = (1 - \eta_{TS}) Q_{TS}(t-1) + (P_{TSC} \eta_{TSC} - P_{TSD} / \eta_{TSD}) \Delta t \tag{13}$$

$$\alpha_{TSC} + \alpha_{TSD} \le 1 \tag{14}$$

$$0 \le P_{TSC} \le \alpha_{TSC} P_{TSC\ max} \tag{15}$$

$$0 \le P_{TSD} \le \alpha_{TSD} P_{TSD\ max} \tag{16}$$

$$Q_{TS\ min} \le Q_{TS}(t) \le Q_{TS\ max} \tag{17}$$

where $Q_{TS}(t)$ and $Q_{TS}(t-1)$ represent the thermal stored by thermal storage unit at time $t$ and time $t-1$. $Q_{TS}(t)$ and $Q_{TS}(t-1)$ are the charging heating and discharge heating at time $t$ and time $t-1$. $\eta_{TS}, \eta_{TSC}, \eta_{TSD}$ are the self-discharge ratio, charging efficiency and discharge efficiency, respectively. $P_{TSC\,max}$ and $P_{TSD\,max}$ are the rated charging thermal and the rated discharge thermal, respectively. $E_{ES\,min}$ and $E_{ES\,max}$ are the lower and upper climbing limits.

*2.2. Objective Function*

The VPP was modelled using the mixed integer linear programming (MILP) method in LINGO software. The objective of the optimization was to maximize the VPP profit, which consists of the incomes of participating power and gas markets, benefits from demand sides, carbon emission fees, and unbalanced penalties. Therefore, the objective function is defined as follows:

$$
\begin{aligned}
\max C_m = & \sum_{t=1}^{T} c_d(t)P_d(t) - \sum_{t=1}^{T} F_{carbon}(t)\eta_{carbon} + \sum_{t=1}^{T} c_l(t)P_l(t) - \\
& \sum_{t=1}^{T} F_{eco}(t)P_{c,a}(t) - c_{co}(t)\sum_{t=1}^{T} P_{TS}(t) - C_b
\end{aligned}
\tag{18}
$$

where $c_d(t)$ is the prices in the electricity market. $\eta_{carbon}$ is the fee of carbon emissions, yuan/t $CO_2$. $c_l(t)$ is the price for VPP's users. $C_{co}(t)$ is the fee of energy storage loss and $C_b$ is the unbalanced penalty cost.

The operating cost is the sum of the purchased energy cost and equipment maintenance cost which are showed as follows:

$$
F_{eco} = C_{op} + C_{en}
\tag{19}
$$

where $F_{eco}$ is the operation cost. $C_{op}$ is the maintenance cost. $C_{en}$ is the purchased energy cost which is shown as follows:

$$
\begin{aligned}
C_{op} = \sum_{t=1}^{T} & [\lambda_{WT}P_{WT}(t) + \lambda_{PV}P_{PV}(t) + \lambda_{GT}P_{GT}(t) + \lambda_{GB}P_{GB}(t) + \\
& \lambda_{EC}P_{EC}(t) + \lambda_{RC}P_{RC}(t) + \lambda_{ES}P_{ES}(t) + \lambda_{TS}P_{TS}(t)]\Delta t
\end{aligned}
\tag{20}
$$

$$
C_{en} = \delta_{gas}\sum_{t=1}^{T}[F_{GT}(t) + F_{GB}(t)]\Delta t + \delta_{el}\sum_{t=1}^{T} P_{grid}(t)\Delta t
\tag{21}
$$

where $P_X(t)$ represents the average power output of X. $\lambda_X$ represents the cost coefficient of operation and maintenance. $\delta_{gas}$ and $\delta_{el}$ are the prices of natural gas and electricity, respectively. $F_{GT}(t)$, $F_{GB}(t)$, $P_{grid}(t)$ indicate the average combustion ratio of natural gas in CHP, the average combustion ratio of natural gas in the gas boiler, and the average input of the power grid, respectively.

The carbon emissions of a VPP could be calculated using the following equation:

$$
F_{carbon} = \sum_{t=1}^{T}\{\lambda_{gas}[F_{GB}(t) + F_{GT}(t)] + \lambda_{el}P_{grid}(t)\}\Delta t
\tag{22}
$$

where $F_{carbon}$ represents the carbon emissions of the VPP, t $CO_2$. $\lambda_{gas}, \lambda_{el}$ are the carbon emission coefficients of natural gas and the power grid, t $CO_2$/MW.

*2.3. Constraint Conditions*

In order to make the energy network safe and stable, the variables in the energy network need to meet certain constraints in the VPP. The energy conservation constraints include

$$
P_{grid} + P_{WT} + P_{PV} + P_{EGT} = P_E + P_{EC} + P_{ESC}(-P_{ESD})
\tag{23}
$$

$$
P_{HGT} + P_{GB} = P_H + P_{RC} + P_{TSC}(-P_{TSD})
\tag{24}
$$

$$\eta_{EC}P_{EC} + \eta_{RC}P_{RC} = P_C \tag{25}$$

where $P_{grid}$ is the power supply of the power gird. $P_E$, $P_H$, $P_C$ are the electric load, thermal load, and cooling load.

All equipment operates between the maximum output and the minimum output to ensure long-term safe operation in VPP. The constraints include

$$0 \leq P_{GT} \leq P_{GT\ max} \tag{26}$$

$$0 \leq P_{GB} \leq P_{GB\ max} \tag{27}$$

$$0 \leq P_{CEC} \leq P_{CEC\ max} \tag{28}$$

$$0 \leq P_{HRC} \leq P_{HRC\ max} \tag{29}$$

$$0 \leq P_{grid} \leq P_{grid\ max} \tag{30}$$

$$0 \leq F_{GB} + F_{GT} \leq F_{max} \tag{31}$$

where $P_{GT\ max}$, $P_{GB\ max}$, $P_{CEC\ max}$, $P_{HRC\ max}$ are the maximum output of the CHP, gas boiler, and refrigeration unit. $P_{grid\ max}$ and $F_{max}$ represent the maximum electric power supplied by the power grid and the maximum ratio of natural gas supplied by the natural gas pipeline.

The energy storage unit constraints include electric, heating, and cooling storage balance constraints which are as follows:

$$E_{ES}(t) = (1 - \eta_{ES})E_{ES}(t-1) + (P_{ESC}\eta_{ESC} - P_{ESD}/\eta_{ESD})\Delta t \tag{32}$$

$$\alpha_{ESC} + \alpha_{ESD} \leq 1 \tag{33}$$

$$0 \leq P_{ESC} \leq \alpha_{ESC}P_{ESC\ max} \tag{34}$$

$$0 \leq P_{ESD} \leq \alpha_{ESD}P_{ESD\ max} \tag{35}$$

$$E_{ES\ min} \leq E_{ES}(t) \leq E_{ES\ max} \tag{36}$$

$$Q_{TS}(t) = (1 - \eta_{TS})Q_{TS}(t-1) + (P_{TSC}\eta_{TSC} - P_{TSD}/\eta_{TSD})\Delta t \tag{37}$$

$$\alpha_{TSC} + \alpha_{TSD} \leq 1 \tag{38}$$

$$0 \leq P_{TSC} \leq \alpha_{TSC}P_{TSC\ max} \tag{39}$$

$$0 \leq P_{TSD} \leq \alpha_{TSD}P_{TSD\ max} \tag{40}$$

$$Q_{TS\ min} \leq Q_{TS}(t) \leq Q_{TS\ max} \tag{41}$$

where $\alpha_{ESC}$, $\alpha_{ESD}$, $\alpha_{TSC}$, $\alpha_{TSD}$ are binary parameters (0–1) which could constrain the energy storage unit so that it could not charge and discharge simultaneously.

## 3. Flexibility Margin Considering Demand Response

Terminal customers are a strong uncertainty, and the load could be divided into the interruptible, adjustable, and sensitive loads [15,33,34]. We have divided the load demand into certainty and uncertainty loads. The certainty load means the invariable load which must be supplied, and the uncertainty loads are variable loads in the flexibility margin. According to the theory of uncertainty, the electricity price is described by the probability distribution. The output of renewable energy is analyzed by weather prediction.

*3.1. Flexibility Margin*

The flexibility of the VPP refers to the degree of balance between the supply and demand of energy. The difference between the power load and the output of photovoltaic power is described as the net load. We can describe the VPP flexibility requirement as being calculated by

$$F_t = P_{netload,t+1} - P_{netload,t} \tag{42}$$

$$P_{netload,t} = P_{load,t} - P_{PV,t} \tag{43}$$

$$\begin{aligned} F_t{}^{up} &= \{F_t \mid F_t > 0\} \\ F_t{}^{down} &= \{F_t \mid F_t < 0\} \end{aligned} \tag{44}$$

where $P_{PV,t}$ is the actual output of photovoltaic power generation at time $t$. $F_t{}^{up}$, $F_t{}^{down}$ are the upward and downward flexibility requirements at time $t$. $P_{netload,t}$, $P_{load,t}$ are the power load and net load at time $t$. $P_{netload,t+1}$ is the net load at time $t + 1$. Moreover, the prediction error of the photovoltaic system output satisfied the normal distribution, which is described as $\Delta P_{PV,t} \sim N(0, \sigma_{PV,t})$.

*3.2. Flexibility Indicators*

The flexibility margin is described as the difference between flexible supply and demand. The direction includes up and down.

$$\begin{cases} F_{ru\_up} = F_{gong,t}^{up} - F_t^{up} \\ F_{ru\_down} = F_{gong,t}^{down} - F_t^{down} \end{cases} \tag{45}$$

where $F_{ru\_up}$, $F_{ru\_down}$ are the upward and downward flexibility margins, respectively.

## 4. Example Analysis

There is a community which has the data showing electricity load, cooling load, heating load, light radiation, and temperature in Beijing. The time scale is one hour. The charging price of electric vehicles refers to the charging standard of Beijing electric vehicles. The valley periods are 23:00–7:00, the usual periods are 8:00–10:00, 16:00–18:00, and 22:00, and the peak periods are 11:00–15:00 and 19:00–21:00, which are shown in Table 1. The electricity price of users is shown in Table 2. The valley periods are 23:00–6:00, 7:00–9:00, 12:00–18:00, 10:00–11:00–11:00, and 19:00–22:00, and the selling price of energy storage equipment to the grid is set at 0.45 yuan/kWh, which is higher than the electricity price of both valleys and lower than the usual price of both.

**Table 1.** Charging price of the electric vehicles.

| Times | Prices (yuan/kWh) |
| --- | --- |
| valley period | 0.3946 |
| usual period | 0.685 |
| peak period | 1.0044 |

**Table 2.** Purchase electricity price of users.

| Times | Prices (yuan/kWh) |
| --- | --- |
| valley period | 0.284 |
| usual period | 0.52 |
| peak period | 0.89 |

Moreover, the carbon emission of the CHP unit is 0.798 t/(MWh), and the carbon trading price is 52.78 yuan/t. The peak power load of the user is 150 kW, the peak cooling load is 201 kW, the peak heating load is 672 kW, and the PV installation capacity is within the range of 0.5~2 times of the power capacity. Considering the charging demand of electric vehicles, the installed capacity of the battery is 4000 Ah, the installed capacity of air conditioning is 220 kW, the installed capacity of ice storage equipment is 500 kWh, the installed capacity of the electric boiler is 70 kW, and the installed capacity of heat storage tank equipment is 120 kWh. The cost parameters of the energy unit are shown in Table 3.

**Table 3.** The cost parameters of the energy unit.

| Technical Equipment | Installation Cost yuan/kW | Running Costs yuan/kWh | Efficiency | | Period (Year) |
|---|---|---|---|---|---|
| | | | Electrical Efficiency | Heating Efficiency | |
| Internal combustion engine | 5000 | 0.072 | 0.4 | 0.45 | 30 |
| Photovoltaic system | 7500 | 0.01 | 0.12 | 0 | 25 |
| Energy storage system | 4000 | 0.0022 | 0.81 | | 15 |

In our model, the prices of different energies are shown in Figure 2. The gas price and photovoltaic feed-in tariff do not change with time. However, the electricity prices in different voltages change at different times. Moreover, the charging load of the electric vehicle benchmark is shown in Figure 3. It changes with a normal distribution.

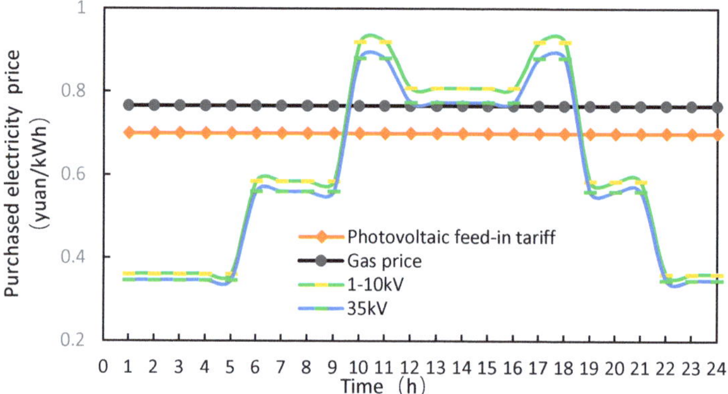


**Figure 2.** Prices of different energy.

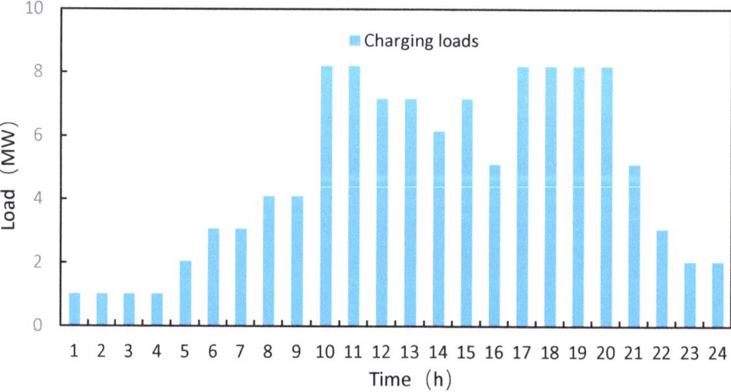


**Figure 3.** Charging load of electric vehicle benchmark.

The typical daily operation is shown in the following figure which is the power balance in the traditional model without a VPP. At the low price, the charging station buys electricity from the power grid, and the CHP system starts and stops twice a day. The photovoltaic output is small in the morning and evening, and at noon, when the photovoltaic output is large, the renewable energy is fully used. There are no power exchanges with the power grid.

The photovoltaic system outputs energy during the day. At the peak time, buying electricity from the grid is not economical. Therefore, it integrates photovoltaic systems into the charging stations, showing a good economy. The gas internal combustion engine in the CHP system is easy to start and stop, which could increase the power safety and stability of charging stations. According to the gas price and the safety and stability requirements of the charging station, it has more benefits with the appropriate gas generator sets. According to the above analysis, the mode of grid-connected and non-connected VPPs is adopted. For the charging station, a 10 MW photovoltaic system, 2 MW CHP unit, and 1 MW energy storage system are arranged to calculate the gas price. The price of the gas is 2 yuan/m$^3$, and the charging cost of the electric vehicle is 1.4 yuan/kWh. The operation strategy is shown in traditional model in Figure 4. As shown in Figure 5, the energy storage station saves energy during the low electricity price at night and discharges during the daytime peak. In the case of the photovoltaic system, when the energy supply of the energy system is higher than the load demand, the energy storage increases the efficient operation of the energy system.

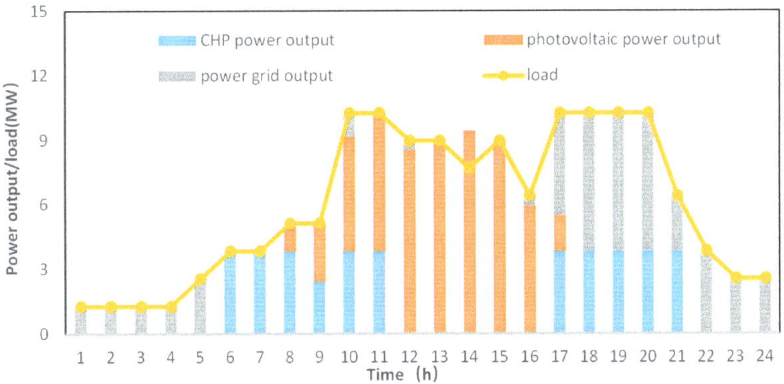

**Figure 4.** Power balance in traditional model without VPP.

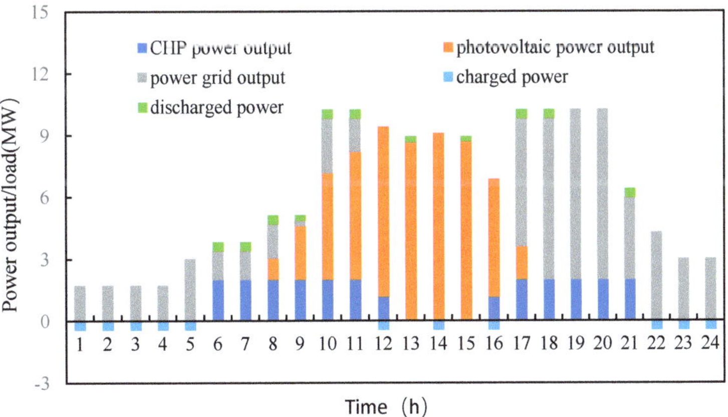

**Figure 5.** Power balance in the model with VPP.

According to the above, the VPP is modelled in Section 2. The typical daily operation situation is shown in Figure 6. Since all the power comes from the photovoltaic, CHP, and energy storage systems, the selected installed capacity must meet the real-time demand of the charging load. The equipment capacities are set relatively high, and the overall investment cost of the system is high. From the perspective of operation, the transmission power exceeds the demand in the low load period. The high output of renewable energy is greater than the load demand at noon. The energy storage system mainly stores renewable energy and releases electricity during the load peak period in the evening.

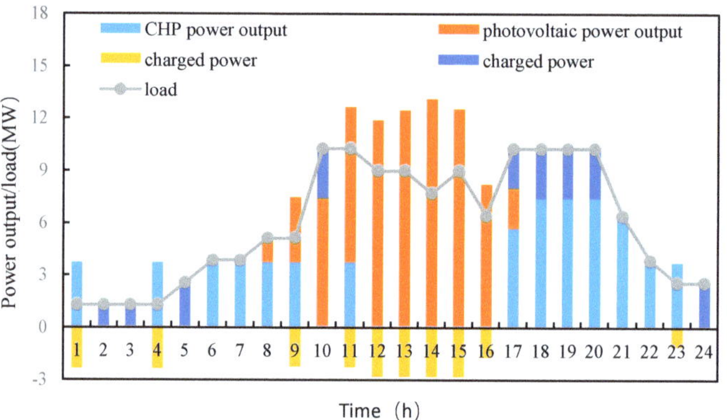


**Figure 6.** Power balance in the model with VPP and electric vehicle.

Figures 4–6 are the different scenarios of the traditional model, VPP model, and the VPP model with an electric vehicle. In Figure 4, there are no power exchanges with the power grid. The power exchanges with the power grid are shown in Figures 5 and 6. Moreover, the huge power exchanges are shown in Figure 6 which illustrates more profits for the VPP operator.

The orderly charging and discharging strategy are adopted in the VPP shown in Figure 7. The power interaction by electric vehicle load is changing, and the load distribution is more reasonable. Peak load filling is carried out, and no new load peak is generated which is conducive to keeping the safe operation of the power grid. The transformer has no overload. In the peak period of electricity consumption, the discharge is conducted by electric vehicle according to the demand of the users. It not only reduces the load rate of the transformer but also improves the income of the users.

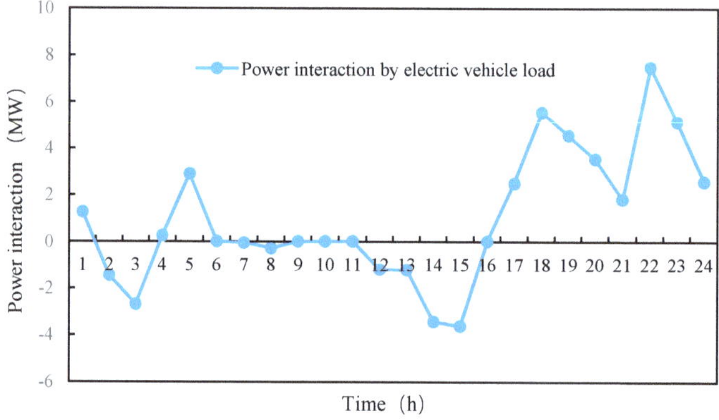


**Figure 7.** Power interaction between power grid and electric vehicle.

Figures 8 and 9 are the benefits of the proposed VPP model. We could give the conclusion that the charging price is the key point for the electric vehicle, which is the flexibility resources of the VPP. The CHP system has more income in the night when the power load is at its peak. Moreover, the flexibility resources based on the flexibility margin have more benefits all day.

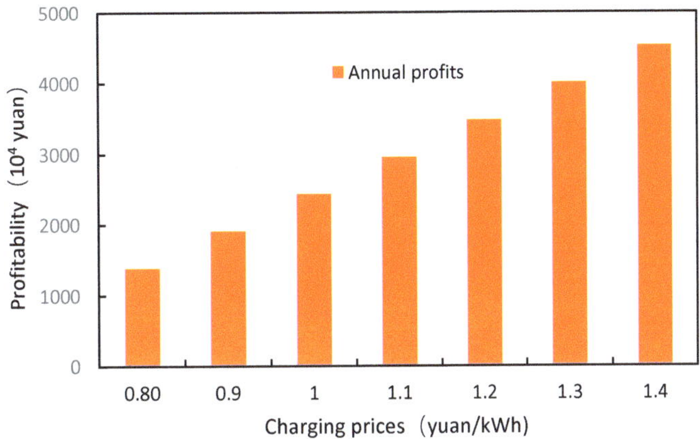


**Figure 8.** The annual profit of VPP with different charging prices.

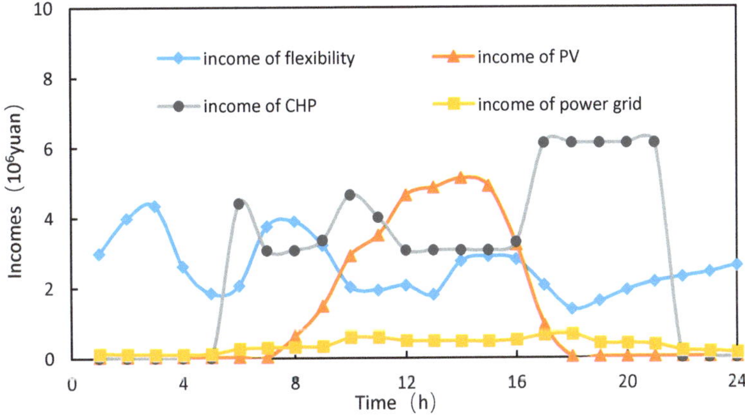


**Figure 9.** The incomes of different participators.

## 5. Conclusions

This paper puts forward the orderly charging and discharging strategy of electric vehicles in a VPP considering the needs and interests of users based on the flexibility margin in the VPP. The numerical results showed that the proposed VPP optimization method reduced the operation cost very well. The strategy could prevent overload, complete load transfer, realize peak shifting and valley filling, and solve the problems of peaks and new peaks caused by disorderly power utilization. Moreover, the VPP strategy proposed in this paper changes the multi-objective function into a single-objective function by optimizing the load model of electric vehicles which could increase the economic efficiency of the VPP. Finally, the orderly charging and discharging strategy of electric vehicles could reduce the charging cost for users participating in the peak-regulating auxiliary services market.

**Author Contributions:** Conceptualization, Y.T.; methodology, Y.Z.; validation, Z.L.; formal analysis, H.F.; investigation, J.W. and T.Z. All authors have read and agreed to the published version of the manuscript.

**Funding:** This work was supported by the National Natural Science Foundation of China (NSFC) (Grant No:51879140), State Key Laboratory of Hydroscience and Hydraulic Engineering (Grant No. 2021-KY-04), Tsinghua-Foshan Innovation Special Fund (TFISF) 2021THFS0209, and Creative Seed Fund of Shanxi Research Institute for Clean Energy, Tsinghua University.

**Conflicts of Interest:** The authors declare no conflict of interest.

## Abbreviations

| | |
|---|---|
| $\alpha_{PV}$ | the power temperature reduction coefficient in the standard state |
| $\eta_E$ | the electric efficiency CHP |
| $\eta_{EC}$ | the utilization efficiency of the electric refrigeration unit |
| $\eta_{ES}$ | the self-discharge ratio |
| $\eta_{ESC}$ | the charging efficiency |
| $\eta_{ESD}$ | the discharge efficiency |
| $\eta_H$ | the thermal efficiency CHP |
| $\eta_{GB}$ | the gas utilization efficiency of the gas boiler |
| $\eta_{loss}$ | the heat loss rate of CHP |
| $\eta_{RC}$ | the refrigeration efficiency |
| $E_{ES\,min}$ | the lower climbing limits (MW/h) |
| $E_{ES\,max}$ | the upper climbing limits (MW/h) |
| $E_{ES}(t)$ | The electric energy stored by the electric energy storage unit at time $t$ (MW) |
| $E_{ES}(t-1)$ | the electric energy stored by the electric energy storage unit at time $t-1$ (MW) |
| $f_{pv}$ | the reduction coefficient |
| $F_{GB}$ | the consumption of natural gas (m$^3$) |
| $F_{GT}$ | the energy of gas combustion (MW) |
| $F_{ru\_up}$ | the upward flexibility margin (MW) |
| $F_{ru\_udown}$ | the downward flexibility margin (MW) |
| $G$ | the illumination intensity of the current position (W/m$^2$) |
| $G_S$ | the illumination intensity of the standard state (W/m$^2$) |
| $P_C$ | the cooling load (MW) |
| $P_E$ | the electric load (MW) |
| $P_{EC}$ | the cooling output of the electric refrigeration unit (MW) |
| $P_{EGT}$ | the electric power output by CHP (MW) |
| $P_{ESC}$ | the charging power output (MW) |
| $P_{ESD}$ | the discharge power output (MW) |
| $P_{ESC\,max}$ | the rated charging power output (MW) |
| $P_{ESD\,max}$ | the rated discharge power output (MW) |
| $P_{grid}$ | the power supply of the power gird (MW) |
| $P_{GB}$ | The thermal output power of the gas boiler (MW) |
| $P_H$ | the thermal load (MW) |
| $P_{HGT}$ | the thermal power output by CHP (MW) |
| $P_{HRC}$ | the cooling output of the absorption refrigeration unit (MW) |
| $P_{PV}$ | the power output of the photovoltaic power system (MW) |
| $P_{PVR}$ | the rated power output in the standard state (MW) |
| $P_{RC}$ | the cooling output and heating input of the absorption refrigeration unit (MW) |
| $T$ | the temperature on the surface of the photovoltaic panel (°C) |
| $Ts$ | the temperature of the photovoltaic surface in the standard state (°C) |

## References

1. Kong, X.; Xiao, J.; Wang, C.; Cui, K.; Jin, Q.; Kong, D. Bi-level multi-time scale scheduling method based on bidding for multi-operator virtual power plant. *Appl. Energy* **2019**, *249*, 178–189. [CrossRef]
2. Wei, C.; Xu, J.; Liao, S.; Sun, Y.; Jiang, Y.; Ke, D.; Zhang, Z.; Wang, J. A bi-level scheduling model for virtual power plants with aggregated thermostatically controlled loads and renewable energy. *Appl. Energy* **2018**, *224*, 659–670. [CrossRef]

3.  Ju, L.; Tan, Z.; Yuan, J.; Tan, Q.; Li, H.; Dong, F. A bi-level stochastic scheduling optimization model for a virtual power plant connected to a wind–photovoltaic–energy storage system considering the uncertainty and demand response. *Appl. Energy* **2016**, *171*, 184–199. [CrossRef]
4.  Riveros, J.Z.; Bruninx, K.; Poncelet, K.; D'haeseleer, W. Bidding strategies for virtual power plants considering CHPs and intermittent renewables. *Energy Convers. Manag.* **2015**, *103*, 408–418. [CrossRef]
5.  Alsaleh, M.; Abdul-Rahim, A.S. Bioenergy industry and the growth of the energy sector in the EU-28 region: Evidence from panel cointegration analysis. *J. Renew. Sustain. Energy* **2018**, *10*, 53–61. [CrossRef]
6.  Alsaleh, M.; Abdul-Rahim, A.S. Bioenergy Intensity and Its Determinants in European Continental Countries: Evidence Using GMM Estimation. *Resources* **2019**, *8*, 43. [CrossRef]
7.  Royapoor, M.; Pazhoohesh, M.; Davison, P.J.; Patsios, C.; Walker, S. Building as a virtual power plant, magnitude and persistence of deferrable loads and human comfort implications. *Energy Build.* **2020**, *213*, 109794. [CrossRef]
8.  Sikorski, T.; Jasiński, M.; Ropuszyńska-Surma, E.; Węglarz, M.; Kaczorowska, D.; Kostyla, P.; Leonowicz, Z.; Lis, R.; Rezmer, J.; Rojewski, W.; et al. A Case Study on Distributed Energy Resources and Energy-Storage Systems in a Virtual Power Plant Concept: Technical Aspects. *Energies* **2020**, *13*, 3086. [CrossRef]
9.  Loßner, M.; Böttger, D.; Bruckner, T. Economic assessment of virtual power plants in the German energy market—A scenario-based and model-supported analysis. *Energy Econ.* **2017**, *62*, 125–138. [CrossRef]
10.  Van Summeren, L.F.; Wieczorek, A.J.; Bombaerts, G.J.; Verbong, G.P. Community energy meets smart grids: Reviewing goals, structure, and roles in Virtual Power Plants in Ireland, Belgium and the Netherlands. *Energy Res. Soc. Sci.* **2020**, *63*, 101415. [CrossRef]
11.  Ullah, Z.; Mokryani, G.; Campean, F.; Hu, Y.F. Comprehensive review of VPPs planning, operation and scheduling considering the uncertainties related to renewable energy sources. *IET Energy Syst. Integr.* **2019**, *1*, 147–157. [CrossRef]
12.  Nosratabadi, S.M.; Hooshmand, R.-A.; Gholipour, E. A comprehensive review on microgrid and virtual power plant concepts employed for distributed energy resources scheduling in power systems. *Renew. Sustain. Energy Rev.* **2017**, *67*, 341–363. [CrossRef]
13.  Arslan, O.; Karasan, O.E. Cost and emission impacts of virtual power plant formation in plug-in hybrid electric vehicle penetrated networks. *Energy* **2013**, *60*, 116–124. [CrossRef]
14.  Ju, L.W.; Tan, Q.L.; Lu, Y.; Tan, Z.F.; Zhang, Y.X.; Tan, Q.K. A CVaR-robust-based multi-objective optimization model and three-stage solution algorithm for a virtual power plant considering uncertainties and carbon emission allowances. *Int. J. Electr. Power Energy Syst.* **2019**, *107*, 628–643. [CrossRef]
15.  Cui, H.; Li, F.; Hu, Q.; Bai, L.; Fang, X. Day-ahead coordinated operation of utility-scale electricity and natural gas networks considering demand response based virtual power plants. *Appl. Energy* **2016**, *176*, 183–195. [CrossRef]
16.  Kasaei, M.J.; Gandomkar, M.; Nikoukar, J. Optimal management of renewable energy sources by virtual power plant. *Renew. Energy* **2017**, *114*, 1180–1188. [CrossRef]
17.  Luo, Z.; Hong, S.; Ding, Y. A data mining-driven incentive-based demand response scheme for a virtual power plant. *Appl. Energy* **2019**, *239*, 549–559. [CrossRef]
18.  Li, Y.; Gao, W.; Ruan, Y. Feasibility of virtual power plants (VPPs) and its efficiency assessment through benefiting both the supply and demand sides in Chongming country, China. *Sustain. Cities Soc.* **2017**, *35*, 544–551. [CrossRef]
19.  Alahyari, A.; Ehsan, M.; Mousavizadeh, M. A hybrid storage-wind virtual power plant (VPP) participation in the electricity markets: A self-scheduling optimization considering price, renewable generation, and electric vehicles uncertainties. *J. Energy Storage* **2019**, *25*, 100812. [CrossRef]
20.  Sousa, T.; Morais, H.; Vale, Z.; Faria, P.; Soares, J. Intelligent Energy Resource Management Considering Vehicle-to-Grid: A Simulated Annealing Approach. *IEEE Trans. Smart Grid* **2012**, *3*, 535–542. [CrossRef]
21.  Daraei, M.; Campana, P.E.; Thorin, E. Power-to-hydrogen storage integrated with rooftop photovoltaic systems and combined heat and power plants. *Appl. Energy* **2020**, *276*, 115499. [CrossRef]
22.  Kolenc, M.; Nemček, P.; Gutschi, C.; Suljanović, N.; Zajc, M. Performance evaluation of a virtual power plant communication system providing ancillary services. *Electr. Power Syst. Res.* **2017**, *149*, 46–54. [CrossRef]
23.  Bloess, A. Modeling of combined heat and power generation in the context of increasing renewable energy penetration. *Appl. Energy* **2020**, *267*, 1–17. [CrossRef]
24.  Arteconi, A.; Mugnini, A.; Polonara, F. Energy flexible buildings: A methodology for rating the flexibility performance of buildings with electric heating and cooling systems. *Appl. Energy* **2019**, *251*, 113387. [CrossRef]
25.  Magdy, F.E.Z.; Ibrahim, D.K.; Sabry, W. Energy management of virtual power plants dependent on electro-economical model. *Ain Shams Eng. J.* **2020**, *11*, 643–649. [CrossRef]
26.  Yu, S.; Fang, F.; Liu, Y.; Liu, J. Uncertainties of virtual power plant: Problems and countermeasures. *Appl. Energy* **2019**, *239*, 454–470. [CrossRef]
27.  Liu, Y.; Shen, Z.; Tang, X.; Lian, H.; Li, J.; Gong, J. Worst-case conditional value-at-risk based bidding strategy for wind-hydro hybrid systems under probability distribution uncertainties. *Appl. Energy* **2019**, *256*, 113918. [CrossRef]
28.  Liang, Z.; Alsafasfeh, Q.; Jin, T.; Pourbabak, H.; Su, W. Risk-Constrained Optimal Energy Management for Virtual Power Plants Considering Correlated Demand Response. *IEEE Trans. Smart Grid* **2019**, *10*, 1577–1587. [CrossRef]
29.  Adu-Kankam, K.O.; Camarinha-Matos, L.M. Towards collaborative Virtual Power Plants: Trends and convergence. *Sustain. Energy Grids Netw.* **2018**, *16*, 217–230. [CrossRef]

30. Robu, V.; Chalkiadakis, G.; Kota, R.; Rogers, A.; Jennings, N.R. Rewarding cooperative virtual power plant formation using scoring rules. *Energy* **2016**, *117*, 19–28. [CrossRef]
31. Guerrero, J.; Gebbran, D.; Mhanna, S.; Chapman, A.C.; Verbic, G. Towards a transactive energy system for integration of distributed energy resources: Home energy management, distributed optimal power flow, and peer-to-peer energy trading. *Renew. Sustain. Energy Rev.* **2020**, *132*, 110000. [CrossRef]
32. Hooshmand, R.A.; Nosratabadi, S.M.; Gholipour, E. Event-based scheduling of industrial technical virtual power plant considering wind and market prices stochastic behaviors—A case study in Iran. *J. Clean. Prod.* **2018**, *172*, 1748–1764. [CrossRef]
33. Obringer, R.; Mukherjee, S.; Nateghi, R. Evaluating the climate sensitivity of coupled electricity-natural gas demand using a multivariate framework. *Appl. Energy* **2020**, *262*, 114419. [CrossRef]
34. Shabanzadeh, M.; Sheikh-El-Eslami, M.-K.; Haghifam, M.-R. An interactive cooperation model for neighboring virtual power plants. *Appl. Energy* **2017**, *200*, 273–289. [CrossRef]

*Article*

# Environment-Friendly Power Scheduling Based on Deep Contextual Reinforcement Learning

Awol Seid Ebrie [1], Chunhyun Paik [2], Yongjoo Chung [3] and Young Jin Kim [4,*]

[1] Major in Industrial Data Science and Engineering, Department of Industrial and Data Engineering, Pukyong National University, Busan 48513, Republic of Korea; es.awol@pukyong.ac.kr
[2] Department of Industrial Management and Big Data Engineering, Dongeui University, Busan 47340, Republic of Korea; chpaik@deu.ac.kr
[3] Department of Global Marketing, Busan University of Foreign Studies, Busan 46234, Republic of Korea; chungyj@bufs.ac.kr
[4] Department of Systems Management and Engineering, Pukyong National University, Busan 48513, Republic of Korea
* Correspondence: youngk@pknu.ac.kr; Tel.: +82-51-629-6486

**Abstract:** A novel approach to power scheduling is introduced, focusing on minimizing both economic and environmental impacts. This method utilizes deep contextual reinforcement learning (RL) within an agent-based simulation environment. Each generating unit is treated as an independent, heterogeneous agent, and the scheduling dynamics are formulated as Markov decision processes (MDPs). The MDPs are then used to train a deep RL model to determine optimal power schedules. The performance of this approach is evaluated across various power systems, including both small-scale and large-scale systems with up to 100 units. The results demonstrate that the proposed method exhibits superior performance and scalability in handling power systems with a larger number of units.

**Keywords:** power scheduling; unit commitment; reinforcement learning; agent-based simulation

**Citation:** Ebrie, A.S.; Paik, C.; Chung, Y.; Kim, Y.J. Environment-Friendly Power Scheduling Based on Deep Contextual Reinforcement Learning. *Energies* **2023**, *16*, 5920. https://doi.org/10.3390/en16165920

Academic Editors: Marialaura Di Somma, Jianxiao Wang and Bing Yan

Received: 16 July 2023
Revised: 6 August 2023
Accepted: 8 August 2023
Published: 10 August 2023

## 1. Introduction

Electrical energy generated from fossil fuels emits significant amounts of greenhouse gases (GHGs), including carbon dioxide ($CO_2$), sulfur dioxide ($SO_2$), and nitrous oxide ($N_2O$). These emissions have detrimental effects on human health and contribute to climate change and global warming. However, the prevailing focus on economic concerns in power generation often results in higher emission levels since economic costs and environmental impacts tend to be inversely related. This effect has been amplified since the implementation of the global emissions trading scheme (ETS) in 2005, aimed at controlling GHGs [1,2]. As a result, power generation based solely on economic costs leads to increased financial penalties for emissions, along with adverse environmental impacts. Therefore, the traditional approach of scheduling power generation based solely on economic costs, which overlooks environmental ETS, is no longer acceptable [2]. Consequently, it becomes imperative to determine an efficient and environmentally friendly power generation schedule that might have lower emission costs, indicating that it generates fewer GHGs or other pollutants during its operation. The primary objective of environmentally friendly power scheduling is to achieve a sustainable cost and emission balance, where economic concerns are considered without compromising environmental sustainability. This means effectively meeting electricity demand while minimizing the environmental impact, especially in terms of greenhouse gas emissions (GHGs) and other pollutants. By adopting this approach, not only does the profitability of power generation increase, but it also leads to reduced emission levels through efficient management and scheduling of generating units [1,2].

Prior research has explored various model-based approaches, such as conventional and dynamic programming and stochastic optimizations, to address the challenges in power

Equations (5)–(9) using classical methods is highly computationally demanding even for a moderate number of units and might also stack in some local optima. As a result, the power scheduling problem remains a strongly NP-hard problem, causing the curse of dimensionality, and the implicit burden of computation has limited the scope of numerical optimization [3]. A model-free RL approach may provide a promising methodological framework for solving the power scheduling problem [5].

## 3. Proposed Methodological Framework

A novel multi-agent deep contextual RL algorithm for power scheduling is proposed by constructing a specialized environment called an "agent-based contextual simulation environment", whose main peculiarities are explained in Section 3.1. Within the environment, the generating units are represented as cooperative types of RL agents [15]. The agents are active in observing the contextual changes in the environment, and they can make independent decisions regarding their commitment status and optimal power outputs. On the other hand, the agents collaborate to satisfy demand, including the reserve availability at each hour (timestep) described in Equations (8) and (9), and the total operation cost of the entire planning horizon (episode) is to be minimized. The power scheduling dynamics from the agent-to-environment interactions follow an agent-based simulation strategy [16] because the agents are heterogeneous, active, and autonomous. These dynamics are simulated in the form of a Markov decision process (MDP), whose key elements are defined below, and then fed as inputs for the deep RL model.

### 3.1. The Power Scheduling Dynamics as an MDP

Since the planning horizon is an hourly divided day, each hour is considered a timestep $t, \forall t$. For each timestep $t$ of an episode, the system will be state $s_t$ consisting of different components: current timestep $t$, minimum capacities ($p_{it}^{min}; \forall i$), and maximum capacities ($p_{it}^{max}; \forall i$) based on the maximum ramp-down ($p_{i*}^{down}; \forall i$) and ramp-up ($p_{i*}^{up}; \forall i$) rates of the units, current operating (online/offline) time durations ($t_{it}; \forall i$) of the units based on the minimum up-time duration $\left(t_{i*}^{up}; \forall i\right)$ and down-time duration $\left(t_{i*}^{down}; \forall i\right)$ of the units, and the demand ($d_t$) to be satisfied. The state $s_t$ at timestep $t$ is defined as $s_t = \left(t, p_t^{min}, p_t^{max}, t_t, d_t\right)$ where $t$ is the current timestep, $p_t^{min}$ is a vector of minimum capacities, $p_t^{max}$ is a vector of maximum capacities, $t_t$ is a vector of current (online/offline) time duration, and $d_t$ is the demand. Overall, the system's state space for the entire episode can be described as $\mathscr{S} = \left(t, P^{min}, P^{max}, T, d\right)$. There are two possible actions (switch-to/stay ON or switch-to/stay OFF) for each of the $n$ agents. This implies, there are a total of $2^n$ action combinations (i.e., unit commitments) in the action space $\mathscr{A}$. Thus, each agent $i$ will decide their optimal action ($a_{it} \in \{0, 1\}$) at timestep $t$ (or in state $s_t$). Then, the decisions of all the $n$ agents together constitute an $n$-dimensional vector $a_t = \{0, 1\}^n \in \mathscr{A}$. This actions vector $a_t$ is the change of the switch (ON/OFF) status $z_t$ of units at period $t$ to $z_{t+1}$ at the next period $t + 1$. Once the agents take actions $a_t \in \mathscr{A}$ in the current state $s_t \in \mathscr{S}$, there is a transition (or probability) function $\mathscr{P}(s_{t+1}|s_t, a_t)$ that leads to the next state $s_{t+1}$. The transition function must satisfy all the constraints.

At each timestep $t$ of the planning horizon, the agents first observe the state $s_t = \left(t, p_t^{min}, p_t^{max}, t_t, z_t, d_t\right) \in \mathscr{S}$. Then, each agent can decide to be either ON or OFF, which would result in $2^n$ combinations of unit commitment found in an action space $\mathscr{A}$. The decisions of all agents constitute the action vector $a_t = \{0, 1\}^n \in \mathscr{A}$. Then, the agents get an aggregate reward $r_t \in \mathbb{R}$ that can lead to the next state $s_{t+1} \in \mathscr{S}$ through a transition (probability) function $\mathscr{P}(s_{t+1} = s'|s_t = s, a_t)$. Therefore, these power scheduling dynamics can be represented as a 4-tuple $(\mathscr{S}, \mathscr{A}, \mathscr{P}, r)$ Markovian decision process where $\mathscr{S}$ is a state space, $\mathscr{A}$ is an action space, $\mathscr{P}$ is a transition (or probability) function, and $r$ is a reward.

### 3.2. Agent-Based Contextual Simulation Environment Algorithm

The structure of agent-based simulation environment is roughly similar to an OpenAI Gym as described below.

**Step 1.** The parameters of supply units are initialized as $\jmath_0 = \left(0, p_0^{min}, p_0^{max}, t_0, z_0, d_0\right)$.

**Step 2.** The minimum and maximum marginal costs of all agents are determined based on the average production cost, Equation (10).

$$\lambda_i^{min} = \alpha_i p_{i*}^{max} + \beta_i + \frac{\delta_i}{p_{i*}^{max}} \text{ and } \lambda_i^{max} = \alpha_i p_{i*}^{min} + \beta_i + \frac{\delta_i}{p_{i*}^{min}}; \forall i \tag{10}$$

**Step 3.** The must-ON $(u_{it}^1)$ or must-OFF $(u_{it}^0)$ agents are identified based on the operating times, Equation (11).

$$u_{it}^1 = 1 \text{ if } 0 < t_{it} < t_{i*}^{up}; \text{ and } u_{it}^0 = 1 \text{ if } - t_{i*}^{down} < t_{it} < 0; \forall i. \tag{11}$$

**Step 4.** The agents execute their action $a_t$ in state $\jmath_t$, and then pass to the next state $\jmath_{t+1}$ through a transition (or probability) function, $\mathcal{P}(\jmath_{t+1}|\jmath_t, a_t)$, satisfying all constraints.

**Step 4.1.** The legality of action $a_{it} \in a_t$ of each agent is confirmed and legalized if there are any violations of the constraints specified in Equation (11), as shown in Equation (12).

$$a_{it} = 1 \text{ if } a_{it} = 0 \,\Big|\, u_{it}^1 = 1; \text{ and } a_{it} = 0 \text{ if } a_{it} = 1 \,\Big|\, u_{it}^0 = 0, \forall i. \tag{12}$$

**Step 4.2.** The aggregate supply capacity of agents is checked for sufficiency in satisfying the demand and future demands when OFF units have not completed their downtime. Then contextual capacity adjustments are made if necessary, and if possible.

- If $\sum_{i=1}^n a_{it} p_{it}^{max} < (1+r)d_t$, then set each $a_{it} = 1|u_{it}^1 = 0$ based on the increasing order of $\lambda_i^{min}$'s of Equation (10) until $\sum_{i=1}^n a_{it} p_{it}^{max} \geq (1+r)d_t$. If the capacity shortage is not fully corrected due to unconstrained OFF units, then $\jmath_t$ is labeled as a terminal state $(\jmath_t^+)$ that would result an incomplete episode $(\mathbb{I}_{[\jmath_t^+]} = 1)$.

- If $\sum_{i=1}^n a_{it} p_{it}^{min} > (1+r)d_t$, then set each $a_{it} = 0|u_{it}^0 = 1$ as per the decreasing order of $\lambda_i^{min}$'s of Equation (10) until $\sum_{i=1}^n a_{it} p_{it}^{min} \leq (1+r)d_t$. If the excess capacity is not yet fully adjusted due to an insufficient number of unconstrained ON units, it results in an incomplete episode $(\mathbb{I}_{[\jmath_t^+]} = 1)$ as the state $\jmath_t$ is terminal $(\jmath_t^+)$.

- If the current capacity does not satisfy future demands, set each $a_{it} = 1\Big|\left(t_{it} \geq t_{i*}^{up}\right)$ as per the decreasing order of $\lambda_i^{min}$'s of Equation (10). The current state $\jmath_t$ is also labeled as terminal $(\jmath_t^+)$ if the future demands cannot be meet while the offline units must still be OFF due to an insufficient number of unconstrained OFF units.

**Step 5.** The total operation cost at timestep $t$ is determined. First, start-up and shutdown costs are obtained based on the action $a_t$. Second, a lambda iteration algorithm is used for solving the optimal power loads $p_{it}$, in Equation (2), which are then used to estimate the emission costs specified in Equation (3). Lastly, the total operation cost is obtained using Equation (13) where $z_{i,t+1} = a_{it}, \forall i$.

$$\mathscr{C}_t = \sum_{i=1}^n \left\{ z_{i,t+1}(c_{it}^{prod} + c_{it}^{emis}) + z_{i,t+1}(1 - z_{it})c_{it}^{ON} + (1 - z_{i,t+1})z_{it}c_i^{OFF} \right\} \tag{13}$$

**Step 6.** The agents get an aggregate reward according to the predefined function given in Equation (14), which is the negative of the normalized total operation cost scaled to 100.

$$r_t = \mathscr{R}(\jmath_t, a_t, \jmath_{t+1}) = \left(1 - \frac{\left(1 - \mathbb{I}_{[\jmath_t^+]}\right)\mathscr{C}_t + \mathbb{I}_{[\jmath_t^+]}\mathscr{C}_t^+ - \mathscr{C}^{min}}{\mathscr{C}^{max} - \mathscr{C}^{min}}\right) \times 100 \tag{14}$$

In the episodic task of RL, incomplete episodes need to be avoided, and large penalties are recommended by [11]. For this purpose, while the cost function in Equation (13) is used for non-terminal states, the cost for terminal states is defined as $\mathscr{C}_t^+ = \mathscr{C}^{max} - \frac{t}{23}$ $\left(\mathscr{C}^{max} - \mathscr{C}^{p^{max}}\right)$ where $\mathscr{C}^{max} = \sum_{i=1}^n \lambda_i^{max} p_{i*}^{max}$ and $\mathscr{C}^{p^{max}}$ is the sum of Equations (2) and (3), assuming $p_{it} = p_{i*}^{max}$. This provides evenly distributed penalties, maintaining the desired proximity to the final timestep.

**Step 7.** If the current state is terminal (i.e., $\mathbb{I}_{[\jmath_t^+]} = 1$) or $t \geq 24$, then go to Step 1 to re-initialize the environment and restart a new episode to state $\jmath_0$. But, if $t < 24$ and $\mathbb{I}_{[\jmath_t^+]} = 0$, the agents pass to the next state $\jmath_{t+1} = \left(t + 1, p_{t+1}^{min}, p_{t+1}^{max}, t_{t+1}, z_{t+1}, d_{t+1}\right)$ where $z_{t+1} = a_t$ is a vector commitment status; $p_{t+1}^{min} = max\left\{p_*^{min}, z_t'z_{t+1}\left(p_t - p_*^{down}\right)\right\}$ is a vector of minimum capacities; $p_{t+1}^{max} = min\left\{p_*^{max}, z_t'z_{t+1}\left(p_t + p_*^{up}\right)\right\} + \mathbb{I}_{[z_t'z_{t+1}=0]}p_*^{max}$ is a vector of maximum capacities; $t_{t+1} = (t_t + 1$ if $z_{t+1} = 1|t_t > 0, -1$ else if $z_{t+1} = 0|t_t > 0, 1$ else if $z_{t+1} = 1|t_t < 0, t_t - 1$ if $z_{t+1} = 0|t_t < 0)$ is a vector of operating time durations.

**Step 8.** Update the must-ON and must-OFF agents for the next timestep $t + 1$ as defined in Equation (11): $u_{i,t+1}^1 = 1$ if $0 < t_{i,t+1} < t_{i*}^{up}$, and $u_{i,t+1}^0 = 1$ if $-t_{i*}^{down} < t_{i,t+1} < 0; \forall i$.

**Step 9.** The execution of $a_t$ in the environment returns the next state ($\jmath_{t+1}$), the reward ($r_t$), indicates whether the state is terminal ($\mathbb{I}_{[\jmath_t^+]}$) and loads dispatch information together with the legally confirmed action ($a_t, p_t$).

It should be noted that action $a_t$ returned in Step 9 is not necessarily the same as the action $a_t$ executed in Step 4 since the environment makes contextual corrections in Step 4.1 and 4.2 using the idea of a contextual search [17]. This agent-based contextual simulation environment, one of the main contributions of this study, can be highly effective in reducing the computing and training time of the multi-agent deep contextual RL model described below.

*3.3. Deep Contextual Reinforcement Learning*

At each timestep $t \in \mathcal{T}$, the agents observe a state $\jmath_t$ from $\mathcal{S}$ and select their respective actions $a_t$ from the action space $\mathcal{A}$ according to a policy $\pi(a_t|\jmath_t)$, where $\pi$ is a mapping from states $\jmath_t$ to actions $a_t$. The agents then receive a reward $r_t$ and proceed to the next state $\jmath_{t+1}$. This process continues until the agents finish the entire episode or reach a terminal state, both of which reinitialize the environment. The agents' goal is to learn a policy $\pi(a_t|\jmath_t)$ that maximizes the long-run cumulative sum of rewards called return, defined as $G_t \doteq \sum_{k=0}^{\infty} \gamma^k r_{t+k}$ where $\gamma$ is a discount rate $\gamma \in [0, 1]$ of the MDP. The expected return of action, $a_t$ in the state $\jmath_t$ can be expressed as an action-value function $Q^{\pi}(\jmath_t, a_t) = \mathbb{E}(G_t|\jmath_t, a_t)$. It can be approximated using a deep Q-network (DQN) which can be applied in a high-dimensional state and/or action space [18]. The action-value function can now be written as $Q(\jmath_t, a_t|\theta)$, where $\theta$ consists of the parameters of the DQN model whose inputs are the power scheduling dynamics simulated from the environment in the form of MDPs. The size of action space $\mathcal{A}$ is $2^n$, which may render an exponential growth in computation. Thus, it is impractical to parameterize the model into $2^n$ output nodes. Instead, it is parameterized into $2n$ output nodes corresponding to the two possible actions (ON/OFF) of each agent. As a result, the model estimates action-values for $2n$ output nodes, and then the decisions of agents made using an exponential decay epsilon-greedy exploration strategy collectively constitute the action vector $a_t$.

In a power scheduling problem, the action in a particular state affects the rewards and a set of future states, which yields serial correlations among the MDPs. In such cases, direct application of DQN may not be efficient, as it might result in unstable and slow learning processes [11]. Employing the notion of experience replay, the autocorrelation among the states may be properly addressed, and the training process can be expedited and stabilized [19]. After storing a transition tuple ($\jmath_t, a_t, r_t, \jmath_{t+1}$) to a replay buffer $\mathbb{B}$, a batch $b$ of experiences ($\jmath, a, r, \jmath'$) is sampled to approximate the action-value function $Q(\jmath, a|\theta)$ and the target network $Q(\jmath', a'|\theta')$, where $\jmath$ and $\jmath'$ are of size ($b \times 2n$), and the sizes of $a$ and $r$ are ($b \times n$) and ($b \times 1$), respectively. The DQN algorithm minimizes the mean-squared loss (i.e., temporal difference) $L_e(\theta)$ defined by

$$L_e(\theta) = \mathbb{E}_{(\jmath, a, r, \jmath') \sim \mathbb{B}}\left[r + \gamma \max_{a'} Q(\jmath', a'|\theta') - Q(\jmath, a|\theta)\right]^2 \tag{15}$$

where $\theta'$ is the parameter of target network, which is periodically updated from the Q-network parameters $\theta$, and $e$ denotes the iteration index.

## 4. Demonstrative Example

The applicability of the proposed deep contextual RL method is demonstrated with the power system investigated in [20]. The test system consists of five units, for which supply and demand profiles and emission parameters can be found in [20]. The deep RL utilizes a feedforward neural network featuring a rectified linear unit (ReLU) activation function on both the hidden and output layers. The learning rate and discount factor were set to 0.0001 and 0.99, respectively, and the Adam's optimizer was used. Employing the popular genetic algorithm, the optimal cost in [20] was \$430,331, summing up the start-up, production, and emission costs of \$3140 (0.7%), \$289,178 (67.2%), and \$138,010 (32.1%), respectively. On the other hand, the proposed method in this study yields an improved optimal operation cost of \$413,122 as shown in Table 1 and Figures 1 and 2, corresponding to \$16,808 (4.0%) lower daily operation cost than the results reported in [20] as compared in Table 2 and Figure 3. The total cost is composed of \$2230 (0.5%) for start-up costs, \$275,962 (66.8%) for production costs, and \$134,931 (32.6%) for emission costs. It is also asserted that the proposed method may be computationally efficient by adopting the experience replay. The scalability of the proposed method may thus be tested with a large-scale power system comprising a large number of generating units. Duplicating the five-unit test system multiple times and scaling the demands proportionately, the proposed method has been applied to obtain the optimal power scheduling scheme of individual units. The optimal costs of duplicated large-scale power systems are summarized in Table 3. It is worth noting that the optimal operating cost of each test system is lower than the scaled optimal operating cost of the original five-unit system. It is implied that the proposed method may easily be extended to render an economically and environmentally better solution for larger-scale power systems.

**Table 1.** Optimal commitments, optimal loads, and available reserve of test power system I using the proposed method.

| Hour (t) | Optimal Commitments | | | | | Optimal Loads (MW) | | | | | $r$ (%) |
|---|---|---|---|---|---|---|---|---|---|---|---|
| | $z_{1t}$ | $z_{2t}$ | $z_{3t}$ | $z_{4t}$ | $z_{5t}$ | $p_{1t}$ | $p_{2t}$ | $p_{3t}$ | $p_{4t}$ | $p_{5t}$ | |
| 1 | 1 | 0 | 0 | 0 | 0 | 400.0 | 0 | 0 | 0 | 0 | 13.8 |
| 2 | 1 | 0 | 1 | 0 | 0 | 426.5 | 0 | 23.5 | 0 | 0 | 30.0 |
| 3 | 1 | 0 | 1 | 0 | 0 | 450.9 | 0 | 29.1 | 0 | 0 | 21.9 |
| 4 | 1 | 0 | 1 | 0 | 0 | 455.0 | 0 | 45.0 | 0 | 0 | 17.0 |
| 5 | 1 | 0 | 1 | 0 | 0 | 455.0 | 0 | 75.0 | 0 | 0 | 10.4 |
| 6 | 1 | 1 | 1 | 0 | 0 | 455.0 | 36.6 | 58.4 | 0 | 0 | 30.0 |
| 7 | 1 | 1 | 1 | 0 | 0 | 455.0 | 52.0 | 73.0 | 0 | 0 | 23.3 |
| 8 | 1 | 1 | 1 | 0 | 0 | 455.0 | 62.3 | 82.7 | 0 | 0 | 19.2 |
| 9 | 1 | 1 | 1 | 0 | 0 | 455.0 | 72.6 | 92.4 | 0 | 0 | 15.3 |
| 10 | 1 | 1 | 1 | 1 | 0 | 455.0 | 77.7 | 97.3 | 20 | 0 | 22.3 |
| 11 | 1 | 1 | 1 | 1 | 0 | 455.0 | 93.1 | 111.9 | 20 | 0 | 16.9 |
| 12 | 1 | 1 | 1 | 1 | 0 | 455.0 | 103.3 | 121.7 | 20 | 0 | 13.6 |
| 13 | 1 | 1 | 1 | 1 | 0 | 455.0 | 77.7 | 97.3 | 20 | 0 | 22.3 |
| 14 | 1 | 1 | 1 | 0 | 0 | 455.0 | 72.5 | 92.5 | 0 | 0 | 15.3 |
| 15 | 1 | 1 | 1 | 0 | 0 | 455.0 | 62.3 | 82.7 | 0 | 0 | 19.2 |
| 16 | 1 | 1 | 1 | 0 | 0 | 455.0 | 36.6 | 58.4 | 0 | 0 | 30.0 |
| 17 | 1 | 0 | 1 | 0 | 0 | 455.0 | 0 | 45.0 | 0 | 0 | 17.0 |
| 18 | 1 | 0 | 1 | 1 | 0 | 455.0 | 0 | 75.0 | 20 | 0 | 20.9 |
| 19 | 1 | 0 | 1 | 1 | 0 | 455.0 | 0 | 125.0 | 20 | 0 | 10.8 |
| 20 | 1 | 0 | 1 | 1 | 1 | 455.0 | 0 | 130.0 | 55 | 10 | 10.8 |
| 21 | 1 | 0 | 1 | 1 | 0 | 455.0 | 0 | 125.0 | 20 | 0 | 10.8 |
| 22 | 1 | 0 | 1 | 1 | 0 | 455.0 | 0 | 75.0 | 20 | 0 | 20.9 |
| 23 | 1 | 0 | 1 | 0 | 0 | 455.0 | 0 | 45.0 | 0 | 0 | 17.0 |
| 24 | 1 | 0 | 1 | 0 | 0 | 426.4 | 0 | 23.6 | 0 | 0 | 30.0 |

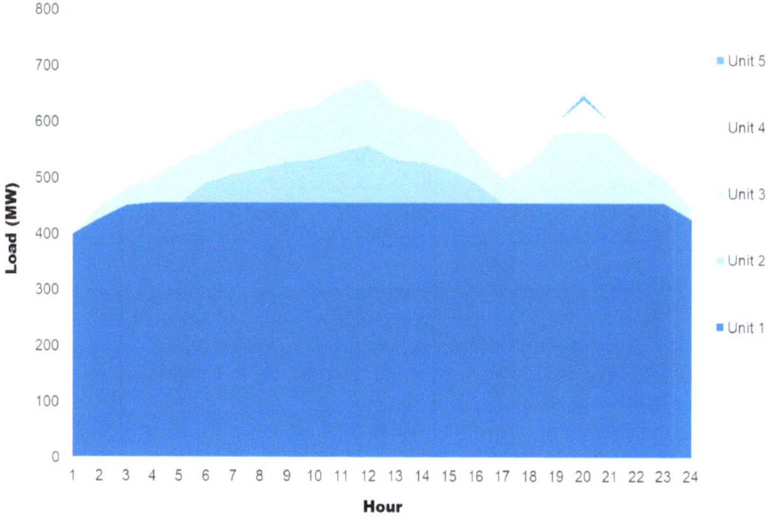

**Figure 1.** Optimal loads of test power system I using the proposed method.

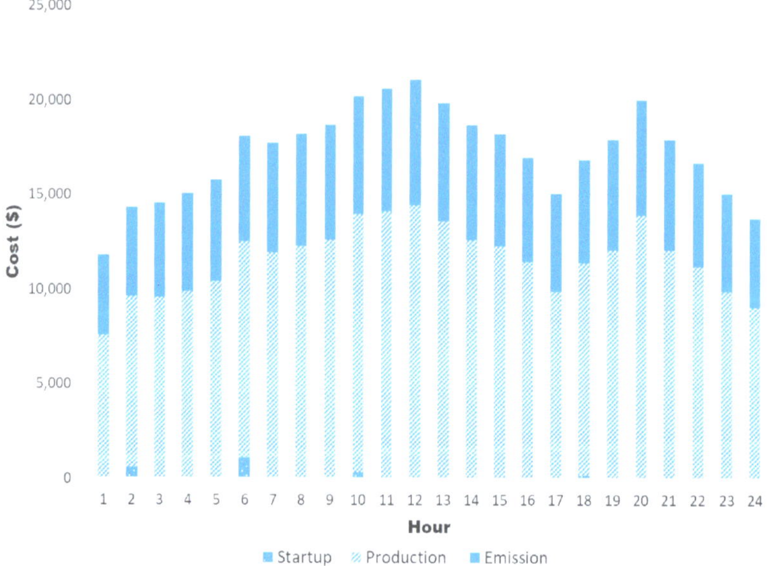

**Figure 2.** Costs of test power system I using the proposed method.

**Table 2.** Comparison of the optimal costs (start-up cost, production cost, emission cost, and total cost) of test power system I between genetic algorithm (GA) [20] and the proposed RL method.

| Hour (t) | Genetic Algorithm [20] | | | | Proposed RL | | | |
|---|---|---|---|---|---|---|---|---|
| | Start-Up | Production | Emission | Total | Start-Up | Production | Emission | Total |
| 1 | 0 | 8466 | 4824 | 13,290 | 0 | 7553 | 4241 | 11,793 |
| 2 | 0 | 8466 | 4828 | 13,294 | 560 | 9061 | 4702 | 14,324 |
| 3 | 60 | 10,564 | 5044 | 15,668 | 0 | 9560 | 5004 | 14,564 |
| 4 | 0 | 10,564 | 5044 | 15,608 | 0 | 9893 | 5170 | 15,063 |
| 5 | 1120 | 11,327 | 5825 | 18,271 | 0 | 10,395 | 5401 | 15,797 |
| 6 | 0 | 11,327 | 5825 | 17,151 | 1100 | 11,427 | 5555 | 18,082 |
| 7 | 0 | 11,327 | 5825 | 17,151 | 0 | 11,930 | 5786 | 17,717 |
| 8 | 60 | 13,425 | 6045 | 19,529 | 0 | 12,267 | 5940 | 18,207 |
| 9 | 0 | 13,425 | 6045 | 19,469 | 0 | 12,604 | 6094 | 18,699 |
| 10 | 340 | 13,523 | 6145 | 20,008 | 340 | 13,591 | 6251 | 20,183 |
| 11 | 30 | 15,621 | 6365 | 22,016 | 0 | 14,099 | 6483 | 20,582 |
| 12 | 0 | 15,621 | 6365 | 21,986 | 0 | 14,439 | 6637 | 21,076 |
| 13 | 0 | 13,523 | 6145 | 19,668 | 0 | 13,591 | 6251 | 19,843 |
| 14 | 30 | 13,425 | 6045 | 19,499 | 0 | 12,604 | 6094 | 18,699 |
| 15 | 1100 | 13,456 | 6045 | 20,601 | 0 | 12,267 | 5940 | 18,207 |
| 16 | 0 | 11,358 | 5825 | 17,182 | 0 | 11,427 | 5555 | 16,982 |
| 17 | 0 | 11,358 | 5825 | 17,182 | 0 | 9893 | 5170 | 15,063 |
| 18 | 0 | 11,358 | 5825 | 17,182 | 170 | 11,213 | 5481 | 16,865 |
| 19 | 340 | 13,554 | 6145 | 20,039 | 0 | 12,059 | 5866 | 17,926 |
| 20 | 0 | 13,554 | 6145 | 19,699 | 60 | 13,862 | 6085 | 20,007 |
| 21 | 0 | 13,554 | 6145 | 19,699 | 0 | 12,059 | 5866 | 17,926 |
| 22 | 0 | 11,358 | 5825 | 17,182 | 0 | 11,213 | 5481 | 16,695 |
| 23 | 60 | 10,564 | 5044 | 15,668 | 0 | 9893 | 5170 | 15,063 |
| 24 | 0 | 8466 | 4824 | 13,290 | 0 | 9061 | 4702 | 13,764 |
| Total | 3140 | 289,178 | 138,013 | 430,331 | 2230 | 275,962 | 134,931 | 413,122 |

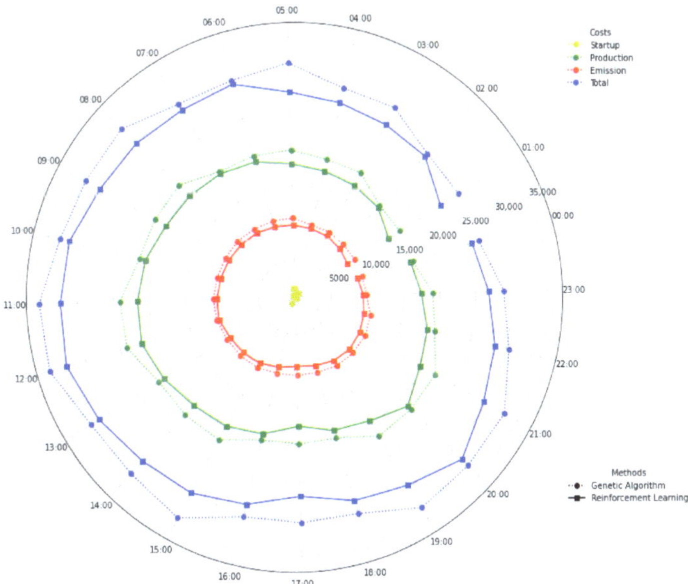


**Figure 3.** Radar plot of the hourly optimal costs using the genetic algorithm (GA) and the proposed reinforcement learning (RL).

**Table 3.** Optimal costs of power production for large-scale systems.

| Number of Units | Cost ($) | | | |
| --- | --- | --- | --- | --- |
| | Start-Up | Production | Emission | Total |
| 10 | 4840 | 545,837 | 270,724 | 821,401 |
| 20 | 9300 | 1,083,979 | 540,476 | 1,633,754 |
| 30 | 14,370 | 1,622,473 | 811,866 | 2,448,710 |
| 40 | 18,980 | 2,160,114 | 1,082,666 | 3,261,760 |
| 50 | 31,660 | 2,744,090 | 1,329,071 | 4,104,821 |
| 80 | 51,320 | 4,369,093 | 2,129,526 | 6,549,939 |
| 100 | 58,960 | 5,456,700 | 2,674,207 | 8,189,867 |

## 5. Concluding Remarks

This article presents a novel RL-based algorithm designed to optimize the economic and environmental cost of power scheduling. The algorithm utilizes a contextually corrective agent-based RL environment, which simulates power scheduling dynamics using the framework of MDP. To evaluate the applicability and performance of the proposed method, the algorithm is tested on different test systems comprising up to 100 generating units. It is demonstrated that the algorithm provides superior solutions and is scalable to handle larger power systems. The potential for incorporating renewable power sources and investigating their impacts further highlights the versatility and applicability of the proposed method in addressing real-world power scheduling challenges.

**Author Contributions:** Conceptualization, A.S.E. and Y.J.K.; methodology, A.S.E. and C.P.; software, A.S.E.; validation, C.P. and Y.C.; formal analysis, A.S.E.; investigation, C.P. and Y.C.; resources, Y.J.K.; data curation, Y.C.; writing—original draft preparation, A.S.E.; writing—review and editing, Y.J.K.; supervision, Y.J.K.; project administration, Y.J.K.; funding acquisition, Y.J.K. All authors have read and agreed to the published version of the manuscript.

**Funding:** This research was supported by the Basic Science Research Program through the National Research Foundation of Korea (NRF), funded by the Ministry of Education (2021R1I1A3047456).

**Data Availability Statement:** Not applicable.

**Conflicts of Interest:** The authors declare no conflict of interest.

## Nomenclature

Indices

| | |
| --- | --- |
| $n$: | Number of units. |
| $m$: | Number of emission types. |
| $\mathfrak{I} = \{1, 2, \ldots, n\}$: | Indices of all units, $i \in \mathfrak{I}$. |
| $\mathscr{M} = \{1, 2, \ldots, m\}$ | Indices of all types of emissions, $h \in \mathscr{M}$. |
| $\mathscr{T} = \{1, 2, \ldots, 24\}$: | Indices of all periods, $t \in \mathscr{T}$. |

Units and Demand Profiles

| | |
| --- | --- |
| $p_{i*}^{max}, p_{i*}^{min}$: | Max, min capacity of unit $i$ (MW). |
| $p_{it}^{max}, p_{it}^{min}$: | Max, min capacity of unit $i$ at period $t$ (MW). |
| $p_{it}$: | Power output of unit $i$ at period $t$ (MW). |
| $t_{i*}^{up}, t_{i*}^{down}$: | Min online, offline duration of unit $i$ (hour). |
| $t_{it}$: | Operating (online/offline) duration of unit $i$ at period $t$ (hour). |
| $t_{it}^{ON}, t_{it}^{OFF}$: | Online (up), offline (down) duration of unit $i$ at period $t$ (hour). |
| $u_{it}^{1}, u_{it}^{0}$: | Indicator if unit $i$ must $-$ ON, must $-$ OFF at time $t$. |
| $d_t$: | Demand at period $t$ (MW). |
| $r$: | Percentage of demand for reserve capacity. |

Objective Function

| | |
|---|---|
| $\mathscr{C}, \mathscr{C}_t$: | Total generation $\cos t$ function of a day and at period $t$. |
| $\alpha_i, \beta_i, \delta_i$: | Quadratic, linear, constant parameters of cost function of unit $i$. |
| $\phi_h, \psi_{ih}$: | Externality cost of emission type $h$ (\$/g), emission factor of unit $i$ for type $h$ (g/MW). |
| $c_{it}^{ON}, c_i^{OFF}$: | Start $-$ up $\cos t$ at period $t$, shutdown $\cos t$ of unit $i$. |

Others

| | |
|---|---|
| $\mathbb{E}$: | Expected value. |
| $\mathbb{I}$: | Indicator function. |

## References


1. Asokan, K.; Ashokkumar, R. Emission controlled Profit based Unit commitment for GENCOs using MPPD Table with ABC algorithm under Competitive Environment. *WSEAS Trans. Syst.* **2014**, *13*, 523–542.
2. Roque, L.; Fontes, D.; Fontes, F. A multi-objective unit commitment problem combining economic and environmental criteria in a metaheuristic approach. In Proceedings of the 4th International Conference on Energy and Environment Research, Porto, Portugal, 17–20 July 2017.
3. Montero, L.; Bello, A.; Reneses, J. A review on the unit commitment problem: Approaches, techniques, and resolution methods. *Energies* **2022**, *15*, 1296. [CrossRef]
4. De Mars, P.; O'Sullivan, A. Applying reinforcement learning and tree search to the unit commitment problem. *Appl. Energy* **2021**, *302*, 117519. [CrossRef]
5. De Mars, P.; O'Sullivan, A. Reinforcement learning and A* search for the unit commitment problem. *Energy AI* **2022**, *9*, 100179. [CrossRef]
6. Jasmin, E.A.; Imthias Ahamed, T.P.; Jagathy Raj, V.P. Reinforcement learning solution for unit commitment problem through pursuit method. In Proceedings of the 2009 International Conference on Advances in Computing, Control, and Telecommunication Technologies, Bangalore, India, 28–29 December 2009.
7. Jasmin, E.A.T.; Remani, T. A function approximation approach to reinforcement learning for solving unit commitment problem with photo voltaic sources. In Proceedings of the 2016 IEEE International Conference on Power Electronics, Drives and Energy Systems, Trivandrum, India, 14–17 December 2016.
8. Li, F.; Qin, J.; Zheng, W. Distributed Q-learning-based online optimization algorithm for unit commitment and dispatch in smart grid. *IEEE Trans. Cybern.* **2019**, *50*, 4146–4156. [CrossRef] [PubMed]
9. Navin, N.; Sharma, R. A fuzzy reinforcement learning approach to thermal unit commitment problem. *Neural Comput. Appl.* **2019**, *31*, 737–750. [CrossRef]
10. Dalal, G.; Mannor, S. Reinforcement learning for the unit commitment problem. In Proceedings of the 2015 IEEE Eindhoven PowerTech, Eindhoven, Netherlands, 29 June–2 July 2015.
11. Qin, J.; Yu, N.; Gao, Y. Solving unit commitment problems with multi-step deep reinforcement learning. In Proceedings of the 2021 IEEE International Conference on Communications, Control, and Computing Technologies for Smart Grids, Aachen, Germany, 25–28 October 2021.
12. Ongsakul, W.; Petcharaks, N. Unit commitment by enhanced adaptive Lagrangian relaxation. *IEEE Trans. Power Syst.* **2004**, *19*, 620–628. [CrossRef]
13. Nemati, M.; Braun, M.; Tenbohlen, S. Optimization of unit commitment and economic dispatch in microgrids based on genetic algorithm and mixed integer linear programming. *Appl. Energy* **2018**, *2018*, 944–963. [CrossRef]
14. Trüby, J. *Thermal Power Plant Economics and Variable Renewable Energies: A Model-Based Case Study for Germany*; International Energy Agency: Paris, France, 2014.
15. Zhang, K.; Yang, Z.; Başar, T. Multi-agent reinforcement learning: A selective overview of theories and algorithms. In *Handbook of Reinforcement Learning and Control*; Vamvoudakis, K.G., Wan, Y., Lewis, F.L., Cansever, D., Eds.; Springer: Berlin/Heidelberg, Germany, 2021; pp. 321–384.
16. Wilensky, U.; Rand, W. *An Introduction to Agent-Based Modeling: Modeling Natural, Social, and Engineered Complex Systems with NetLogo*; The MIT Press: London, UK, 2015.
17. Sutton, R.; Barto, A. *Reinforcement Learning: An Introduction*; The MIT Press: London, UK, 2018.
18. Matzliach, B.; Ben-Gal, I.; Kagan, E. Detection of static and mobile targets by an autonomous agent with deep Q-learning abilities. *Entropy* **2022**, *24*, 1168. [CrossRef] [PubMed]

19. Adam, S.; Busoniu, L.; Babuska, R. Experience replay for real-time reinforcement learning control. *IEEE Trans. Syst. Man Cybern. Part C Appl. Rev.* **2012**, *42*, 201–212. [CrossRef]
20. Yildirim, M.; Özcan, M. Unit commitment problem with emission cost constraints by using genetic algorithm. *Gazi Univ. J. Sci.* **2022**, *35*, 957–967.

*Article*

# Forecast of Operational Downtime of the Generating Units for Sediment Cleaning in the Water Intakes: A Case of the Jirau Hydropower Plant

Lenio Prado, Jr. [1,*], Marcelo Fonseca [2], José V. Bernardes, Jr. [3], Mateus G. Santos [1], Edson C. Bortoni [3] and Guilherme S. Bastos [1]

[1] Systems Engineering and Information Technology Institute, Itajubá Federal University, Itajubá 37500-903, Brazil; mateus.gabriel@unifei.edu.br (M.G.S.); sousa@unifei.edu.br (G.S.B.)
[2] JIRAU ENERGIA, Distrito de Jaci-Paraná, Porto Velho 76840-000, Brazil; marcelo.fonseca@jirauenergia.com.br
[3] Electric and Energy Systems Institute, Itajubá Federal University, Itajubá 37500-903, Brazil; jusevitor@unifei.edu.br (J.V.B.J.); bortoni@unifei.edu.br (E.C.B.)
* Correspondence: lenio@unifei.edu.br

**Abstract:** Hydropower plants (HPP) in the Amazon basin suffer from issues caused by trees and sediments carried by the river. The Jirau HPP, located in the occidental Amazon basin, is directly affected by high sediment transportation. These materials accumulate in the water intakes and obstruct the trash racks installed in the intake system to prevent the entry of materials. As a result, head losses negatively impact the efficiency of the generating units and the power production capacity. The HPP operation team must monitor these losses and take action timely to clear the intakes. One of the possible actions is to stop the GU to let the sediment settle down. Therefore, intelligent methods are required to predict the downtime for sediment settling and restoring operational functionality. Thus, this work proposes a technique that utilizes hidden Markov models and Bayesian networks to predict the fifty Jirau generation units' downtime, thereby reducing their inactive time and providing methodologies for establishing operating rules. The model is based on accurate operational data extracted from the hydropower plant, which ensures greater fidelity to the daily operational reality of the plant. The results demonstrate the model's effectiveness and indicate the extent of the impact on downtime under varying sediment levels and when neighboring units are generating or inactive.

**Keywords:** Bayesian networks; correlation techniques; hidden Markov models; hydropower generation units operational downtime; sediment decantation

**Citation:** Prado, L., Jr.; Fonseca, M.; Bernardes, J.V., Jr.; Santos, M.G.; Bortoni, E.C.; Bastos, G.S. Forecast of Operational Downtime of the Generating Units for Sediment Cleaning in the Water Intakes: A Case of the Jirau Hydropower Plant. *Energies* **2023**, *16*, 6354. https://doi.org/10.3390/en16176354

Academic Editors: Marialaura Di Somma, Jianxiao Wang and Bing Yan

Received: 27 July 2023
Revised: 18 August 2023
Accepted: 22 August 2023
Published: 1 September 2023

## 1. Introduction

Hydropower plants (HPP) offer a convenient solution for meeting energy demands, taking advantage of renewable water resources [1]. Moreover, the hydropower plant's operation is closely tied to efficiently using available water resources [2].

The HPPs operation can be categorized as run-of-river or reservoir systems, using single or multiple reservoirs—operating independently or in cascade [3]. The run-of-river HPPs have minimal water storage volumes, and, consequently, in opposition to the reservoir systems HPP, the water resources cannot be stored and must flow along the course of the river.

Accurately determining the availability of generation units (GUs) is essential for optimizing energy generation using water resources [4,5]. Furthermore, establishing operating rules based on the GUs' operational status is crucial for effectively allocating generation resources [6].

Several factors can impact the availability of GUs, including maintenance operations aimed at preventing issues and correcting failures that may reduce the productive capacity of the GUs or render them unusable [7].

In specific basins, like the Madeira River basin, a significant amount of sediment is transported downstream, including elements such as trees, branches, algae and debris. These substances can accumulate on trash racks, potentially leading to adverse effects on power generation efficiency [8,9].

The Jirau HPP installed in this basin is directly affected by the sediment accumulation problem. During the flood season, the volume of material transported by the river is extremely high. Over time, the sediment accumulates in the water intakes and gradually reduces the GUs efficiency, making the unit unavailable for power generation.

Efforts are made to address or mitigate the impacts of such accumulation. One of these approaches is to clear the trash racks using cleaning claws to remove a substantial portion of the accumulated sediment. Another method is stopping the GU for a certain period, allowing the sediment to settle and reducing the obstruction of the trash racks.

However, cleaning using the claws is only performed occasionally due to the operational cost and effort involved. Therefore, plant operators more frequently opt to stop the GUs for decanting. The challenge lies in determining the optimal downtime required for the decanting process to be effective. The stoppage time is also influenced by neighboring GUs, directly affecting the duration needed to reduce the sediment volume.

This work presents an innovative technique using hidden Markov models and Bayesian networks to estimate the ideal stopping time of the GUs, ensuring that the decantation process achieves its objective within the shortest possible downtime. Considering the vast amount of information associated with each GU, correlation techniques were employed to select the most relevant parameters for analysis. The study also explored the impact of neighboring GUs' operational states, whether active or inactive, on the downtime of the GU under investigation.

Using hidden Markov models enables the prediction of expected obstruction levels after a GU is stopped for cleaning. Bayesian networks contribute to achieving more accurate results by considering the influence of neighboring GUs. Moreover, historical data utilization facilitates the development of models that align more closely with the operational realities of the HPP.

With the resulting models, it is possible to estimate the required time for sediment decantation more accurately, taking into account the specific conditions of the plant. Furthermore, the proposed technique can be employed to develop improved operational rules, thereby enhancing the pre-operation process through a systematized process and leading to operational benefits.

The main contributions of this study are as follows:

- Development of a hidden Markov model to estimate the required downtime for GUs;
- Modeling of a Bayesian network to calculate conditional probabilities for estimating the necessary decanting downtime under various scenarios;
- Use of correlation techniques to reduce the number of analyzed variables while maintaining the quality and relevance of the information;
- Investigation of the influence of neighboring GUs on sediment movement when they are stopped;
- Improvements in pre-operation processes enhancing GUs performance.

The work is structured as follows: Section 2 presents the problem definition. Section 3 presents related works. Section 4 presents the proposed solution. Section 5 shows the discussions and results obtained, and Section 6 concludes the work.

## 2. Problem Definition

Hydropower plants installed in the Amazon basin suffer from problems caused by trees and sediments transported by the river [10]. These sediments reach the HPP operation area and, over time, accumulate in the GUs water intake trash racks (to prevent the entry of materials). The sediment accumulation causes a pressure loss at the turbine inlet, reducing the available net head. Another issue stemming from this accumulation of materials is the proportional increase in force exerted on the trash racks.

Usually, there is a safety limit for the force exerted on the trash racks, and this limit is represented by the HPP operation teams by the head loss in meters. Exceeding this limit poses safety risks as the trash racks may break, adversely affecting the plant.

The Jirau HPP is directly affected by sediment-related problems, and the case study presented in this work was carried out in this plant. It is installed in the Madeira River basin, one of the main sub-basins of the Amazon basin, covering an area of over 1.3 million km$^2$. Figure 1 presents the hydrography of the Madeira River basin, where it is possible to observe the main tributaries of this basin: the Guaporé, Mamoré, Beni, Abunã and Madre de Dios rivers, with the headwaters located in Brazil, Bolivia and Peru [11,12].

The Amazon basin is the largest in the world, with around 7 million m$^2$, covering seven countries in South America. The legal territory of the Amazon is divided into the occident and orient Amazon basin [13].

The occident Amazon basin has approximately 2,400,000 km$^2$. Its most important rivers are the Solimões and Madeira Rivers, and they share essential characteristics such as large dimensions, large flows, low slopes and significant variations in level and flow from droughts to flood periods [14].

**Figure 1.** Hydrography of the Madeira River basin where the Jirau HPP is located.

Following the Brazilian government's determination, the Madeira River basin has a tiny passing reservoir at the HPP dam with low regulation capacity. However, this reservoir has a high flow speed to carry many materials, especially trees and sediments [15,16]. The volumes of precipitation received by the Madeira River vary between 500 and 5000 mm per year. Figure 2 presents the average monthly flows of Madeira River [10]. The government also regulates the levels for the damn operation throughout the different flow seasons throughout the year.

The Jirau HPP is approximately 120 km from Porto Velho, the capital of Rondônia, Brazil. Figure 3 shows the annual temperature history of Porto Velho, which varies between 21 and 34 degrees Celsius and is rarely lower than 18 or higher than 36 degrees.

The plant consists of 50 bulb-type turbines with an installed capacity of 75 MW each, totaling an installed capacity of 3,750 MW. The plant has a dam that stretches over 7875 m. An important aspect of the Jirau HPP is its run-of-river reservoir type without storage capacity, which means all inflow must be either used for generation or released downstream [17].

Jirau is the fourth-largest hydropower plant in Brazil in terms of installed capacity and the largest in the world in terms of the number of GUs [18]. However, the immense structure of the HPP presents significant operational challenges. Furthermore, the construction of the plant was carried out with precautions to preserve the biodiversity and natural characteristics of the river by allowing the passage of transported trees.

**Figure 2.** Average monthly flows of Madeira River at Porto Velho. Adapted from [10].

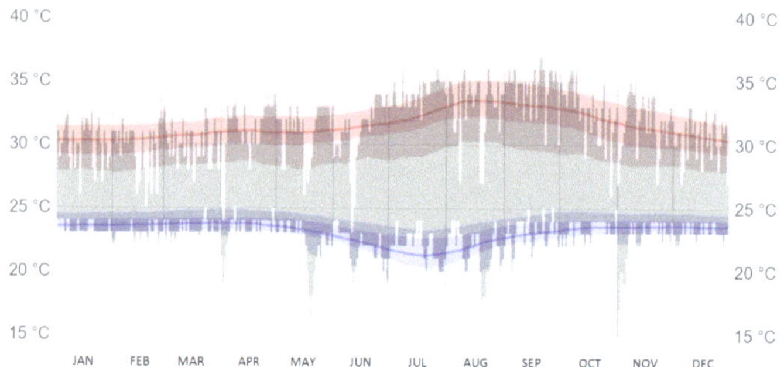

**Figure 3.** Porto Velho annual temperature history: temperature in the region ranges between 21 and 34 degrees Celsius. Gray bars: daily range of recorded temperatures; red and blue lines: daily highs and lows, respectively.

A complex set of equipment, referred to as auxiliary services, is essential for the proper functioning, operation and power generation of the HPP GUs. These auxiliary services encompass refrigeration systems, pressure control mechanisms, protection and control systems, temperature sensors and more. Figure 4 provides an aerial view, illustrating the physical dimensions of the HPP.

The HPP infrastructure of sensors, relays, actuators and systems totals more than 100,000 information points. All data collected from these components within the plant are directed to a SCADA system and stored in a relational database. The AVEVA Osisoft PI System software was acquired a few years ago to enable fast historical data access and reliable information retrieval. In addition, it also provided a solution for managing the large volume of data collected from GUs and their auxiliary services.

The PI System is a comprehensive software portfolio designed for collecting, storing, visualizing, analyzing and sharing operational data internally and externally. It includes a temporal database for information storage using tags with time stamps. Additionally, it offers a tool for effectively organizing the existing assets, displaying them in a tree format that aligns with the company's organizational structure.

**Figure 4.** Aerial view of HPP Jirau. Right margin with 28 UGs on the left of the image. Left margin with 22 UGs on the right of the image.

Due to the characteristics of the Madeira River, whose beds carry a great deal of material that usually leads to head losses in the water intakes of the GUs, stops are performed to allow the settling of accumulated sediment. During the GU operation, level measurements are performed before and after the trash racks, and the difference between these measurements represents the head loss in meters.

Regression analysis using the power information and the observed head of the GU is carried out to determine the flow rate of each unit. This regression is based on the GU hill curve. The obtained regression flow value and the previously recorded head loss information in meters are used to calculate $K$, representing the trash racks' obstruction factor. An increase in the $K$ factor indicates a greater accumulation of sediments deposited on the trash racks. The formula for $K$ is expressed in (1),

$$\Delta h_{gra} = K_{gra} \cdot \frac{\left(\frac{q_i}{2}\right)^2}{2.A_{gra}^2 \cdot g} \tag{1}$$

where:

$\Delta h_{gra}$ = Head loss on the rack [m];
$K_{gra}$ = Dimensionless coefficient relative to the rack;
$A_{gra}$ = Trash rack cross-sectional area;
$g$ = Gravity;
$q_i$ = turbine flow in unit $i$ [m$^3$/s].

According to studies conducted at the HPP, the trash racks installed at the water intake can support a maximum head loss of up to 1.5 m. As a result, operators typically continue operating the GU until the obstruction level reaches this maximum safety limit. Any further accumulation could break the trash racks, causing damage to the GUs and their associated systems, resulting in significant financial losses.

An operation rule was devised for the HPP to deal with the obstruction level issue, where the electrical power dispatch is reduced whenever the head loss reaches the safety limit. This reduction consequently reduces the flow, which causes a decrease in the head loss in meters as it is proportional to the flow. This process is repeated when the limit is reached until the dispatch exceeds the minimum operating limits of the GU. When the maximum obstruction level is reached, cleaning the trash racks or stopping the GU for sediment settling is necessary.

Therefore, accurately estimating the downtime is extremely important for the plant as it enables proper planning and utilization of the GU. Given the large amount of existing data related to the operation of the plant, the status of the GUs and their auxiliary systems, an automated and innovative method has been developed to predict the GU downtime and thus systematize a random process that was carried out based on the operator knowledge and experience. The technique allows the establishment of more consistent operating rules that align with the plant's actual conditions. It makes it possible to provide reliable data to the pre-operation team, enabling better allocation of resources for efficient power generation.

In addition to the downtime forecasting, the work's main contribution lies in the fact that the proposed analysis also considers the impact of neighboring GUs activity in addition to time-series data, instead of only considering the time-series as in other methods, such as ARIMA, dynamic regressions, state space models, etc.

The HMM usage in this work is intended to infer the future level of obstruction in the trash racks after some elapsed time, considering that the previous UG obstruction level is known. As HMMs do not capture the influence factors on the future clogging level, BN models are used to map and capture the neighboring GUs relationship in the decantation process. The integrated approach trumps the separate use of techniques. The HMM considers the temporal sequences of obstruction levels, while the BNs incorporate the modeling of uncertainties. This results in enriched forecasting modeling integrating the two techniques.

## 3. Related Work

Although some works deal with the sediments subject, to the best of our knowledge, these studies do not address the impacts caused by sediments on the GUs' operational performance. This fact makes it difficult to compare the proposed work with the literature.

As examples of work dealing with sediments-related problems, we cite [9] that explores the cost-effective aspects of the sediment abrasion effect, the possible disturbances, ecological relevance and the sediment bypass systems. The work of [19] analyses the reduction in the downstream sediment caused by installing hydropower dams, impacting one of the world's largest freshwater fisheries, which supports 17 million livelihoods. Furthermore, in [20], reservoirs' water storage capacity decrease is studied, resulting from the human activities and climatic changes that accelerate soil erosion and increase reservoir sedimentation.

### 3.1. Hidden Markov Models

The hidden Markov model (HMM) is a stochastic process in which states are hidden or not directly observable. They can only be inferred through the sequence of symbols produced by an underlying stochastic process. This probabilistic modeling technique is commonly used to handle uncertainty by representing a system as a Markov process with hidden states that are not explicitly visible [21,22].

A triple $\pi$, $A$ and $B$ typically represents the HMM, consisting of an initial probability vector over states $\pi$, a transition probability matrix $A$ that defines the set of possible states and an emission probabilities matrix $B$ that represents the observation probability distribution over the hidden states [21,22].

A review is completed exploring various applications of HMMs in the context of energy production and associated problems.

One such application, presented in [23], proposes an approach for situation analysis and anomaly detection using a hierarchy of hidden semi-Markov models. The methodology models the expected behavior of a system to detect contextual anomalies in SCADA systems, aiming to predict and prevent potential risks of attacks that could disrupt or damage water supply or power grid systems structures.

Modeling and forecasting electricity prices are proposed in [24]. The technique is based on input–output HMM. It considers the uncertainty of some involved variables, such as competitor behavior in the energy market, power source availability, water inflows, system energy demand and related costs. The market states are modeled as hidden states, and a conditional probability transition matrix is used to estimate probabilities when a new market session is opened. Finally, the paper reviews other electricity price models, and related works utilizing each type of model are presented, highlighting the strengths and weaknesses of each approach.

Integrating intermittent renewable energy sources into the power grid presents new challenges. To tackle these challenges, [25] conducted a study that focuses on modeling the power output of a wind farm. The author used discrete HMMs and inferred the model parameters from available data. By incorporating measurement data from multiple turbines and capturing the interdependencies between their outputs, the developed models successfully replicated crucial features of wind farm power output with high accuracy.

Non-intrusive load monitoring (NILM) is a technique used to identify appliance consumption at a disaggregated level. In [26], a hierarchical HMM framework is proposed to model home appliances and anticipate load characteristics at low voltage levels and distinct power consumption profiles in devices with multiple built-in operational modes. In addition, models were also built using dynamic Bayesian network representation. An expectation–maximization approach using the forward–backward algorithm was applied in the HMM fitting process. Tests related to the estimation of energy disaggregation showed that the proposed solution using HMM and a dynamic Bayesian network could effectively handle the modeling of appliances with multiple functional modes.

Islanding, a problem faced by power system engineers in smart grids, is addressed in [27] using an HMM-based algorithm approach to predict the probability of islanding events. The underlying process maps standard or faulty cases as a sequence of states. The HMM can help detect these states despite them not being directly observable but follow a pattern. Phasor measurements from the smart grid are used, and statistical analysis is conducted to determine the HMM parameters and tests were conducted in an IEEE nine-bus system. A trained artificial neural network provides HMM emission probabilities, enabling the prediction of islanding events based on posterior probabilities.

Reliability analysis of phasor measurement units is presented in [28] as another application of HMMs. The proposed methodology computes the transient probability, allowing for better monitoring systems during transient states. This ability enables faster and more effective restorative initiatives, providing a reliable method for operating, monitoring and controlling wide-area measurement systems.

The oil-immersed power transformers fault diagnosis using dissolved gases analysis is properly and commonly used. This technique is used with HMMs to estimate the health state of power transformers to infer operation failures in the work of [29]. In addition, a dynamic fault prediction technique is proposed where a Gaussian mixture model is used as a clustering method to extract health state features from datasets with 1600 days of operation. The HMM transition probability was calculated and analyzed to relate different health states. The results showed the proposed solution's effectiveness in predicting fault in a condition-based operation.

An alternative method for faults classification is proposed in [30], utilizing an HMM algorithm to process electrical signals in multivariate time series. A comparative analysis between the proposed technique and artificial neural network, support vector machine, K-nearest neighbor and random forest is presented. When considering a significance level of $\alpha = 5\%$, the results indicated that only the artificial neural network (ANN) and random forest (RF) classifiers achieved results comparable to the HMM algorithm. Compared to other classifiers, the presented algorithm significantly reduced computational costs, with processing time reduced by over 90%.

The massive integration of plug-in electric vehicles into the power distribution networks directly affected the planning, control and operation processes. To contribute to understanding the power needs of this kind of vehicle, the work of [31] presents an analytical approach for modeling PEV travel behaviors and charging demand. Monte Carlo simulation was employed considering the temporal travel purposes and state of charge of vehicles. The Markov model and HMM formulated the probabilistic correlation between multiple PEV states and state of charge ranges. The technique was tested using an IEEE 53-bus test network with field data, with results demonstrating the benefits of the proposed modeling.

### 3.2. Bayesian Networks

Bayesian networks (BNs) have emerged as a powerful technique to address uncertainty problems in scenarios characterized by randomness, indeterminism or lack of predictability [32].

Specifically, within the context of energy generation and its associated tasks, several studies have presented approaches using BNs to address maintenance-related issues [33], stakeholder decision support [34], watershed management [35] and solar plant failure detection [36], among others.

This section provides an overview of BNs and their applications related to power production and associated problems.

The work of [37] employed statistical approaches to analyze runoff and sediment characteristics in China's Three Gorges Reservoir (TGP). The study utilizes cumulative anomaly analysis, Fisher-ordered clustering and maximum entropy spectral analysis to study variations and forecast flow and sedimentary load using hydrological series of several decades. The ARIMA model is used to build the prediction model over the monthly average runoff and sediment inflow. The findings indicate a decreasing trend in runoff and sediment, with notable changes observed in 1991 and 2001.

Another case study combining BNs, neural networks and a multiagent system is presented in [38] to support and improve the automatic control of solar power plants. The BN model provides probabilistic values to aid operators in making informed decisions regarding remote control of the solar power plant, offering an optimized solution through distributed artificial intelligence technologies in industrial control systems for facilities based on solar photovoltaic energy sources.

In [39], a dynamic BN model is proposed for predicting the generation reserve size in renewable energy environments. This technique considers factors such as the availability of conventional generator capacity, weather conditions and market prices. Additionally, a new dynamic metric for calculating the reliability level of the power grid is introduced, serving as a real-time stochastic decision support tool. The approach is validated using seven years of historical data from the Australian Energy Market Operator, demonstrating improved accuracy in forecasting the risk of involuntary load shedding.

An agent-based model utilizing BNs is proposed in [40] to address the problem of short-term strategic bidding in a generation company's power pool. The agents employ probabilistic models based on dynamic BNs and online learning algorithms to train the model and estimate optimal bidding strategies, leveraging incomplete public information to infer the future state of the market correctly. The model was tested on two different time scales: hour ahead and day ahead. According to the results, the agents predicted the market equilibrium in advance with acceptable errors using incomplete information data.

Furthermore, a BN-based approach is applied in [41] to predict wind power ramp events, employing an imprecise conditional probability estimation method. The proposed solution utilizes the maximum weight spanning tree, a greedy search method to fit the observed data with the highest degree, and a modified version of the Dirichlet model to estimate the network parameters. Given the meteorological conditions, the proposed solution is meant to detect the possibility of a random ramp event, quantifying the uncertainty of the event. The method's effectiveness is demonstrated through tests using three-year operational data from a real wind farm.

Additionally, a BN is employed in the work of [42] for fault detection in power transformers. The model analyzes dissolved gases in oil, using concentration ratios of specific gases to identify normal deterioration and electrical and thermal failures. The solution used a historical database in the learning process, and, compared to data in the literature, the BN presented a high degree of reliability.

Lastly, the study in [43] discusses a BN-based approach for estimating faulty sections in transmission power systems within blackout areas. Three BN models are proposed, capable of testing the faultiness of components using uncertain or incomplete data and knowledge about power system diagnosis. The model uses a similar error backpropagation algorithm employed in artificial neural networks, with priors requiring domain experts' knowledge and network structure modeling.

In conclusion, the applications and case studies presented highlight the versatility and effectiveness of BNs in addressing uncertainty and decision making challenges in various power-related domains. Using BNs allows for improved complex systems analysis, prediction and control.

### 3.3. Analysis Summary

After a comprehensive review of the existing literature across three key subjects: HMMs, BNs and sediment transportation downstream, to the best of our knowledge, no prior work has sought to integrate these distinct topics in analyzing the impacts on the operational efficiency in HPPs caused by riverbed material transportation.

Notably, the critical challenge of sediment transportation prevails in the Amazon basin, where two of Brazil's largest HPPs are installed: Jirau and Santo Antonio. These plants hold the respective ranks of the fourth and fifth largest HPPs regarding installed capacity within the country. Although our contribution does not introduce new techniques, the innovation lies in the amalgamation of BNs and HMMs harnessed to address a pressing predicament within the Brazilian power sector. Such a fusion of methodologies offers a solution to an imperative issue, underscoring the novelty and significance of our work.

## 4. Methods

### 4.1. Data Selection

Due to the large amount of equipment-related data, using all available information in any forecasting technique is practically infeasible. This limitation arises from the time required for data processing and the computational resources consumed in performing such tasks. Pearson's correlation technique is used to identify the degree of dependence between the analyzed variables. This approach aims to reduce the data required while representing the relevant attributes of interest.

The correlation coefficient quantifies the relationship between variables, with values ranging between $-1$, indicating a strong negative relationship, 1, indicating a strong positive relationship and zero indicating no relationship. Pearson's correlation coefficient formula is expressed in (2).

$$r = \frac{\sum_{i=1}^{n}(x_i - \bar{x})(y_i - \bar{y})}{\sqrt{\sum_{i=1}^{n}(x_i - \bar{x})^2}\sqrt{\sum_{i=1}^{n}(y_i - \bar{y})^2}} \tag{2}$$

This work uses data collected from the 50 GUs at Jirau HPP. The database encompasses three months, from November 2021 to January 2022, with a sampling interval of 10 min. Figure 5 illustrates the dispatch power and efficiency attributes derived from the data set.

In the first analysis attempt, numerous attributes of the GUs were used. However, the strong correlation between the attributes reduced the set to a few elements, which can still represent the necessary information. The resulting attributes from the analysis were net head, dispatch power, efficiency, calculated flow and racks loss (head loss on the trash racks), in which the last attribute represents the generation power loss related to sediment in the water intakes.

**Figure 5.** Monthly efficiency and dispatch power readings extracted from the dataset.

Applying the Pearson correlation formula to the data set, the heat map correlation shown in Figure 6 is obtained. It is possible to visualize a strong correlation between net head and current power and calculated flow and current power and to notice that the relationship between net head and calculated flow is weak. However, the calculated flow strongly correlates with the rack loss.

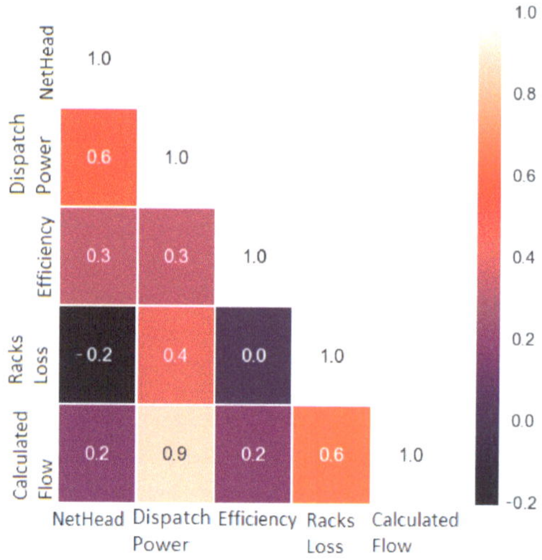

**Figure 6.** Attributes correlation heat map.

*4.2. HMM—Hidden Markov Models*

The HMM is applied to predict the GUs obstruction factor after a specific time stopped, given the head loss observed when the GU was stopped. As a result, it is possible to detect a relationship dependency between the GU obstruction level and the required decanting time. The application of HMM modeling allows the extraction of this relationship.

Once the relevant attributes are identified, the HMM is developed by creating the initial probability vector, the transition probability matrix and the emission probability matrix. The HMM states are mapped using the $K$ factor, and four value ranges are defined for their use in the model. These ranges include the cleanest range S1, where K reanges from 0 to $1^{-6}$, the S2 range from $1^{-6}$ to $4^{-6}$, the S3 range from $4^{-6}$ to $5^{-6}$ and the most obstructed range S4 for values above $5^{-6}$.

During the operating period of the GU, the $K$ factor and the head loss in meters can be calculated. Once the obstruction level on the trash racks reaches the maximum supported value, the operator stops the unit for decanting.

Four intervals are created to map the necessary GU downtime: interval H1 from 1 to 4 downtime hours, interval H2 from 4 to 8, interval H3 from 8 to 12 and interval H4 above 12 h.

Given the obstruction level at the stop, the HMM presents the GU probabilities of being in each mapped obstruction level interval over time.

It is important to emphasize that the probability vector, the transition and emission matrices were derived from HPP historical data so that the results from the model reflect the reality of the GU downtime for decanting.

The historical data are divided into two sets, one representing the training set and the other the test set. Separation is necessary to evaluate the model using a different group from the one used in training, thus avoiding data overfitting.

In the HPP, it is impossible to determine the current level of GU obstruction after decanting for a few hours. Therefore, the HMM is used in this scenario to estimate the GU obstruction level through probabilities to determine whether it is possible to restart the GU operation.

For the model creation, the following information is inferred from historical data: the prior or initial probability vector, the transition probability and emission probability matrices.

The initial probability vector denotes the probability of the GU being in a specific initial obstruction state, serving to determine the most probable initial state for the GU. The initial probabilities are defined based on the ratio between the number of decanting stops and the obstruction level when the GU was stopped. Consequently, the obtained initial state distribution vector, presented in Table 1, confirms the expected observation that decanting stops for the GU are more frequent when the obstruction level is higher.

**Table 1.** Initial state distribution vector.

|         | S1   | S2   | S3   | S4   |
|---------|------|------|------|------|
| $\pi =$ | 0.08 | 0.12 | 0.30 | 0.50 |

The transition probability matrix represents the likelihood of the GU transitioning from one state to another. As the GU downtime increases, there is a higher probability of transitioning from a more obstructed state to a lesser one. Since the current obstruction level is not directly observable, it is considered a hidden state. The resulting transition probability matrix can be seen in Table 2.

**Table 2.** Initial probability matrix.

|       |    | S1   | S2   | S3   | S4   |
|-------|----|------|------|------|------|
|       | S1 | 0.80 | 0.11 | 0.08 | 0.01 |
|       | S2 | 0.48 | 0.44 | 0.01 | 0.07 |
| $A =$ | S3 | 0.43 | 0.42 | 0.10 | 0.05 |
|       | S4 | 0.15 | 0.44 | 0.06 | 0.34 |

On the other hand, the emission probability matrix corresponds to the observed information related to the GU's current obstruction level. This information is the elapsed decanting time since the GU was stopped. As time progresses, GU's trash racks become cleaner, which aligns with expectations. Thus, the HMM model utilizes the probability vector and matrices to estimate the GU's obstruction level after a random time. Table 3 presents the resulting emission probability matrix.

The HMM implementation is carried out using the Pomegranate, a Python package for probabilistic models [44]. The model construction involved the three entities: $\pi$, $A$ and $B$. Once the model is trained, given a sequence of observations $O$, the model determines a score for the observed sequence using the so-called forward algorithm, or $\alpha$-pass, using the dynamic programming concept.

**Table 3.** Emission probability matrix.

|  |  | **H1** | **H2** | **H3** | **H4** | **H5** | **H6** |
|---|---|---|---|---|---|---|---|
|  | S1 | 0.11 | 0.066 | 0.198 | 0.077 | 0.022 | 0.527 |
| $B =$ | S2 | 0.088 | 0.099 | 0.198 | 0.187 | 0.033 | 0.396 |
|  | S3 | 0.053 | 0.105 | 0.105 | 0.211 | 0.053 | 0.474 |
|  | S4 | 0.279 | 0.131 | 0.158 | 0.153 | 0.049 | 0.23 |

After obtaining the score for the observed sequence, the next step is to reveal the most probable sequence of states given the presented observations. Given the GU stop elapsed time, the *Viterbi* algorithm is used to expose the hidden states, representing the actual $K$ factor. The *Viterbi* algorithm generates the most likely sequence of hidden states for a given list of observations, using dynamic programming to generate the output sequence recursively.

*4.3. Bayesian Networks*

Bayesian networks offer the possibility of representing a domain problem through a graphical structure composed of nodes comprising a set of random domain variables. The arcs connect the nodes through pairs, meaning the direct dependence of the variables. The conditional probability distribution of each associated node governs the strength of the relationship between the variables.

Using BNs in this work makes it possible to represent the variables directly affecting the required downtime to decrease the GU obstruction level.

The following variables are considered in the BN modeling: $K$ factor before the unit stops, an indication if the left, the right or both neighboring GUs are in operation during the analyzed GU stopped time, and the power at which the neighboring units were operating if it is the case.

The resulting BN diagram, designed to reflect information about the GUs, is shown in Figure 7.

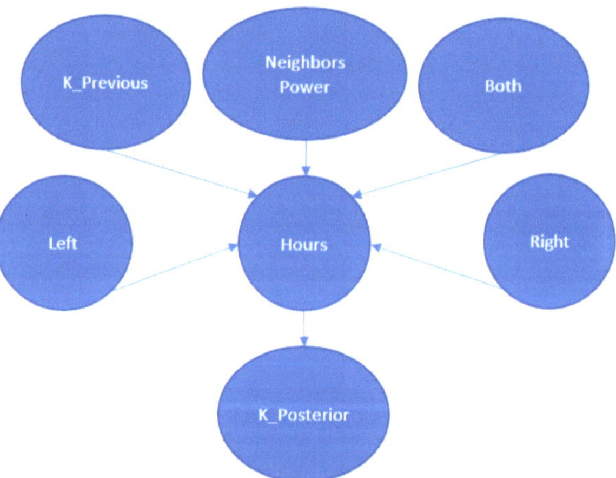


**Figure 7.** Bayesian network diagram.

The $K$ factor is discretized using the same four obstruction level intervals defined in Section 4.2. The BN model can be queried by one or more variables, obtaining the conditional probability according to the provided inputs.

To account for the neighboring operating time while the analyzed GU is stopped, all the intervals of each working neighboring are added up to obtain the relationship between the neighboring working time by the analyzed GU stopped time.

For example, if the analyzed GU is stopped for 12 h and the right neighbor operates for 3 h, then the neighbor operates for 25% of the analyzed GU stopped time. Therefore, the operating time of the neighboring GUs was divided into four ranges, the range T1 up to 25% of the time, T2 from 25% to 50%, range T3 from 50% to 75% and range T4 from 75% to 100%.

The average is used to compute the neighboring operating power during left or right GU usage hours. If both neighbors are operational, the average hourly power each neighbor generated is used. Finally, the same four intervals shown in Section 4.2 are used to infer the necessary GU downtime information.

The conditional probabilities distributions (CPD) related to each model variable are obtained through parameter learning using the provided data and model structure. The maximum likelihood estimation (MLE) algorithm is used in this work for CPD extracting using a data set [45,46].

The Bayesian model and CPDs make inferences using several scenarios to validate the proposed technique and compare the obtained model results with the plant operational data.

## 5. Results and Discussions

### 5.1. Results

After elaborating on the proposed models using the techniques presented, study cases are performed to verify their performance. In this section, the obtained results are provided below.

The obtained results through the HMM application for GU 2 are outlined in Table 4 and provide a comprehensive representation of the probabilities associated with specific hidden states. These hidden states correspond to varying levels of obstruction caused by the accumulation of sediments. The likelihood of the GU being in distinct hidden states can be ascertained by analyzing the data within each observed hourly interval.

**Table 4.** HMM results for GU 2: state probability after given time elapsed. Columns represent obstruction levels, and lines represent the time elapsed.

|  | S1 | S2 | S3 | S4 |
|---|---|---|---|---|
| H1 | 48.4 | 38.5 | 42.1 | 41.0 |
| H2 | 26.9 | 35.2 | 31.6 | 31.1 |
| H3 | 17.2 | 18.7 | 15.8 | 13.1 |
| H4 | 7.5 | 7.6 | 10.5 | 14.8 |

Whenever the GU is in state S4, which means it has the highest level of obstruction, the transition probability towards a less obstructed state becomes evident only after the time interval H3. This outcome aligns with the plant's operational practice when they usually keep the unit offline for extended periods when the obstruction level reaches a higher degree.

Alternatively, if the same GU currently has the obstruction level S2 and the GU remains inactive for the time interval H3, there is a significantly higher probability (18.7%) of it remaining in that state. The GU transition to a state cleaner than S2 demands a more extended downtime due to the characteristics of the sedimentation process, in which denser materials take more time to settle.

At the S1 obstruction level, the unit can remain in the same state, or, sometimes, the evolution is identified to a higher obstruction level, changing to S2. This event may occur due to the operation of neighboring GUs, which contributes to the movement of sediments, migrating material to the stopped GU.

In an ideal scenario for the HPP operation, a GU at the highest obstruction level should remain stopped until it is at the lowest dirt level, when it can return to activity. A study case was performed to obtain the probability of the GU migrating from the initial state S4 to S1 as its final state. The result is presented in Table 5.

**Table 5.** Probability of the GU 2 migrating from initial state S4 to S1 as its final state.

|  | S4 → S1 |
|---|---|
| H1 | 7.80% |
| H2 | 18.25% |
| H3 | 24.99% |
| H4 | 30.20% |

It is possible to notice that the provided scenario is more likely to occur only after the H3, with a probability of 24.99%, and it is most probably, with 30.20%, at H4. As expected, it is not common to reach S1 starting from S4 after the H1 or H2 periods, corresponding to 1 to 4 h or 4 to 8 h intervals.

To analyze the differences in downtime between the different GUs, Table 6 presents information relating to the level of obstruction and the downtime for GUs 1 and 3, respectively.

**Table 6.** HMM results for GUs 1 and 3: state probability after time elapsed. Columns represent obstruction levels, and lines represent the time elapsed.

|  | GU 1 | | | | GU 3 | | | |
|---|---|---|---|---|---|---|---|---|
|  | **S1** | **S2** | **S3** | **S4** | **S1** | **S2** | **S3** | **S4** |
| H1 | 62.0 | 34.1 | 27.3 | 28.1 | 59.0 | 37.5 | 28.9 | 31.1 |
| H2 | 32.4 | 24.5 | 26.9 | 25.9 | 35.2 | 26.4 | 26.4 | 28.1 |
| H3 | 3.7 | 22.9 | 25.6 | 23.4 | 4.1 | 15.1 | 23.9 | 20.9 |
| H4 | 1.9 | 18.5 | 20.2 | 22.6 | 1.7 | 21 | 20.8 | 19.9 |

The following considerations can be conducted through data analysis and using the same study case performed on GU 2 to obtain the probability of the GU migrating from the initial state S4 to S1 as its final state:

The probability of remaining in state S4 shows a balanced dispersion over time for GU 1. This pattern indicates situations where the GU transitions to a cleaner state even within the H1 interval. Conversely, there are cases where a significant time lapse, such as H4, is required for the transition. This variance can be attributed to GU 1's proximity to the dam's ravine, potentially contributing to sediment accumulation in specific circumstances.

Concerning GU 3, the likelihood of persisting in state S4 during the H1 and H2 intervals is higher, registering values of 31.1% and 28.1%, respectively. The prevailing trend is for GU 3 to transition to a cleaner state only after the H3 interval.

The dissimilar behaviors observed between GUs 2 and 3 can be attributed to the following factors: GU 1 absorbs sediment from the riverbank and can consequently transfer sediment to GU 2, explaining why GU 2 shifts to a cleaner state only after a more extended downtime. Conversely, GU 3 is unaffected by the same issue due to its greater distance from GU 1, illustrating the impact of neighboring GUs on the sediment decantation process.

Probability outcomes for GUs 31 and 32 are shown in Table 7.

**Table 7.** HMM results for GUs 31 and 32: state probability after time elapsed. Columns represent obstruction levels, and lines represent the time elapsed.

|  | GU 31 | | | | GU 32 | | | |
|---|---|---|---|---|---|---|---|---|
|  | **S1** | **S2** | **S3** | **S4** | **S1** | **S2** | **S3** | **S4** |
| H1 | 35.4 | 38.5 | 42.1 | 41 | 46.9 | 44.3 | 45.7 | 43.2 |
| H2 | 26.9 | 35.2 | 31.6 | 31.1 | 32.7 | 31.9 | 36.1 | 34 |
| H3 | 27.2 | 18.7 | 15.8 | 13.1 | 15.2 | 17.6 | 11.1 | 12.7 |
| H4 | 10.5 | 7.6 | 10.5 | 14.8 | 5.2 | 6.2 | 7.1 | 10.1 |

The behaviors of GUs 31 and 32 differ from those presented for GUs 1 to 3. This difference can be attributed to these GUs being on different margins, separated by kilometers, and to the curvature of the river displayed on the left margin where these GUs are installed. It is possible to observe a certain similarity between the probabilities for GUs 31 and 32, with slight variations in the required time to change between states. Generally, there is migration between states only after the H3 interval, which can be associated with the type of material accumulated in the trash racks.

When the generation units (GUs) exhibit a notably high degree of obstruction, a prevalent trend emerges: substantial clearance of the trash racks occurs only after prolonged GUs downtime. Specifically, if the GU experiences a brief stop time upon resumption of operational activity, a considerable amount of material is expected to obstruct the trash racks persistently.

Observations reveal that units positioned near the riverbank experience a notably higher sediment accumulation, leading to a more pronounced obstruction of the trash racks. Adjacent GUs also experience a residual effect from this sediment accumulation, albeit with a lesser impact.

It is essential to highlight that the HMM technique cannot consider whether neighboring GUs are in operation, nor does it account for the GUs operating power or the time it was generating.

For this reason, BNs are used to consider the factors that directly affect the operation and consequently alter the sediment flow during the GU stop time. Separated BNs are created for each GU to reflect the specificities of each one.

Below are presented the obtained results for different types of BNs queries. For example, Table 8 shows the CPDs for GU 2.

**Table 8.** BN results for GU 2: state probability after given time elapsed. Columns represent time elapsed, and lines represent obstruction levels.

|    | H1    | H2    | H3    | H4    |
|----|-------|-------|-------|-------|
| S1 | 0.172 | 0.269 | 0.075 | 0.484 |
| S2 | 0.187 | 0.385 | 0.077 | 0.352 |
| S3 | 0.158 | 0.316 | 0.105 | 0.421 |
| S4 | 0.41  | 0.311 | 0.131 | 0.148 |

Utilizing models derived from BN offers a significant advantage due to their inherent query capabilities. Queries involving any model attributes mapped within the network can be completed, thereby facilitating the prediction of posterior values. Specifically, these models empower the prediction of the obstruction level for each distinct downtime interval.

This predictive capacity enhances the ability to forecast and anticipate the progression of obstruction levels during various operational downtimes. In essence, BNs allow for a comprehensive exploration of the network's attributes, enabling the generation of valuable insights into the system's expected behavior over time.

Using the BN network shown in Figure 7, it is possible to estimate the resulting obstruction level using a given scenario to verify which parameters influence the decantation process the most.

Below are the values entered for the BN parameters.

- A = [K_Previous: S4]
- B = [K_Previous: S4, Right: T3]
- C = [K_Previous: S4, Right: T3, Left: T1]
- D = [K_Previous: S4, Both: T4]

Data referring to UG 2 were used. In all scenarios, it is considered that the UG is at the highest level of obstruction, S4.

Scenario A is parameterized only with this S4 obstruction information.

Scenario B is configured with the additional 'Right' information, whose defined value is T3, which comprises the value ranging from 50% to 75% of the time.

The information 'Right', 'Left' or 'Both' refers to the percentage of time that the adjacent unit operated when the GU was stopped for decantation. In this case of scenario B, the analyzed GU is the 2, and the 'Right' neighbor GU is the 3.

Scenario C is configured with the same value for the 'Right' parameter: T3. Additionally, the 'Left' information is set to T1, which comprises the value up to 25% of the time.

Scenario D is configured with the value for the 'Both' parameter equal to T4, which means that both the 'Right' and 'Left' GUs, in this case, 3 and 4, respectively, operated for the time interval comprising the 75% to 100% of time in which the GU 2 was stopped for decantation.

The results are presented in Table 9.

**Table 9.** BN results for GU 2: probabilities obtained for scenarios A, B, C and D, using S4 as the initial obstruction level.

| K Posterior | A | B | C | D |
| --- | --- | --- | --- | --- |
| S1 | 0.0935 | 0.0803 | 0.4008 | 0.484 |
| S2 | 0.3883 | 0.3232 | 0.2703 | 0.352 |
| S3 | 0.3297 | 0.3077 | 0.2594 | 0.421 |
| S4 | 0.1885 | 0.2888 | 0.0695 | 0.148 |

In scenario A, the probabilities that GU 2, stopped at the worst obstruction level, will resume operation at levels S2 and S3 are approximately 38.8% and 32.9%, respectively. In scenario B, these values are close, 32.3% and 30.7%, respectively. The operation of the neighboring GU, in this case, GU 3, did not significantly impact the obstruction level of GU 2.

In scenario C, the most favorable cleaning results were obtained, with a 40% probability that the UG would return at the cleanest level of obstruction: S1. The probable explanation for such behavior may be that the operation of the left GU, in this case, GU 1, has pulled the sediment from GU 2, migrating the GU more quickly to a lower level of obstruction.

Finally, in scenario D, both neighbors were in operation for the entire time GU 2 was stopped. The results demonstrate a more uniformly distributed probability between levels S1, S2 and S3, with values of 48.4%, 35.2% and 42.1%, respectively.

The results show that the decantation process when the GU is stopped is significantly influenced by the neighboring GUs. This relationship changes depending on the time the neighboring GUs were operating and the level of dirt when the GU stopped.

The BN was parameterized to present modeling outputs for each final obstruction level when resuming GU operation for all available downtimes to enable a more comprehensive view of data, including more complete probability results. The obtained results are shown in Table 10.

Given that the GU was stopped at the higher obstruction level S4, the following behavior can be observed in the table for many scenarios: after stopping time H1, the highest probability is that the GU resumes operation still at level S4. For time H3, the restart must occur at level S2, and, finally, the stop for time H4 increases the probability of resuming at level S1.

Only on the H2 stop time interval does this pattern not hold. Instead of resuming at level S3, the GU remains at level S4, demonstrating that stopping the GU for short intervals does not influence the level of obstruction so strongly.

*5.2. Discussions*

The main feature of HMMs is their suitability for use with sequential data, where the order of observations is essential. In this work, the HMMs captured temporal dependencies and transitions between different states, which evidenced the relation between the obstruction level and elapsed time when GUs are stopped for decanting. The flexibility of HMMs enabled usage with time-series input data, while levels of obstruction are mapped as states in the model.

**Table 10.** BN results for GU 2: for each scenario A through D, and for each time interval H1 through H4, the probabilities of the GU returning to operation at dirt levels S1 through S4 are presented, using S4 as the level of initial obstruction.

| Hours | K Posterior | A | B | C | D |
|---|---|---|---|---|---|
| H1 | S1 | 0.056 | 0.033 | 0.029 | 0.039 |
| H1 | S2 | 0.056 | 0.033 | 0.029 | 0.039 |
| H1 | S3 | 0.139 | 0.083 | 0.072 | 0.097 |
| H1 | S4 | 0.167 | 0.100 | 0.087 | 0.117 |
| H2 | S1 | 0.019 | 0.019 | 0.017 | 0.019 |
| H2 | S2 | 0.019 | 0.019 | 0.017 | 0.019 |
| H2 | S3 | 0.094 | 0.096 | 0.083 | 0.096 |
| H2 | S4 | 0.113 | 0.115 | 0.099 | 0.115 |
| H3 | S1 | 0.019 | 0.028 | 0.024 | 0.023 |
| H3 | S2 | 0.096 | 0.139 | 0.120 | 0.114 |
| H3 | S3 | 0.019 | 0.028 | 0.024 | 0.023 |
| H3 | S4 | 0.038 | 0.056 | 0.048 | 0.046 |
| H4 | S1 | 0.095 | 0.143 | 0.200 | 0.145 |
| H4 | S2 | 0.049 | 0.079 | 0.021 | 0.081 |
| H4 | S3 | 0.010 | 0.010 | 0.105 | 0.010 |
| H4 | S4 | 0.012 | 0.018 | 0.025 | 0.018 |

The two main advantages of HMMs are related to the probabilistic modeling and the incorporation of hidden states. The first feature captures the uncertainties, which fits the objectives of this work: map the ratio of accumulated sediment and the required downtime to settle this material. The second maps the unobserved obstruction levels underlying processes as hidden states, enabling the estimation of the sediment settlement according to the elapsed time.

On the other hand, the limitations of HMMs in this model are, once the transition probabilities are influenced by the neighboring GU, the Markov property is directly affected and may not hold. The Markov property can also struggle to capture long-range dependencies effectively.

Another limitation is related to the fixed state space: the number of hidden states was determined in advance and may not represent the best possible scenario. Choosing the appropriate number of states was challenging since it could affect model performance. A significant effort was required to ensure that the selected state space reflects the best option.

The results showed that the expressiveness of the HMM technique alone is limited since the models might not effectively capture complex relationships between variables. The training complexity represents a time bottleneck since a new training cycle must be performed with each new parameter and state mapping variation.

The Bayesian networks' advantages rely on the fact that BNs provided a natural and intuitive way to model uncertainties and dependencies in data obtained from the real HPP operation. As the BN is a probabilistic framework, it was possible to infer even when some variables appear unrelated.

The causal inference of BNs allowed for understanding how the neighboring GU usage affected other variables. That feature was valuable for decision making about using the GUs while the next ones are stopped for sediment settling. With the help of problem domain experts of the Jirau HPP, it was possible to validate prior beliefs and causal relationships obtained by the resulting modeling.

BNs allowed exploratory analysis and efficient inference related to sediment behavior, which helped to uncover hidden patterns that were not immediately apparent from the raw data and compute probabilities of different scenarios, providing query evidence. As the real operational data from the HPP were available, it was possible to use the BN to learn the conditional probabilities parameters from these data, which makes the model consistent with the plant reality.

Bayesian networks present some limitations, such as the heavy dependence on the graph structure. It was challenging to correct specifying the design, which required domain expertise since the model might not effectively capture the true relationships without expert input.

The needed training time was a bottleneck to realize parameter variations during the modeling because the high number of GUs represents a computational complexity problem. Finally, the correlation and causation relationships represented a challenge because assuming causation based on correlation could sometimes be dangerous.

A strength of the presented work is the union of the HMM with the BN, which made it possible to take advantage of the main characteristics of each of the techniques. The model's robustness allows probability information extraction. It brings to light details related to the behavior of the obstruction in the trash racks, including the neighboring GU impact in the sediment settling behavior.

Because it is a pioneering work, which addresses the problem of obstruction of trash racks with consequent impact on the operation of hydroelectric plants, studies still need to be conducted for contextual reference and comparison. This sediment issue is specific to HPPs located within the Amazon basin. In the case of the Jirau plant, the challenge of high sediment transport rates arises only during particular periods of the year. For this reason, only data referring to flood seasons are used.

## 6. Conclusions

This paper proposes techniques to estimate the ideal stopping time of the GUs in the Jirau HPP using hidden Markov models and Bayesian networks as inference methods. Field operational data are used to obtain the presented models. The results demonstrate consistency with daily plant operation, allowing the use of the model in the operator's decision making, thus helping to operate the high number of existing GUs in the Jirau HPP.

An essential advantage of the presented methodology is that it allows systematized and data-based means to model information inferring, enabling more consistent operating rules at the HPP.

As the Jirau is a run-of-river plant and does not allow water resources storage, the presented proposal in this work offers methods that enable using a robust model in the plant operation planning under several possible scenarios, extracting the resulting probability under each perspective.

This innovative proposal aims to bring greater clarity and robustness to the operating rules extraction for HPPs whose accumulation of materials on the trash rack can negatively influence daily operations, especially those in the Amazon basin.

The applied methodology in this work can be used in other HPPs, both for operating rules extraction and for HMM and BN modeling. In addition, the resulting model can help to identify factors that alter the GUs operational efficiency, providing tools and methods to operate the plant efficiently.

Although the generated models use HPP operating data, the training was offline. As future work, it is proposed to integrate plant data for online model training, presenting real-time information to assist operators in decision making and minimizing GUs downtime.

**Author Contributions:** Writing—original draft, L.P.J.; Writing—review & editing, J.V.B.J., M.G.S. and E.C.B.; Supervision, M.F.; Project administration, G.S.B. All authors have read and agreed to the published version of the manuscript.

**Funding:** This research was funded by Energia Sustentável do Brasil S.A.

**Data Availability Statement:** The access to the data underlying the findings of this study is not available due to privacy considerations and in accordance with company operational policies.

**Acknowledgments:** This work was developed during the execution of a project regulated by ANEEL and developed within the scope of the Jirau Energia R&D Program under the corporate name 'Energia Sustentável do Brasil S.A' (PD06631-0011/2020). The authors thank Jirau Energia, Capes and IF Sul de Minas for their continued support.

**Conflicts of Interest:** The authors declare no conflict of interest.

## References


1. Contreras, E.; Herrero, J.; Crochemore, L.; Pechlivanidis, I.; Photiadou, C.; Aguilar, C.; Polo, M.J. Advances in the definition of needs and specifications for a climate service tool aimed at small hydropower plants' operation and management. *Energies* **2020**, *13*, 1827. [CrossRef]
2. Berga, L. The role of hydropower in climate change mitigation and adaptation: A review. *Engineering* **2016**, *2*, 313–318. [CrossRef]
3. Jiang, Z.; Song, P.; Liao, X. Optimization of year-end water level of multi-year regulating reservoir in cascade hydropower system considering the inflow frequency difference. *Energies* **2020**, *13*, 5345. [CrossRef]
4. Sibtain, M.; Li, X.; Azam, M.I.; Bashir, H. Applicability of a three-stage hybrid model by employing a two-stage signal decomposition approach and a deep learning methodology for runoff forecasting at Swat River catchment, Pakistan. *Pol. J. Environ. Stud.* **2021**, *30*, 369–384. [CrossRef]
5. Betti, A.; Crisostomi, E.; Paolinelli, G.; Piazzi, A.; Ruffini, F.; Tucci, M. Condition monitoring and predictive maintenance methodologies for hydropower plants equipment. *Renew. Energy* **2021**, *171*, 246–253. [CrossRef]
6. Feng, Z.k.; Niu, W.j.; Zhang, R.; Wang, S.; Cheng, C.t. Operation rule derivation of hydropower reservoir by k-means clustering method and extreme learning machine based on particle swarm optimization. *J. Hydrol.* **2019**, *576*, 229–238. [CrossRef]
7. Thaeer Hammid, A.; Awad, O.I.; Sulaiman, M.H.; Gunasekaran, S.S.; Mostafa, S.A.; Manoj Kumar, N.; Khalaf, B.A.; Al-Jawhar, Y.A.; Abdulhasan, R.A. A review of optimization algorithms in solving hydro generation scheduling problems. *Energies* **2020**, *13*, 2787. [CrossRef]
8. Singh, V.K.; Singal, S.K. Operation of hydro power plants—A review. *Renew. Sustain. Energy Rev.* **2017**, *69*, 610–619. [CrossRef]
9. Hauer, C.; Wagner, B.; Aigner, J.; Holzapfel, P.; Flödl, P.; Liedermann, M.; Tritthart, M.; Sindelar, C.; Pulg, U.; Klösch, M.; et al. State of the art, shortcomings and future challenges for a sustainable sediment management in hydropower: A review. *Renew. Sustain. Energy Rev.* **2018**, *98*, 40–55. [CrossRef]
10. Tucci, C.E. *Análises dos Estudos Ambientais dos Empreendimentos do rio Madeira*; Instituto Brasileiro de Meio Ambiente–IBAMA: Brasilia, Brazil, 2007.
11. Carpio, J.M. Hidrologia e sedimentos. In *Águas Turvas: Alertas Sobre as Conseqüências de Barrar o Maior Afluente do Amazonas/Glenn Switkes*; International Rivers: Oakland, CA, USA, 2008.
12. Castro, N.P.d. *Avaliação de Indicadores de Alteração Hidrológica na Bacia Hidrográfica do rio Madeira: Grandes Obras Hidráulicas, Sedimentos e os Possíveis Impactos na Dinâmica Fluvial*; International Rivers: São Paulo, Brazil, 2019.
13. Frappart, F.; Papa, F.; da Silva, J.S.; Ramillien, G.; Prigent, C.; Seyler, F.; Calmant, S. Surface freshwater storage and dynamics in the Amazon basin during the 2005 exceptional drought. *Environ. Res. Lett.* **2012**, *7*, 044010. [CrossRef]
14. Hemming, J. *Change in the Amazon Basin: The Frontier after a Decade of Colonisation*; Manchester University Press: Manchester, UK, 1985; Volume 2.
15. Santos, R.E.; Pinto-Coelho, R.M.; Fonseca, R.; Simões, N.R.; Zanchi, F.B. The decline of fisheries on the Madeira River, Brazil: The high cost of the hydroelectric dams in the Amazon Basin. *Fish. Manag. Ecol.* **2018**, *25*, 380–391. [CrossRef]
16. Sorí, R.; Marengo, J.A.; Nieto, R.; Drumond, A.; Gimeno, L. The atmospheric branch of the hydrological cycle over the Negro and Madeira river basins in the Amazon Region. *Water* **2018**, *10*, 738. [CrossRef]
17. Jirau Energia. *Conheça a UHE—Jirau Energia.* 2022. Available online: https://www.jirauenergia.com.br/conheca-a-uhe/ (accessed on 21 August 2023).
18. Operador Nacional do Sistema Elétrico. *ONS—Manual de Procedimentos da Operação.* 2022. Available online: http://www.ons.org.br/ (accessed on 21 August 2023).
19. Schmitt, R.J.; Bizzi, S.; Castelletti, A.; Opperman, J.; Kondolf, G.M. Planning dam portfolios for low sediment trapping shows limits for sustainable hydropower in the Mekong. *Sci. Adv.* **2019**, *5*, eaaw2175. [CrossRef] [PubMed]
20. Amasi, A.; Wynants, M.; Blake, W.; Mtei, K. Drivers, impacts and mitigation of increased sedimentation in the hydropower reservoirs of east Africa. *Land* **2021**, *10*, 638. [CrossRef]
21. Rabiner, L.R. A tutorial on hidden Markov models and selected applications in speech recognition. *Proc. IEEE* **1989**, *77*, 257–286. [CrossRef]
22. Stamp, M. *A Revealing Introduction to Hidden Markov Models*; Department of Computer Science, San Jose State University: San Jose, CA, USA, 2004; pp. 26–56.
23. Zohrevand, Z.; Glasser, U.; Shahir, H.Y.; Tayebi, M.A.; Costanzo, R. Hidden Markov based anomaly detection for water supply systems. In Proceedings of the 2016 IEEE International Conference on Big Data (Big Data), IEEE, Washington, DC, USA, 5–8 December 2016; pp. 1551–1560.

24. González, A.M.; Roque, A.S.; García-González, J. Modeling and forecasting electricity prices with input/output hidden Markov models. *IEEE Trans. Power Syst.* **2005**, *20*, 13–24.
25. Bhaumik, D.; Crommelin, D.; Kapodistria, S.; Zwart, B. Hidden Markov models for wind farm power output. *IEEE Trans. Sustain. Energy* **2018**, *10*, 533–539. [CrossRef]
26. Kong, W.; Dong, Z.Y.; Hill, D.J.; Ma, J.; Zhao, J.; Luo, F. A hierarchical hidden Markov model framework for home appliance modeling. *IEEE Trans. Smart Grid* **2016**, *9*, 3079–3090. [CrossRef]
27. Kumar, D.; Bhowmik, P.S. Hidden markov model based islanding prediction in smart grids. *IEEE Syst. J.* **2019**, *13*, 4181–4189. [CrossRef]
28. Murthy, C.; Mishra, A.; Ghosh, D.; Roy, D.S.; Mohanta, D.K. Reliability analysis of phasor measurement unit using hidden Markov model. *IEEE Syst. J.* **2014**, *8*, 1293–1301. [CrossRef]
29. Jiang, J.; Chen, R.; Chen, M.; Wang, W.; Zhang, C. Dynamic fault prediction of power transformers based on hidden Markov model of dissolved gases analysis. *IEEE Trans. Power Deliv.* **2019**, *34*, 1393–1400. [CrossRef]
30. Freire, J.C.A.; Castro, A.R.G.; Homci, M.S.; Meiguins, B.S.; De Morais, J.M. Transmission line fault classification using hidden Markov models. *IEEE Access* **2019**, *7*, 113499–113510. [CrossRef]
31. Sun, S.; Yang, Q.; Yan, W. A novel Markov-based temporal-SoC analysis for characterizing PEV charging demand. *IEEE Trans. Ind. Inform.* **2017**, *14*, 156–166. [CrossRef]
32. Pearl, J. *Probabilistic Reasoning in Intelligent Systems: Networks of Plausible Inference*; Morgan kaufmann: Burlington, MA, USA 1988.
33. Leoni, L.; BahooToroody, A.; De Carlo, F.; Paltrinieri, N. Developing a risk-based maintenance model for a Natural Gas Regulating and Metering Station using Bayesian Network. *J. Loss Prev. Process Ind.* **2019**, *57*, 17–24. [CrossRef]
34. Barton, D.N.; Kuikka, S.; Varis, O.; Uusitalo, L.; Henriksen, H.J.; Borsuk, M.; de la Hera, A.; Farmani, R.; Johnson, S.; Linnell, J.D. Bayesian networks in environmental and resource management. *Integr. Environ. Assess. Manag.* **2012**, *8*, 418–429. [CrossRef] [PubMed]
35. Borsuk, M.; Clemen, R.; Maguire, L.; Reckhow, K. Stakeholder values and scientific modeling in the Neuse River watershed. *Group Decis. Negot.* **2001**, *10*, 355–373. [CrossRef]
36. Coleman, A.; Zalewski, J. Intelligent fault detection and diagnostics in solar plants. In Proceedings of the 6th IEEE International Conference on Intelligent Data Acquisition and Advanced Computing Systems, IEEE, Prague, Czech Republic, 15–17 September 2011; Volume 2, pp. 948–953.
37. Hao, C.F.; Qiu, J.; Li, F.F. Methodology for analyzing and predicting the runoff and sediment into a reservoir. *Water* **2017**, *9*, 440. [CrossRef]
38. Oviedo, D.; Romero-Ternero, M.d.C.; Hernández, M.; Sivianes, F.; Carrasco, A.; Escudero, J. Multiple intelligences in a MultiAgent System applied to telecontrol. *Expert Syst. Appl.* **2014**, *41*, 6688–6700. [CrossRef]
39. Fahiman, F.; Disano, S.; Erfani, S.M.; Mancarella, P.; Leckie, C. Data-driven dynamic probabilistic reserve sizing based on dynamic Bayesian belief networks. *IEEE Trans. Power Syst.* **2018**, *34*, 2281–2291. [CrossRef]
40. Dehghanpour, K.; Nehrir, M.H.; Sheppard, J.W.; Kelly, N.C. Agent-based modeling in electrical energy markets using dynamic Bayesian networks. *IEEE Trans. Power Syst.* **2016**, *31*, 4744–4754. [CrossRef]
41. Zhao, Y.; Zhu, W.; Yang, M.; Wang, M. Bayesian Network Based Imprecise Probability Estimation Method for Wind Power Ramp Events. *J. Mod. Power Syst. Clean Energy* **2020**, *9*, 1510–1519. [CrossRef]
42. Carita, A.J.Q.; Leite, L.C.; Medeiros, A.P.P.; Barros, R. Bayesian networks applied to failure diagnosis in power transformer. *IEEE Lat. Am. Trans.* **2013**, *11*, 1075–1082. [CrossRef]
43. Yongli, Z.; Limin, H.; Jinling, L. Bayesian networks-based approach for power systems fault diagnosis. *IEEE Trans. Power Deliv.* **2006**, *21*, 634–639. [CrossRef]
44. Jacob Schreiber. Pomegranate—A Python Package for Probabilistic Models. 2023. Available online: https://pomegranate.readthedocs.io/ (accessed on 21 August 2023).
45. Kjaerulff, U.B.; Madsen, A.L. Bayesian networks and influence diagrams. *Springer Sci. Bus. Media* **2008**, *200*, 114.
46. Koller, D.; Friedman, N. *Probabilistic Graphical Models: Principles and Techniques*; MIT Press: Cambridge, MA, USA, 2009.

*Article*

# Electric Vehicle Fleet Management for a Prosumer Building with Renewable Generation

**Matteo Fresia and Stefano Bracco ***

Department of Electrical, Electronic, Telecommunication Engineering and Naval Architecture, University of Genoa, 16145 Genova, Italy; matteo.fresia@edu.unige.it
* Correspondence: stefano.bracco@unige.it

**Abstract:** The integration of renewable energy systems in buildings leads to a reduction in energy bills for end users and a reduction in the carbon footprint of such buildings, usually referred to as prosumers. In addition, the installation of charging points for the electric vehicles of people working or living in these buildings can further improve the energy efficiency of the whole system if innovative technologies, such as vehicle-to-building (V2B) technologies, are implemented. The aim of this paper is to present an Energy Management System (EMS) based on mathematical programming that has been developed to optimally manage a prosumer building equipped with photovoltaics, a micro wind turbine and several charging points for electric vehicles. Capabilities curves of renewable power plant inverters are modelled within the EMS, as well as the possibility to apply power curtailment and V2B. The use of V2B technology reduces the amount of electricity purchased from the public grid, while the use of smart inverters for the power plants allows zero reactive power to be drawn from the grid. Levelized cost of electricity (LCOE) is used to quantify curtailment costs, while penalties on reactive power absorption from the distribution network are evaluated in accordance with the current regulatory framework. Specifically, the model is applied to a prosumer building owned by the postal service in a large city in Italy. The paper reports the main results of the study and proposes a sensitivity analysis on the number of charging stations and vehicles, as well as on the consideration of different typical days characterized by different load and generation profiles. This paper also investigates how errors in forecasting energy production from renewable sources impact the optimal operation of the whole system.

**Keywords:** prosumer building; vehicle-to-building; solar energy; wind energy; reactive power management; curtailment

**Citation:** Fresia, M.; Bracco, S. Electric Vehicle Fleet Management for a Prosumer Building with Renewable Generation. *Energies* **2023**, *16*, 7213. https://doi.org/10.3390/en16207213

Academic Editors: Marialaura Di Somma, Jianxiao Wang and Bing Yan

Received: 31 July 2023
Revised: 2 October 2023
Accepted: 20 October 2023
Published: 23 October 2023

## 1. Introduction

The building sector is a major consumer of energy, accounting for approximately 40% of final energy consumption and 36% of Greenhouse Gas (GHG) emissions in the European Union (EU). Specifically, the electricity consumption constitutes 35% of the energy use in buildings [1,2].

To address this issue, governments have implemented various measures in recent years to reduce energy consumption in buildings and promote the adoption of Renewable Energy Sources (RESs). The European Commission has introduced the so-called "Fit for 55" package, which aims to reduce GHG emissions by at least 55% by 2030 (compared to 1990 levels) and increase the market share of renewable energy by up to 40%. Specifically, within the building sector, the target is to achieve a 49% share of RES energy consumption [3].

In addition, private transportation accounts for 15% of $CO_2$ emissions in the EU. The "Fit for 55" package proposes a gradual reduction plan for $CO_2$ emissions from private vehicles, with a goal of achieving a 100% reduction by 2035. This implies that all new cars and vans entering the market after 2035 should be zero-emission vehicles, leading to reduced pollution and improved air quality, especially in densely populated areas [3]. National governments will play a crucial role in this transition by promoting the adoption of Electric Vehicles (EVs) and RESs, starting with public facilities and buildings such as schools, universities, and government offices [4].

RES technologies will play a leading role in boosting this transition: in particular, in dealing with the building sector, small-scale applications of RESs will be needed. The most suitable technologies for this application are Photovoltaic (PV) units and Wind Turbines (WTs). PV units are one of the major sources of distributed renewable energy and are usually involved in domestic rooftop installations, allowing the exploitation of previously unused spaces [5]. This trend is favoured due to the continuous decrease in PV installation prices and the incentives issued by governments.

Small-size WTs are typically positioned in close proximity to facilities in order to minimize potential wind disturbances caused by buildings but can also be installed on the rooftops of high-rise buildings to exploit the high wind speed. In urban areas, wind speed estimation is very important in order to evaluate the suitability of WT installation [6].

However, due to their inherent unpredictability, RESs are unable to instantly follow the profile of the demand. In order to cope with the imbalance between generation and load, Energy Storage Systems (ESSs) such as Battery Energy Storage Systems (BESSs) are commonly installed, enhancing the flexibility of the system. Anyway, in smaller-scale applications such as buildings, the integration of BESSs can significantly impact both installation and operational costs.

Within this regulatory and technical framework, so-called "Prosumption" is gaining increasing importance. "Prosumption" is the combination of both the consumption and the production of energy, and those who are involved in these activities are called "Prosumers". In addition to consuming their own renewable energy surplus, users have the option to sell it to the external power grid and receive compensation from the market operator. Prosumers are simultaneously consumers, producers and sellers of renewable energy [7].

Small-scale prosumers are residential facilities with PV units and BESS or PV units and EVs. Larger-scale applications of prosumption are represented by public institutions and small/medium enterprises that may exploit larger-size facilities, like larger PV units, WTs and BESSs and a fleet of EVs.

Besides the economic advantage deriving from the sale of surplus electricity to the external network, other forms of revenue may come from the provision of ancillary balancing services to the distribution system operator, like Frequency Containment Reserve (FCR) and automatic Frequency Restoration Reserve (aFRR), well suited for integrated PV-battery prosumers, as shown in [8] and references therein. Prosumers also offer distributed energy capacity, as highlighted in [9].

RESs are exploited not only to generate active power but also to satisfy the reactive power request of the load, thanks to the reactive power exchange capability of the inverters; in addition, by adjusting their reactive power injection/absorption, they provide voltage regulation at the node at which they are connected, even during night [10,11].

EVs are progressively substituting traditional combustion engine vehicles. In order to fully exploit the advantage of this technology, EVs can be connected to buildings, enabling the so-called Vehicle-to-Building (V2B) functionality: in this application, EVs do not behave only like a load but, when in idle mode, their battery can provide power to the building or absorb surplus power from the building, thanks to the bidirectionality of the power flow. V2B represents an alternative to non-mobile BESSs, also in Local Energy Communities (LECs) [12].

V2B applications have several positive effects on the vehicle–building system: they contribute to the dumping of the oscillation and unpredictability of the production of RESs [13], allowing the time shifting of energy and increasing the level of RES self-consumption for the user [14]; peak shaving is possible with V2B, reducing the grid peak generation power [15]. In addition, EVs can be used as a backup source in case of power outages [16]. Moreover, acting as storage systems, they can be used to reduce RES curtailment [17] and to supply energy to buildings in residential districts [18] where they stay parked for a long time. Enabling V2B technologies for EV fleets opens the door for the participation of EVs in primary frequency regulation, acting as spinning resources that are able to comply with national grid codes in terms of effectiveness and promptness [17,18].

The complexity of a system composed of a building and one or several EVs, with the possibility of performing V2B, means that an Energy Management System (EMS) is required in order to optimize the power flows between all the units.

Several examples of EMS applied to V2B can be found in the literature. In [19], a building EMS is proposed in order to integrate EVs with the aim of levelling the peak load demand to the off-peak hours. In [20] and in therein references, several EMSs and advanced control strategies are proposed, focusing in particular on the integration of buildings and EVs. In [21], the authors focus on optimal strategies for the smart charging of EVs in facilities that couple unpredictable RESs and infrastructure for electric mobility, while in [22], the authors investigated how the number of EVs in commercial buildings impacts electricity bills and how they can act as storage systems in buildings operating in the island mode. In [23], the proposed control strategy is designed to enable the effective interaction of a PV system, stationary batteries, and an EV within a prosumer installation. Assuming that the storage system works according to a fixed charging and discharging schedule, the proposed algorithm controls the operation of the EV battery, taking into account trip data introduced by the driver in order for the EV battery to reach the planned level of state of charge before the time of driving. In [24], the authors propose heuristic vehicle-to-home charging strategies with the goal of increasing self-sufficiency, vehicle availability and traction battery lifetime in different scenarios characterized by different EV driver behaviours. An EMS designed to optimize demand response in a prosumer building is described in [25], where the EV fleet is modelled considering stochastic characteristics, and PV production is modelled under uncertainty using actual data collected via smart meters. In [26], an EMS designed for the microgrids of building prosumers is described, considering both active and reactive power exchange.

An EMS for a public building with RES generation from PV and WT units and an EV for mail delivery is proposed in [27]. The present paper represents an extension of the model proposed in [27]. The main objective of this paper is to define an EMS that ensures the optimal operation and scheduling of a postal service-owned building located in a large city in Italy. The building is connected to the medium-voltage distribution network and is also fed by two small-size RES units: a PV unit and a WT unit; it acts like a prosumer of electricity, thanks to the bidirectional connection to the network. Several Electric Delivery Vehicles (EDVs) are allocated to the facility for mail transportation; each vehicle has a dedicated charging station. Innovative aspects of the model are represented through the modelling of the capability curves of the RES inverters within the EMS to optimally manage reactive power flows, as well as via the introduction of costs related to RES curtailment and reactive power absorption in the objective function. By running the EMS over an entire year with an hourly time resolution, the study aims to evaluate the impact of the number of EVs on the operation of the building, both from an energy and an economic point of view. Three scenarios are analysed, considering the number of EDVs equal to 10, 50 and 100. A sensitivity analysis is developed by varying the price of sale and purchase of electricity, as well as the number of Equivalent Operating Hours (EOHs) of the renewable power plants.

This paper is structured as follows: Section 2 describes the whole system, providing essential data and assumptions that have been made by the authors, and presents a comprehensive description of the optimization mathematical model that has been set up. Section 3 shows the results of the study and discusses them by comparing the three scenarios. Some concluding remarks, together with a discussion of potential future developments, are provided in Section 4.

## 2. Materials and Methods

### 2.1. System Description and Input Data

The developed mathematical model is an EMS, which makes it possible to apply optimal management strategies to the daily operation of a prosumer building equipped with renewable energy systems (mainly PV and WTs) and charging points for EVs. RES power plants are connected to the AC network through smart inverters able to manage both active and reactive powers, while EV charging stations are of the V2B type. As shown in Figure 1, the whole system is modelled as a microgrid connected to the medium-voltage distribution network, and each EV has a dedicated charging station. The EMS is based on a Mixed-Integer Linear Programming (MILP) model, having a time horizon consisting of $T$ time intervals ($t = 1 \dots T$), each one with a duration equal to $\Delta$.

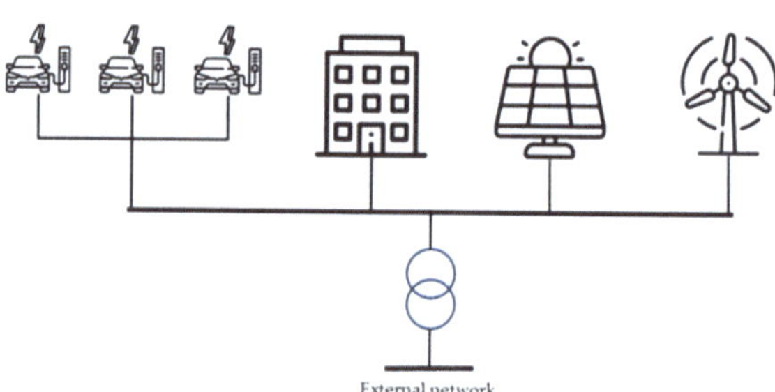


**Figure 1.** Electric system scheme.

The main input data of the model are:

- The number of RES power plants, indicated by $J$;
- The number of EVS, which coincides with the number of charging points, denoted as $N$;
- The size $A_j^{RES}$ [kVA] of the inverter associated with the $j$-th RES plant;
- The estimated average power $P_j^{RES,av}$ [kW] that can be generated by the $j$-th RES plant at time $t$;
- the curtailment cost $c_j^{RES,curt}$ [EUR/kWh] of the $j$-th RES plant, represented by its Levelized Cost of Electricity (LCOE);
- The rated capacity $C_n^{EV}$ [kWh] of the battery installed inside the $n$-th EV, together with its minimum state of charge $SOC_n^{EV,min}$ [%];
- The average energy consumption $F_n^{EV}$ [kWh/km] of the $n$-th EV;
- The transportation demand $D_{n,t}^{EV}$ of the $n$-th EV in the time interval $t$, measured in [km];
- Information on the presence of the vehicles at the facility, as expressed by the factor $y_{n,t}^{EV}$, which is equal to 1 when the $n$-th EV can be connected to its charging point and equal to 0 when the vehicle is not present;
- The minimum power $P_n^{EV,ch,min}$ [kW] that can be delivered to the $n$-th EV;
- The maximum power $P_n^{EV,ch,max}$ [kW] that can be delivered to the $n$-th EV;
- The minimum power $P_n^{EV,dch,min}$ [kW] that can be supplied by the $n$-th EV when it is operated in V2B mode;
- The maximum power $P_n^{EV,dch,max}$ [kW] that can be supplied by the $n$-th EV when it is operated in V2B mode;
- The charging ($\eta^{EV,ch}$) and discharging ($\eta^{EV,dch}$) efficiencies of EVs;
- The electrical load profiles of the building, in terms of active ($P_t^L$) [kW] and inductive reactive power ($Q_t^L$) [kVAr];
- The size $A^{grid}$ [kVA] of the transformer which connects the microgrid to the medium voltage distribution network;

- The active energy purchase and selling prices, respectively, $p_t^{grid}$ [EUR/kWh] and $r_t^{grid}$ [EUR/kWh];
- The penalty $q_t^{grid}$ [EUR/kVArh] on reactive energy absorbed from the distribution network, as set by the authority.

It is assumed that every RES power plant can exchange both active and reactive power in accordance with its capability curve, which is characterized by the shape of a semi-circle in the first two quadrants of the plane, with the reactive power on the horizontal axis and the active power on the vertical axis. Charging points for EVs work at a unitary power factor. Reactive power can be exchanged by the microgrid with the distribution network, thus enabling the transformer to operate in the four quadrants of the reactive/active power plane. The building does not present manageable loads apart from the EV charging points, whose scheduling can be optimized via the EMS. As further described in Section 2.2, the main goal of the EMS is that of managing RES power plants and EV charging points in order to minimize global Net Costs (NCs) and reduce the curtailment of RES sources. Among the costs, one is represented by the penalties related to the reactive power absorbed from the distribution network; to minimize this term, the role of inverters of RES plants in providing inductive reactive power is investigated.

### 2.2. Mathematical Model

In the mathematical model, RES power plants, EV charging points and the transformer are modelled by a set of linear constraints given by equalities or inequalities correlating with the decision variables, both integer and continuous.

The main decision variables that refer to the operation of $j$-th RES power plant at time $t$ are:

- $P_{j,t}^{RES,out}$ [kW]: generated active power;
- $P_{j,t}^{RES,curt}$ [kW]: curtailed active power;
- $Q_{j,t}^{RES,in}$ [kVAr]: inductive reactive power absorbed from the microgrid;
- $Q_{j,t}^{RES,out}$ [kVAr]: inductive reactive power supplied to the microgrid.

The relative constraints are defined from (1) to (2). The energy balances (1) set that the active power available from each RES plant is given by the sum between the generated power and the curtailed one, while constraints (2) guarantee that the maximum curtailed power is the available one. Constraints from (3) to (7) are defined to linearize the circular capability curve of inverters in the active power/reactive power plane, as shown in Figure 2a. The operating points lay in the ochre-colored area. In particular, constraints (3) and (4) fix the upper bounds of the inductive reactive power (absorbed and supplied) as a function of the rated apparent power $A_j^{RES}$, whereas constraints (6) and (7) represent the two tangents to the capability curve, respectively, in the first and in the second quadrants at the points of coordinates $\left(\frac{\sqrt{2}}{2} \cdot A_j^{RES}, \frac{\sqrt{2}}{2} \cdot A_j^{RES}\right)$ and $\left(-\frac{\sqrt{2}}{2} \cdot A_j^{RES}, \frac{\sqrt{2}}{2} \cdot A_j^{RES}\right)$. The set of binary variables $y_{j,t}^{RES,in}$ and $y_{j,t}^{RES,out}$ is introduced to avoid the simultaneous absorption and supply of inductive reactive power at time $t$, as ensured by constraints (5).

$$P_{j,t}^{RES,av} - P_{j,t}^{RES,out} - P_{j,t}^{RES,curt} = 0 \quad \forall j = 1 \ldots J, \ \forall t = 1 \ldots T \tag{1}$$

$$0 \leq P_{j,t}^{RES,curt} \leq P_{j,t}^{RES,av} \quad \forall j = 1 \ldots J, \ \forall t = 1 \ldots T \tag{2}$$

$$0 \leq Q_{j,t}^{RES,in} \leq A_j^{RES} \cdot y_{j,t}^{RES,in} \quad \forall j = 1 \ldots J, \ \forall t = 1 \ldots T \tag{3}$$

$$0 \leq Q_{j,t}^{RES,out} \leq A_j^{RES} \cdot y_{j,t}^{RES,out} \quad \forall j = 1 \ldots J, \ \forall t = 1 \ldots T \tag{4}$$

$$y_{j,t}^{RES,in} + y_{j,t}^{RES,out} \leq 1 \;\; \forall j = 1\ldots J, \; \forall t = 1\ldots T \tag{5}$$

$$P_{j,t}^{RES,out} \leq -Q_{j,t}^{RES,out} + \sqrt{2}\cdot A_j^{RES} \;\; \forall j = 1\ldots J, \; \forall t = 1\ldots T \tag{6}$$

$$P_{j,t}^{RES,out} \leq -Q_{j,t}^{RES,in} + \sqrt{2}\cdot A_j^{RES} \;\; \forall j = 1\ldots J, \; \forall t = 1\ldots T \tag{7}$$

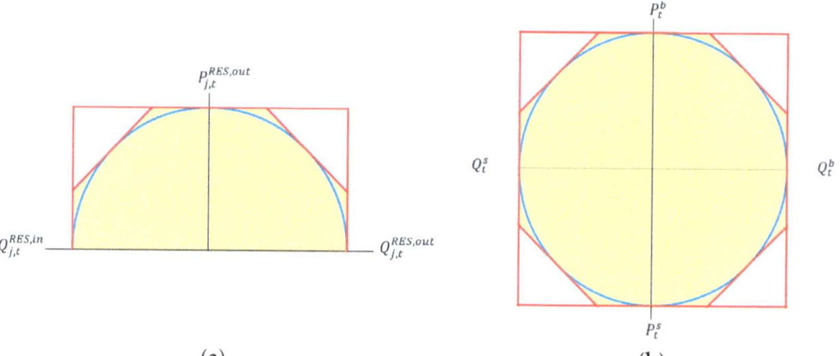


**Figure 2.** Capability curve linearization for RES plants (**a**) and distribution network coupling (**b**).

The decision variables that describe the active and reactive power exchanged between the microgrid and the distribution network at time *t* are:

- $P_t^b$ [kW]: active power withdrawn from the distribution network;
- $P_t^s$ [kW]: active power injected into the distribution network;
- $Q_t^b$ [kVAr]: inductive reactive power absorbed from the distribution network.
- $Q_t^s$ [kVAr]: inductive reactive power provided to the distribution network.

Moreover, a set of auxiliary binary variables are needed to avoid the simultaneous absorption and injection of power from/into the distribution network. Specifically, $x_t^b$ and $x_t^s$ are used for active power, while $y_t^b$ and $y_t^s$ for reactive power. The constraints from (8) to (17) model the interaction between the microgrid and the distribution network and are based on the linearization of the transformer capability curve, as reported in Figure 2b. In particular, constraints (8) and (9) limit the maximum active power that can be exchanged, while constraints (11) and (12) do the same for the reactive power. Constraints (10) and (13) ensure the non-simultaneity between absorptions and withdrawals, while constraints (14) to (17) define the limits imposed by the four oblique segments reported in Figure 2b.

$$0 \leq P_t^b \leq A^{grid}\cdot x_t^b \;\; \forall t = 1\ldots T \tag{8}$$

$$0 \leq P_t^s \leq A^{grid}\cdot x_t^s \;\; \forall t = 1\ldots T \tag{9}$$

$$x_t^b + x_t^s \leq 1 \;\; \forall t = 1\ldots T \tag{10}$$

$$0 \leq Q_t^b \leq A^{grid}\cdot y_t^b \;\; \forall t = 1\ldots T \tag{11}$$

$$0 \leq Q_t^s \leq A^{grid}\cdot y_t^s \;\; \forall t = 1\ldots T \tag{12}$$

$$y_t^b + y_t^s \leq 1 \;\; \forall t = 1\ldots T \tag{13}$$

$$P_t^b \leq -Q_t^b + \sqrt{2} \cdot A^{grid} \quad \forall t = 1 \dots T \tag{14}$$

$$P_t^b \leq -Q_t^s + \sqrt{2} \cdot A^{grid} \quad \forall t = 1 \dots T \tag{15}$$

$$P_t^s \leq -Q_t^b + \sqrt{2} \cdot A^{grid} \quad \forall t = 1 \dots T \tag{16}$$

$$P_t^s \leq -Q_t^s + \sqrt{2} \cdot A^{grid} \quad \forall t = 1 \dots T \tag{17}$$

As far as electric mobility is concerned, the decision variables that refer to the operation of EV charging stations at time $t$ are:

- $P_{n,t}^{EV,ch}$ [kW]: active power supplied to the $n$-th EV;
- $P_{n,t}^{EV,dch}$ [kW]: active power provided by the $n$-th EV when operated in V2B mode;
- $E_{n,t}^{EV}$ [kWh]: energy content of the battery in the $n$-th EV.

The constraints on $P_{n,t}^{EV,ch}$ and $P_{n,t}^{EV,dch}$ can be defined as follows:

$$P_{n,t}^{EV,ch} \geq P_n^{EV,ch,min} \cdot x_{n,t}^{EV,ch} \quad \forall n = 1 \dots N, \forall t = 1 \dots T \tag{18}$$

$$P_{n,t}^{EV,ch} \leq P_n^{EV,ch,max} \cdot x_{n,t}^{EV,ch} \quad \forall n = 1 \dots N, \forall t = 1 \dots T \tag{19}$$

$$P_{n,t}^{EV,dch} \geq P_n^{EV,dch,min} \cdot x_{n,t}^{EV,dch} \quad \forall n = 1 \dots N, \forall t = 1 \dots T \tag{20}$$

$$P_{n,t}^{EV,dch} \leq P_n^{EV,dch,max} \cdot x_{n,t}^{EV,dch} \quad \forall n = 1 \dots N, \forall t = 1 \dots T \tag{21}$$

where the binary variables $x_{n,t}^{EV,ch}$ and $x_{n,t}^{EV,dch}$ become equal to 1 when the $n$-th vehicle is charging or discharging at time $t$, respectively. Obviously, as defined in (22), each vehicle can be charged or discharged only when present at the facility ($y_{n,t}^{EV} = 1$).

$$x_{n,t}^{EV,ch} + x_{n,t}^{EV,dch} \leq y_{n,t}^{EV} \quad \forall n = 1 \dots N, \forall t = 1 \dots T \tag{22}$$

Then, it is necessary to set lower and upper bounds for $E_{n,t}^{EV}$, as shown by constraints (23) and (24).

$$E_{n,t}^{EV} \geq 0.01 \cdot SOC_n^{EV,min} \cdot C_n^{EV} \quad \forall n = 1 \dots N, \forall t = 1 \dots T \tag{23}$$

$$E_{n,t}^{EV} \leq C_n^{EV} \quad \forall n = 1 \dots N, \forall t = 1 \dots T \tag{24}$$

The energy balance of the battery installed in the $n$-th EV can be written as:

$$E_{n,t+1}^{EV} = E_{n,t}^{EV} + \Delta \cdot \left( P_{n,t}^{EV,ch} \cdot \eta^{EV,ch} - \frac{P_{n,t}^{EV,dch}}{\eta^{EV,dch}} \right) - F_n^{EV} \cdot D_{n,t}^{EV} \tag{25}$$
$$\forall n = 1 \dots N, \forall t = 1 \dots T - 1$$

where the quantity of $E_{n,1}^{EV}$ is assumed to be known for all vehicles. It is important to say that the vehicles can be charged only when using the dedicated charging points at the facility.

The remaining set of constraints, (26) and (27), are introduced to represent the active and reactive power balances of the whole system, and are studied with a single bus model given the limited extent of the prosumer power grid.

$$P_t^b + \sum_{j=1}^{J} P_{j,t}^{RES,out} + \sum_{n=1}^{N} P_{n,t}^{EV,dch} = P_t^L + P_t^s + \sum_{n=1}^{N} P_{n,t}^{EV,ch} \quad \forall t = 1 \ldots T \tag{26}$$

$$Q_t^b + \sum_{j=1}^{J} Q_{j,t}^{RES,out} = Q_t^L + Q_t^s + \sum_{j=1}^{J} Q_{j,t}^{RES,in} \quad \forall t = 1 \ldots T \tag{27}$$

The objective function (*Obj*) of the optimization model consists of the minimization of total NCs, which is evaluated as follows:

$$Obj = \Delta \cdot \sum_{t=1}^{T} \left[ p_t^{grid} \cdot P_t^b + q_t^{grid} \cdot Q_t^b + \alpha \cdot \sum_{j=1}^{J} \left( c_j^{RES,curt} \cdot P_{j,t}^{RES,curt} \right) - r_t^{grid} \cdot P_t^s \right] \tag{28}$$

where the multiplication factor $\alpha$ can be chosen as desired to give more or less weight to RES curtailment costs.

## 3. Results

This section presents the results of this study. It is divided into four subsections: the first part presents the numerical values that have been chosen for the input quantities; the second one shows a comparison between the three scenarios, as defined by varying the number of EVs owned by the company; the third one shows some results for four typical days, as defined according to the combination of working/weekend days and to the high or low generation coming from RESs, with a number of EDVs equal to 50; and finally, the last subsection presents the results of the cost-sensitivity analysis, carried out considering 50 EDVs.

### 3.1. Assumptions

In this section, the outcomes of the optimization performed throughout the entire year are presented, assuming a time interval of one hour ($\Delta = 1$ h). Before showing the results, it is necessary to describe some numerical input data to facilitate the reader's understanding.

The EMS has been implemented using the Matlab R2022b/Yalmip (R20210331 release) [28] environment and solved by calling the Gurobi solver.

The considered generation technologies include a PV plant featuring an inverter with a power rating of 29 kVA and a vertical axis WT equipped with an inverter with a power rating of 14 kVA.

The profiles of PV active power available generation and the wind speed profile, from which the WT active power available generation profile is obtained, have been downloaded from the PVGIS Online Tool database [29]. In this scenario, the EOHs of the PV plant are equal to 1348 h, whilst for the WT plant, they are equal to 1000 h.

Possible forecasting errors related to the production of RES power plants are taken into account by means of an hourly random correction coefficient, which is used to scale up and down the hourly RES productions derived from PVGIS. According to this variability, four scenarios have been defined in addition to the base one, according to the different combinations of the different EOHs of the plants. These scenarios are outlined in Table 1.

The EDVs considered for the analysis are E-NV200 models manufactured by Nissan (Yokohama, Japan) and equipped with a battery capacity equal to 40 kWh. The vehicles have been divided into two categories in terms of different average transportation demand and availability at the facility: Category I (EDVs-I) and Category II (EDV-II). EDVs-I represents 40% of the overall number of EDVs, whilst EDVs-II accounts for 60% of the fleet. The daily average transportation demand is presented in Table 2 for the two categories of vehicles.

**Table 1.** Uncertainty scenarios.

|  | **High WT EOH** | **Low WT EOH** |
|---|---|---|
| **High PV EOH** | Scenario I<br>PV: 1415 h<br>WT: 1102 h | Scenario III<br>PV: 1293 h<br>WT: 1102 h |
| **Low PV EOH** | Scenario II<br>PV: 1415 h<br>WT: 916 h | Scenario IV<br>PV: 1293 h<br>WT: 916 h |

**Table 2.** Transportation demand for EDV categories.

|  | **Working Days** | **Holidays** | **Preholiday Days** |
|---|---|---|---|
| EDV-I | 14.5 km | 0 km | 6.25 km |
| EDV-II | 21.75 km | 0 km | 9.375 km |

EDVs-I are available at the facility throughout the holidays; on working days, they are unavailable from 8 a.m. to 12 p.m. and from 1 p.m. to 5 p.m.; on pre-holiday days, EDVs-I are not available from 8 a.m. to 12 p.m. The availability of EDVs-II shifted twelve hours forward in time when compared to EDVs-I because they are supposed to operate mainly during the night.

The transportation demand has been assumed based on the typical work behaviour of a postman in an Italian countryside area. The technical data of the EDV model have been derived from the manufacturer's datasheet.

The load profile of the building, both for active and reactive power, is determined by scaling a real measured load profile of a building located in the Savona Campus of the University of Genoa. It is adjusted to reflect the smaller size of the postal service building and its specific location.

The aforementioned data are summarized in Table 3.

**Table 3.** Load and generation data.

|  | **Annual Energy** |
|---|---|
| Available from PV [kWh] | 39,102 |
| Available from WT [kWh] | 13,996 |
| Active power load [kWh] | 72,646 |
| Reactive power load [KVArh] | 72,157 |

To evaluate the costs reported in the objective function of the optimization model, some input data have to provided:

- $p_t^{grid}$ has been set equal to 0.40 [EUR/kWh] during peak hours and equal to 0.27 [EUR/kWh] during off-peak hours, pre-holiday days and holidays;
- $q_t^{grid}$ has been set equal to 0.0027 [EUR/kVArh] for the whole year;
- $r_t^{grid}$ has been set equal to 0.20 [EUR/kWh] for the whole year.

Curtailment costs for the RESs have been considered equal to the LCOE for those sources: for PV, a value of 0.128 [EUR/kWh] has been chosen, whilst for WTs, a value of 0.60 [EUR/kWh] has been assumed [30].

### 3.2. EMS Results: Sensitivity on the Number of EDVs

This subsection presents the main energy and economic results related to the sensitivity analysis carried out by varying the number of EDVs ($N$) from 10, passing through to 50 and then up to 100. The annual results are summarized in Table 4.

The capability curves of the inverters connected to the PV and to the WT unit are shown in Figures 3 and 4, where $Q^{PV}$ and $Q^{WT}$, respectively, represent $Q_t^{PV,out}$ and $Q_t^{WT,out}$, while

$P^{PV}$ and $P^{WT}$ refer to $P_t^{PV,out}$ and $P_t^{WT,out}$. Both the actual curve and the linearized one are depicted in the figures. As previously mentioned, the linearized curve has been modelled in order to be able to use linear programming models. As it is evident from the pictures, for a few hours during the whole year, the inverters are saturated or overloaded. The two figures do not report the second quadrant since no inductive reactive power is absorbed by RES plants in the case study. The provision of reactive power by RES allows zero reactive energy withdrawal from the distribution network in all the examined scenarios. Due to the fact the RES plants almost never produce the maximum active power, it is not necessary to apply curtailment to provide reactive power within the constraints of the linearized capability curve. As far as the injection of active power into the distribution network is concerned, surplus generation from RES only occurs in the scenario with 10 vehicles, whereas in the other two scenarios, this surplus is used to charge the most numerous vehicles.

**Table 4.** Annual results.

| | [-] | $N_{EDV} = 10$ | $N_{EDV} = 50$ | $N_{EDV} = 100$ |
|---|---|---|---|---|
| PV active energy generation | [kWh] | | 39,102 | |
| PV curtailed energy | [kWh] | | 0 | |
| PV reactive energy generation | [kVArh] | | 44,654 | |
| WT active energy generation | [kWh] | | 13,996 | |
| WT curtailed energy | [kWh] | | 0 | |
| WT reactive energy generation | [kVArh] | | 27,503 | |
| Bought active energy | [kWh] | 34,209 | 73,585 | 124,630 |
| Sold active energy | [kWh] | 14,604 | 0 | 0 |
| Bought reactive energy | [kVArh] | | 0 | |
| Energy charged to EDVs | [kWh] | 21,700 | 62,536 | 113,581 |
| Energy discharged from EDVs | [kWh] | | 84,985 | |
| NCs | [EUR] | 8944 | 19,868 | 33,650 |

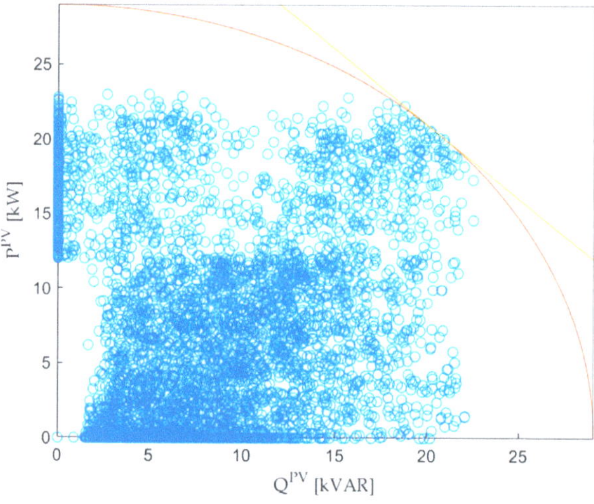


**Figure 3.** PV inverter capability curve. Blue circles represent the hourly operating points of the inverter. The red circular line represents the non-linearized capability curve of the inverter. The yellow line represents the sloped portion of the linearized capability curve of the inverter.

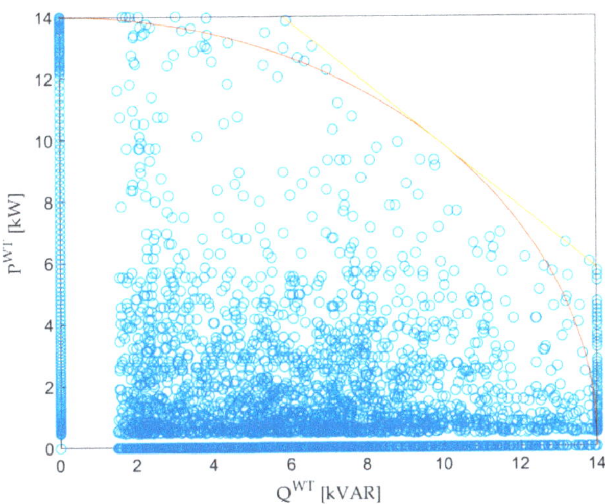

**Figure 4.** WT inverter capability curve. Blue circles represent the hourly operating points of the inverter. The red circular line represents the non-linearized capability curve of the inverter. The yellow line represents the sloped portion of the linearized capability curve of the inverter.

### 3.3. Comparison of EMS Results Taking into Account Forecasting Uncertainty

The present subsection presents the main results of the EMS over the whole year when varying the number of EOHs of the two RES power plants in accordance with the correction coefficient introduced in order to take into account the forecasting uncertainty of RES production. The number of EDVs has been considered equal to 50.

The main results are reported in Table 5.

**Table 5.** Annual results for different scenarios.

|  | [-] | Scenario I | Scenario II | Scenario III | Scenario IV |
|---|---|---|---|---|---|
| PV active energy generation | [kWh] | 41,046 | 41,046 | 37,507 | 37,507 |
| PV curtailed energy | [kWh] | 0 | 0 | 0 | 0 |
| PV reactive energy generation | [kVArh] | 43,601 | 42,850 | 45,753 | 44,837 |
| WT active energy generation | [kWh] | 15,424 | 12,820 | 15,424 | 12,820 |
| WT curtailed energy | [kWh] | 0 | 0 | 0 | 0 |
| WT reactive energy generation | [kVArh] | 28,556 | 29,307 | 26,402 | 27,321 |
| Bought active energy | [kWh] | 70,034 | 72,737 | 73,817 | 76,527 |
| Sold active energy | [kWh] | 0 | 0 | 0 | 0 |
| Bought reactive energy | [kVArh] | 0 | 0 | 0 | 0 |
| Energy charged to EDVs | [kWh] | 61,845 | 62,228 | 62,783 | 63,191 |
| Energy discharged from EDVs | [kWh] | 79,872 | 82,708 | 86,813 | 89,828 |
| NCs | [EUR] | 18,909 | 19,639 | 19,931 | 20,662 |

According to the previously defined scenarios, the RES production is alternatively increased and reduced when compared to the historical data considered in the base scenario. Scenario I is characterized by increased RES production when compared to the base scenario, leading to a reduced energy exchange between the EDVs and the building; Scenario I is also characterized by the lowest NCs. Scenario IV presents reduced RES production to emulate the possible over-estimation of RES production in the base scenario; in this configuration, the NCs are the highest among all the scenarios, and the exchange of active energy between the EDVs and the building is large because the V2B facility has to compensate for the lack of renewable energy. For the same reason, the amount of energy bought from the network is the highest as well. Scenario II and Scenario III represent intermediate scenarios, with moderate values of NCs, V2B energy exchange and energy purchased from the network. In all scenarios, active energy is never sold to the network due to the large transportation demand of EDVs.

It is important to highlight that the aim of the present study is to simulate the annual operation of the building in order to investigate the benefits deriving from RES generation and V2B techniques: for this reason, the proposed EMS is not designed as a real-time EMS; instead, it is developed in order to be run for the 8760 h of a whole year in order to make global technical and economic considerations. As evident from the results, possible uncertainty in the forecasting of RES generation has a limited impact on the NCs ($-4.8\%$ in Scenario I and $+4\%$ in Scenario IV). Even variation in the energy exchange between the building and EDVs is limited: in Scenario I, V2B flow is reduced by $-6\%$, and the B2V flow is reduced by $-1.1\%$; in Scenario IV, V2B exchange is increased by $+5.4\%$ while the B2V exchange is increased by $+1.05\%$.

### 3.4. Comparison of EMS Results for Different Typical Days

The present subsection shows the daily results of the EMS for four typical days, defined according to the combination of two criteria: working day/holiday and high RES production/low RES production. For this analysis, the number of EDVs has been considered equal to 50. Regarding EOHs, the base scenario with no uncertainty has been selected.

The days that have been selected according to the aforementioned criteria are summarized in Table 6, highlighting the energy production of the RES sources.

**Table 6.** Typical days.

|  | **Working Days** | **Holidays** |
|---|---|---|
| **Low RES production** | Wednesday in December<br>19.56 kWh<br>Figure 5 | Sunday in October<br>45.11 kWh<br>Figure 6 |
| **High RES production** | Friday in June<br>379.73 kWh<br>Figure 7 | Sunday in September<br>366.73 kWh<br>Figure 8 |

Figures 5–8 show the optimal active power profiles determined using the EMS for the selected days. In these figures, the power associated with EDVs is considered positive when the vehicles are in discharge mode, whilst the charging of the EDVs is considered as a load and, therefore, associated with negative values.

During the working day with low RES penetration (see Figure 5), it is evident that in order to satisfy the transportation demand of the EDVs, power has to be bought from the network. The purchase of electricity from the network is very significant, especially during the night and early in the morning, for two main reasons: one is that the PV unit is not working and the WT is delivering a very low power; the second reason is that EDVs-II (mainly, but also EDVs-I in some hours) are used as "mobile" BESSs: once they have reached the facility after having fulfilled their duty, they are recharged in order to exploit the low off-peak electricity price, and then they are discharged during the day, to avoid buying electricity from the network. When the RES production is very low, a part of the purchased energy is used to charge the EDVs-I so that they are able to satisfy their transportation demand during the day.

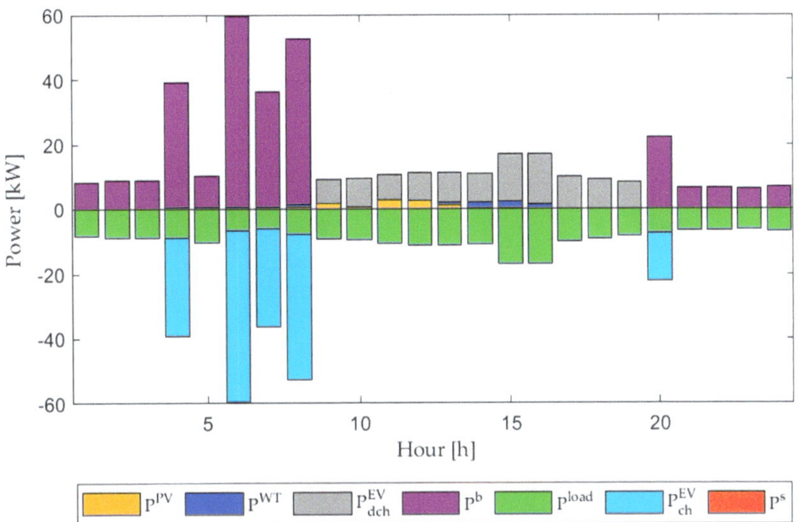

Figure 5. Active power profiles of the 4 December.

During holidays with low RES production (see Figure 6), a large amount of energy must again be purchased from the network, using EDVs as storage systems and exploiting their full availability at the facility throughout the day. Since during holidays, the transportation demand is set to 0, the purchased energy is lower when compared to working days.

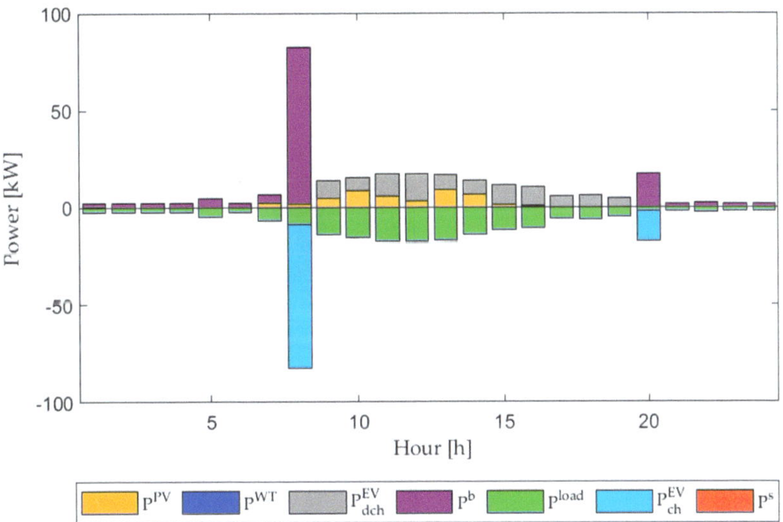

Figure 6. Active power profiles of the 6 October.

During working days with high RES penetration (see Figure 7), it is evident that a much lower power has to be purchased from the network. Thanks to the high RES generation, the EDVs are not discharged during the day, but they are (globally) continuously charged in order to satisfy the transportation demand. The only hours when a more significant amount of power is taken from the network are during the night in order to allow EDVs-II to carry out their service.

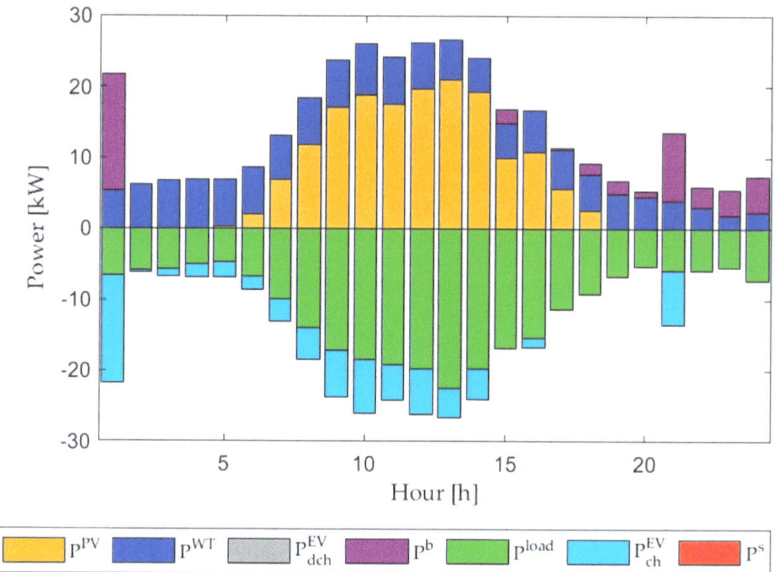

**Figure 7.** Active power profiles of the 7 June.

Finally, during holidays with high RES penetration (see Figure 8), some energy is bought from the network at the beginning of the day and stored in the EDVs to be used for the services of the following day. During the day, the EDVs are continuously charged in order to avoid the curtailment of RESs, to which a minimal, but still existing, cost is associated.

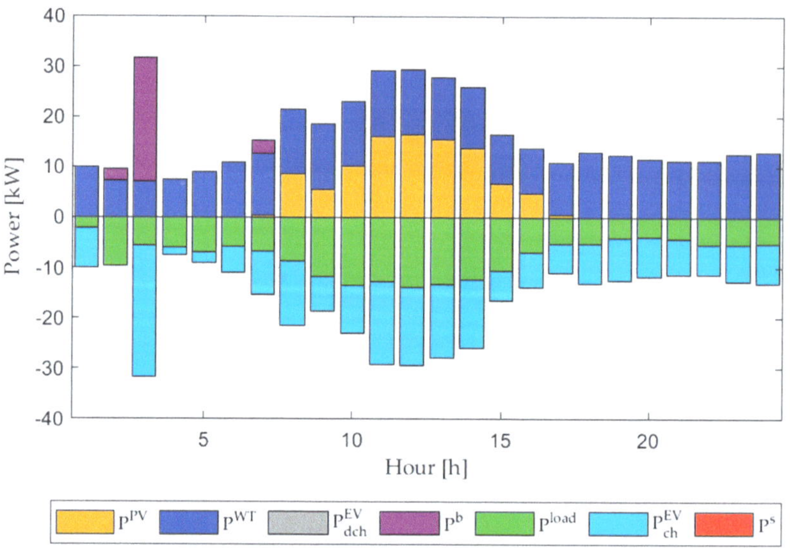

**Figure 8.** Active power profiles of the 29 September.

There is never an excess of electricity sold to the network; this is due to the large number of EDVs that, when exploiting B2V service, charge their batteries in order to use the stored energy to satisfy transportation demands.

V2B services are convenient and have a significant role when RES penetration is low; therefore, they can be used primarily during the winter in order to avoid the purchase of electricity from the network, especially during peak hours, when the cost of energy is higher.

## 4. Conclusions

The purpose of the present paper was to develop an EMS for the efficient management of a prosumer building connected to the distribution network. The building is owned by the postal service and is equipped with two RESs, namely a PV system and a WT unit, and is equipped with a large fleet of EDVs for postal delivery that can be charged at dedicated charging points installed at the facility. The primary goal of the EMS is to minimize the overall costs associated with the operation of the building, such as those related to the exchange of active and reactive power between the distribution network and the building over the whole year.

Three sensitivity analyses were developed. The first analysis aimed to evaluate the impact of the number of EDVs on the operation of the building while also comparing the annual energy quantities and the NCs. The fleet of EDVs was considered to be composed of either 10, 50 or 100 vehicles. The second analysis was performed in order to evaluate the impact of RES generation forecasting errors on the operation of the building; when considering a number of EDVs equal to 50, the PV and WT generations were alternatively scaled up and down in order to emulate possible underestimations and overestimations in terms of RES forecasting. Then, a third analysis was carried out to analyse the behaviour of the building and the optimal active power profiles with a number of EDVs equal to 50 when considering four different typical days, as identified according to RES generation (high or low) and the considered day (working day or holiday). Moreover, the impact of errors in forecasting renewable energy production on the EMS results was also investigated.

This paper showed how increasing the dimension of the EDV fleet affected energy and cost, highlighting the fact that V2B applications are necessary in order to fully exploit the consumption of RES-derived energy.

Future developments could involve the modification of the load profile of the building and the behaviour of the EDVs in terms of availability and transportation demand in order to investigate different scenarios and different types of users. In addition, to expand the proposed EMS, other technical implementations could also be included, such as the possibility for EVs to exchange reactive power with the building and the implementation of a model of the building's electric network in order to take power losses into account.

**Author Contributions:** Conceptualization, S.B. and M.F.; methodology, S.B. and M.F.; software, S.B. and M.F.; validation, M.F.; data curation, M.F.; writing, S.B. and M.F.; project administration, S.B.; funding acquisition, S.B. All authors have read and agreed to the published version of the manuscript.

**Funding:** This research was funded by the National Recovery and Resilience Plan, Mission 4, Component 2, Investment 1.4 "Strengthening research facilities and creating national R&D facilities on some Key Enabling Technologies" funded by the European Union—NextGenerationEU. Code CN000023—Title "Sustainable Mobility Center (National Center for Sustainable Mobility—CNMS)".

**Data Availability Statement:** Data cannot be disclosed due to confidentiality reasons.

**Conflicts of Interest:** The authors declare no conflict of interest.

## References

1. IEA. *Buildings*; IEA: Paris, France, 2022.
2. Renovation Wave: Creating Green Buildings for the Future. Available online: https://europa.eu/!bG4m7p (accessed on 16 June 2023).
3. Fit for 55: Why the EU Is Toughening CO2 Emission Standards for Cars and Vans. Available online: https://europa.eu/!w39R7x (accessed on 16 June 2023).

4.  Fiaschi, D.; Bandinelli, R.; Conti, S. A case study for energy issues of public buildings and utilities in a small municipality: Investigation of possible improvements and integration with renewables. *Appl. Energy* **2012**, *97*, 101–114. [CrossRef]
5.  Alshahrani, A.; Omer, S.; Su, Y.; Mohamed, E.; Alotaibi, S. The technical challenges facing the integration of small-scale and large-scale PV systems into the grid: A critical review. *Electronics* **2019**, *8*, 1443. [CrossRef]
6.  Walker, S.L. Building mounted wind turbines and their suitability for the urban scale—A review of methods of estimating urban wind resource. *Energy Build.* **2011**, *43*, 1852–1862. [CrossRef]
7.  Gautier, A.; Jacqmin, J.; Poudou, J.-C. The prosumers and the grid. *J. Regul. Econ.* **2018**, *53*, 100–126. [CrossRef]
8.  Lopez, A.; Ogayar, B.; Hernández, J.; Sutil, F. Survey and assessment of technical and economic features for the provision of frequency control services by household-prosumers. *Energy Policy* **2020**, *146*, 111739. [CrossRef]
9.  Parag, Y.; Sovacool, B.K. Electricity market design for the prosumer era. *Nat. Energy* **2016**, *1*, 16032. [CrossRef]
10. Lavi, Y.; Apt, J. Using PV inverters for voltage support at night can lower grid costs. *Energy Rep.* **2022**, *8*, 6347–6354. [CrossRef]
11. Talkington, S.; Grijalva, S.; Reno, M.J.; Azzolini, J.A. Solar PV inverter reactive power disaggregation and control setting estimation. *IEEE Trans. Power Syst.* **2022**, *37*, 4773–4784. [CrossRef]
12. Piazza, G.; Bracco, S.; Delfino, F.; Siri, S. Optimal design of electric mobility services for a Local Energy Community. *Sustain. Energy Grids Netw.* **2021**, *26*, 100440. [CrossRef]
13. Ioakimidis, C.; Thomas, D.; Rycerski, P.; Genikomsakis, K. Peak shaving and valley filling of power consumption profile in non-residential buildings using an electric vehicle parking lot. *Energy* **2018**, *148*, 148–158. [CrossRef]
14. Barone, G.; Buonomano, A.; Forzano, C.; Giuzio, G.F.; Palombo, A. Increasing self-consumption of renewable energy through the Building to Vehicle to Building approach applied to multiple users connected in a virtual micro-grid. *Renew. Energy* **2020**, *159*, 1165–1176. [CrossRef]
15. Arias, N.B.; Hashemi, S.; Andersen, P.B.; Træholt, C.; Romero, R. Distribution system services provided by electric vehicles: Recent status, challenges, and future prospects. *IEEE Trans. Intell. Transp. Syst.* **2019**, *20*, 4277–4296. [CrossRef]
16. Borge-Diez, D.; Icaza, D.; Açıkkalp, E.; Amaris, H. Combined vehicle to building (V2B) and vehicle to home (V2H) strategy to increase electric vehicle market share. *Energy* **2021**, *237*, 121608. [CrossRef]
17. Martinenas, S.; Marinelli, M.; Andersen, P.B.; Træholt, C. Implementation and demonstration of grid frequency support by V2G enabled electric vehicle. In Proceedings of the 2014 49th International Universities Power Engineering Conference (UPEC), Cluj-Napoca, Romania, 2–5 September 2014; pp. 1–6.
18. Meng, J.; Mu, Y.; Jia, H.; Wu, J.; Yu, X.; Qu, B. Dynamic frequency response from electric vehicles considering travelling behavior in the Great Britain power system. *Appl. Energy* **2016**, *162*, 966–979. [CrossRef]
19. Aziz, M.; Oda, T.; Kashiwagi, T. Extended utilization of electric vehicles and their re-used batteries to support the building energy management system. *Energy Procedia* **2015**, *75*, 1938–1943. [CrossRef]
20. Zhou, Y.; Cao, S.; Hensen, J.L.; Lund, P.D. Energy integration and interaction between buildings and vehicles: A state-of-the-art review. *Renew. Sustain. Energy Rev.* **2019**, *114*, 109337. [CrossRef]
21. Zheng, S.; Huang, G.; Lai, A.C. Coordinated energy management for commercial prosumers integrated with distributed stationary storages and EV fleets. *Energy Build.* **2023**, *282*, 112773. [CrossRef]
22. He, Z.; Khazaei, J.; Freihaut, J.D. Optimal integration of Vehicle to Building (V2B) and Building to Vehicle (B2V) technologies for commercial buildings. *Sustain. Energy Grids Netw.* **2022**, *32*, 100921. [CrossRef]
23. Kelm, P.; Mieński, R.; Wasiak, I. Energy management in a prosumer installation using hybrid systems combining EV and stationary storages and renewable power sources. *Appl. Sci.* **2021**, *11*, 5003. [CrossRef]
24. Rücker, F.; Schoeneberger, I.; Wilmschen, T.; Sperling, D.; Haberschusz, D.; Figgener, J.; Sauer, D.U. Self-sufficiency and charger constraints of prosumer households with vehicle-to-home strategies. *Appl. Energy* **2022**, *317*, 119060. [CrossRef]
25. Thomas, D.; Deblecker, O.; Ioakimidis, C. Optimal operation of an energy management system for a grid-connected smart building considering photovoltaics' uncertainty and stochastic electric vehicles' driving schedule. *Appl. Energy* **2018**, *210*, 1188–1206. [CrossRef]
26. Farinis, G.K.; Kanellos, F.D. Integrated energy management system for Microgrids of building prosumers. *Electr. Power Syst. Res.* **2021**, *198*, 107357. [CrossRef]
27. Bracco, S.; Fresia, M. Energy Management System for the Optimal Operation of a Grid-Connected Building with Renewables and an Electric Delivery Vehicle. In Proceedings of the IEEE EUROCON 2023—20th International Conference on Smart Technologies, Torino, Italy, 6–8 July 2023; pp. 472–477.
28. Lofberg, J. YALMIP: A toolbox for modeling and optimization in MATLAB. In Proceedings of the 2004 IEEE International Conference on Robotics and Automation (IEEE Cat. No. 04CH37508), Taipei, Taiwan, 2–4 September 2004; pp. 284–289.
29. Šúri, M.; Huld, T.; Cebecauer, T.; Dunlop, E.D. Geographic aspects of photovoltaics in Europe: Contribution of the PVGIS web site. *IEEE J. Sel. Top. Appl. Earth Obs. Remote Sens.* **2008**, *1*, 34–41. [CrossRef]
30. de Simón-Martín, M.; Bracco, S.; Piazza, G.; Pagnini, L.C.; González-Martínez, A.; Delfino, F. Application to Real Case Studies. In *Levelized Cost of Energy in Sustainable Energy Communities: A Systematic Approach for Multi-Vector Energy Systems*; Springer International Publishing: Cham, Switzerland, 2022; pp. 77–120. [CrossRef]

*Article*

# Assessment of Data Capture Conditions Effect on Reverse Electrodialysis Process Using a DC Electronic Load


Jesus Nahum Hernandez-Perez [1,2], Marco Antonio Hernández-Nochebuena [3], Jéssica González-Scott [1], Rosa de Guadalupe González-Huerta [1], José Luis Reyes-Rodríguez [1] and Alfredo Ortiz [2,*]

[1] Instituto Politécnico Nacional—ESIQIE, Laboratorio de Electroquímica, UPALM, Mexico City 07738, Mexico; jesus-nahum.hernandez@alumnos.unican.es (J.N.H.-P.); jgonzalezs1704@alumno.ipn.mx (J.G.-S.); rgonzalezh@ipn.mx (R.d.G.G.-H.); jlreyes@ipn.mx (J.L.R.-R.)

[2] Department of Chemical and Biomolecular Engineering, Universidad de Cantabria, Av. Los Castros 46, 39005 Santander, Spain

[3] Instituto Mexicano del Transporte, Pedro Escobedo 76703, Mexico; marco.hernandez@imt.mx

* Correspondence: alfredo.ortizsainz@unican.es; Tel.: +34-942-200-870



**Abstract:** Reverse electrodialysis (RED), an emerging membrane-based technology, harnesses salinity gradient energy for sustainable power generation. Accurate characterization of electrical parameters in RED stacks is crucial to monitoring its performance and exploring possible applications. In this study, a DC electronic load module (DCELM) is implemented in a constant current condition (CC mode) for characterization of lab scale RED process, using a RED prototype in-house designed and manufactured (RU1), at different data capture setups (DCS), on which the total number of steps for data capture ($NS$) and the number of measurements per step ($\rho$) are the parameters that were modified to study their effect on obtained electrical parameters in RED. $NS$ of 10, 50, and 100 and $\rho$ of 10 and 20 were used with this purpose. The accuracy of resulting current and voltage steps can be enhanced by increasing $NS$ and $]\rho$ values, and according to obtained results, the higher accuracy of resulting output current and voltage steps, with low uncertainty of the average output steps (AOS) inside the operational region of power curve, was obtained using a DCS of $NS = 100$ and $\rho = 20$. The developed DCELM is a low-cost alternative to commercial electronic load devices, and the proposed methodology in this study represents an adaptative and optimizable CC mode characterization of RED process. The results obtained in this study suggest that data capture conditions have a direct influence of RED performance, and the accuracy of electrical parameters can be improved by optimizing the DCS parameters, according to the required specifications and the scale of RED prototypes.

**Keywords:** salinity gradient energy; reverse electrodialysis; red characterization; power production



**Citation:** Hernandez-Perez, J.N.; Hernández-Nochebuena, M.A.; González-Scott, J.; González-Huerta, R.d.G.; Reyes-Rodríguez, J.L.; Ortiz, A. Assessment of Data Capture Conditions Effect on Reverse Electrodialysis Process Using a DC Electronic Load. *Energies* **2023**, *16*, 7282. https://doi.org/10.3390/en16217282

Academic Editor: Svetlozar G. Velizarov

Received: 10 October 2023
Revised: 20 October 2023
Accepted: 23 October 2023
Published: 26 October 2023




## 1. Introduction

Humanity is currently searching for alternative energy sources to fossil fuels, to reduce $CO_2$ emissions and other greenhouse gases that are producing the climate change phenomenon [1,2]. Among several renewable energies sources, salinity gradient energy (SGE) is widely spread along littorals and coasts all over the world, wherever a freshwater stream is discharged into the ocean. The amount of thermodynamic energy released from the mixing process ($\Delta G_{mix}$) is proportional to the concentration gradient and the chemical potential difference between the initial state and the mixed state of the solutions [3,4]. SGE has remarkable advantages regarding other renewable energies, some of which are related to its availability in both diurnal and nocturnal periods over the year if the employed feed waters become from natural sources [5,6], the possibility of regulate the amount of power generated with process parameters without $CO_2$ emissions and the fact that it can be obtained from residual streams of other process, such as reverse osmosis brines or municipal waste water treatment plants [7,8]. However, only a fraction of theoretical potential of rivers and oceans can be recovered in a practical form, which is known as world technical potential and its value is around 0.2–1 TW, according to the employed technique for energy harvest and the concentration of feed solutions [3,9].

Reverse electrodialysis (RED) is an emerging technology that allows converting some part of the $\Delta G_{mix}$ of two water streams (with a different salt concentration) into electricity, through an ion exchange membranes (IEM) stack and redox reactions. Figure 1 shows a schematic representation of a RED unit, which is composed of a membrane stack and the electrode system [10–12].

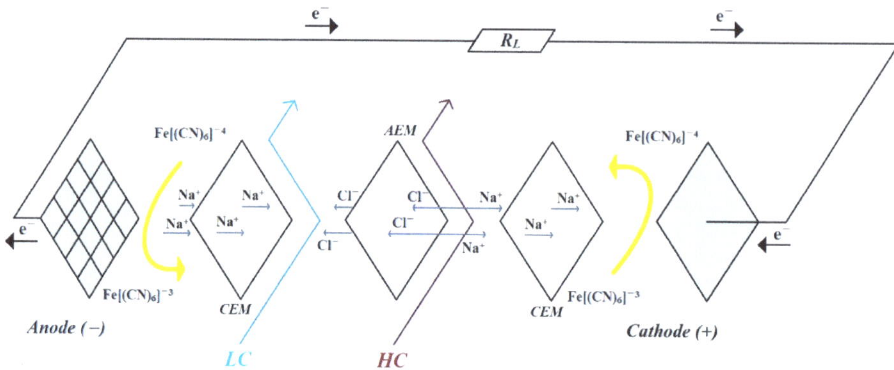


**Figure 1.** Schematic diagram of a RED unit.

In RED, the cation exchange membrane (CEM) and the anion exchange membrane (AEM) are stacked alternately between two electrodes (anode and cathode), and the intermembrane space is flooded with high concentration (HC) and low concentration (LC) solutions, which flow in sequence one after the other. By this, a membrane cell pair (or simply just a cell) is formed.

A RED stack normally uses a large number of cells (around dozens to hundreds) [5,11] and when an external load ($R_L$) is connected to the RED unit, the electrode system transforms the available electromotive force into electricity, using charge transfer process on the electrodes surfaces [11,13–15]. As soon as ions move through the membrane stack, they encounter the electrode rinse solution (ERS) at the electrode compartment, which is recirculated across the end plates of the RED unit and is normally composed of a highly reversible redox pair [16,17]. As a result, electrons can be transferred from the anode to the cathode across the equivalent circuit formed (a representation of equivalent circuit is provided in Figure S1).

*RED Process Parameters and Characterization*

To obtain RED process performance curves, several $R_L$ values must be used to determine the maximum power output that the system can generate to cover an electrical service or application. The Nernst equation can be used to obtain the theoretical electromotive force per cell ($E_{Cell}$) and of the whole stack ($OCV_{Theo}$) [3,5,10,18]. Equation (A1) of Appendix A shows the Nernst equation and its parameters.

The amount of energy that can be obtained from a RED unit is determined by the output voltage (which is the actual potential that the RED device delivers to $R_L$), at a determined current demand condition, and the internal resistance ($R_i$) of the RED unit. In principle, $R_i$ is composed of the ohmic and non-ohmic components on the membrane stack resistance ($R_{Stack}$) and the electrode system resistance ($R_{Elec}$) [5,11,13]. Normally $R_{Stack}$ represents the main contribution to $R_i$, while $R_{Elec}$ is not critical if the proper electrode material is selected and if the stack is composed of several cells [5,17]. Equations (A3) and (A4) of Appendix A describe the $R_i$ components.

In RED, several variables have a direct influence on $OCV_{Theo}$, $R_i$, and (consequently) the obtained power density ($P_d$). The input parameters can be classified into those that depend on "feed parameters" and those that depend on "RED unit parameters" [5,11,18,19]. Figure 2a shows a schematic representation of the RED process parameters classification.

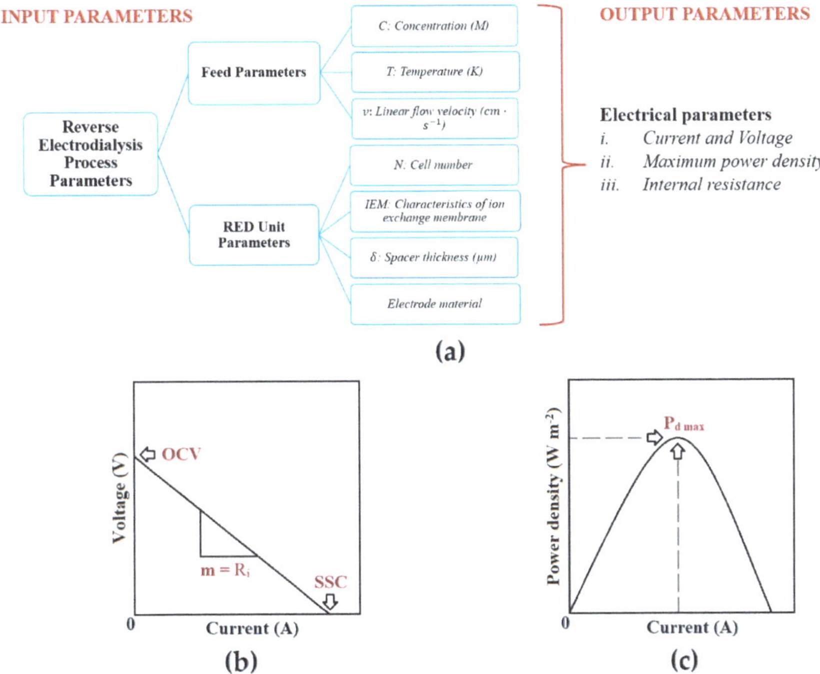


**Figure 2.** (**a**) Considered parameters in reverse electrodialysis process; (**b**) voltage vs. current relationship (polarization curve) and (**c**) power density vs. current relationship (power curve) obtained in a typical RED process.

The feed parameters describe the properties of feed solutions (concentration, temperature, and flow regime), while RED unit parameters are referred to the characteristics of the RED stack and the electrode system (cell number ($N$), properties of IEM used to conform the stack (ion exchange capacity, permselectivity, intrinsic resistance, etc.), the intermembrane space ($\delta$), and the electrode material) [5,11,18]. The linear flow velocity ($v$) can be defined as the fluid velocity inside one flow compartment formed by the spacer and is related to the residence time and hydrodynamic behavior of the feed solutions. Based on the stack dimensions, this parameter can be obtained for determined volumetric flow conditions [11]. Equations (A12) and (A13) of Appendix A describes how to calculate $v$. The proper selection of the electrode material assures that the electron transfer process at the surface will be reversible at the lowest overpotential possible [13,17]. Although the membrane properties are normally considered as the limiting parameters of a RED unit, the rest of the components also have a relevant effect on $E_{Stack}$ and $R_i$ on a different level. In the case of a RED operation, the apparent permselectivity ($\bar{\alpha}$) can be considered for both CEM and AEM as an average parameter that describes in a general form the ability of the membranes to allow the pass to counter-ions and inhibit it for co-ions [5,20], as can be seen in Equation (A14) of Appendix A.

The output parameters can be referred to as electrical parameters. They describe the current–voltage response generated by the RED unit at different $R_L$ conditions. The slope of the output voltage ($U$) vs. output current ($I$) relationship (also called the polarization curve) represents an approximation of $R_i$, as can be seen in Figure 2b. The open circuit voltage (OCV) condition occurs when the $R_L$ value is very high concerning $R_i$, so electrons cannot be transferred from the anode to the cathode across the circuit. The short circuit current (SCC) is a condition where $R_L$ is close to zero, so $I$ have a maximum intensity value while $U$ drops close to zero. The $P_d$ can be defined as the product of current times voltage, divided by the membrane active area (as is described in Equations (A10) and (A11) of Appendix A), and can be represented as a power curve (Figure 2c) by plotting $P_d$ vs. $I$. Thus, the top of the curve represents the maximum power density ($P_{d\,max}$) that can be generated under a determined input parameters condition.

To determine the electrical parameters of RED devices several methods are reported in the literature. Table 1 describes some examples of the mentioned characterization methods for RED devices reported in the literature. In all these methods, the purpose is to determine polarization curves, which can be used to obtain $R_i$ and experimental open circuit voltage ($OCV_{Stack}$), power curves which can be used to define the operation zone of the RED unit, and the maximum power density ($P_{d\,max}$) that can be produce.

**Table 1.** Characterization methods for RED process reported in the literature.

| Instruments Employed for RED Characterization | Data Capture Conditions | Ref. |
|---|---|---|
| Rheostat<br>Voltage: Multimeter UNI-T (UT71D))<br>Current: Fluke 45 Dual Display Multimeter | • OCV condition: 5 min<br>• Time for stabilization until data capture: 120 s | [21] |
| Rheostat<br>Voltage and Current: Multimeter Amprobe AM-520 | • Resistance Range: 0.5–100 $\Omega$ | [22] |
| Rheostat<br>Voltage: Datalogger (LabVIEW™, National Instruments)<br>Current: external Amperemeter | • Resistance Range: 0–22 $\Omega$<br>• OCV condition: ~5 min<br>• 1 h of operation from OCV until OCV/2<br>• Data capture at a frequency of 1 Hz | [23] |
| Rheostat: Five-decade resistance box (COPRICO)<br>Voltage: 3 1/2 Digital multimeter (Veleman, DVM760)<br>Current: 6 1/2 Digital multimeter (Agilent, 34422A) | • OCV, Ri and Pd max were determined by means of linear regression and parabolic correlation | [24] |
| Potentiostat/Galvanostat (HAB-151). Recorded data were processed using a data logging system (midi LOGGER GL200, GRAPHTEC Co.) | • Current range: From OCV until SCC<br>• Scan rate: 0.4 mA s$^{-1}$ | [25] |
| Potentiostat/Galvanostat (Ivium Technologies) in the galvanostatic mode | • Current range: From OCV until SCC<br>• Scan rate: 2 mA s$^{-1}$ | [26] |
| Potentiostat/Galvanostat Metrohm Autolab PGSTAT302N equipped with an FRA module | • Current range: 0–40 A m$^{-2}$<br>• Current steps: 17<br>• Time per step: 40 s | [27] |
| Potentiostat/Galvanostat (Ivium Technologies) using a chronopotentiometric method | • Current range: 0–40 A m$^{-2}$<br>• Current steps: 20<br>• Time per step: 30 s | [28] |
| Electronic load in CC mode (Chroma Systems Solutions 63103A) | • OCV condition: ~5 min<br>• Current rate: 0.025 A<br>• Values were obtained until the system reach a steady-state | [10] |
| Electronic load in CC mode (Chroma Systems Solutions 63103A) | • Current range: 0–1 A<br>• Values were obtained until the system reach a steady-state | [29] |

**Table 1.** *Cont.*

| Instruments Employed for RED Characterization | Data Capture Conditions | Ref. |
|---|---|---|
| Electronic load in CC mode (Chroma Systems Solutions 63103A) | • Values were obtained until the system reach a steady-state | [8] |
| Electronic load in CC mode (Chroma Systems Solutions 63103A) | • OCV condition: ~5 min <br> • Current rate: 0.025 A <br> • Values were obtained until the system reach a steady-state | [19] |
| Electronic load (PLZ 164 W, Kikusui electronics corp.) in I–V mode and CC mode | • Current range: From OCV until SCC <br> • I–V mode, Scan rate: 8.1 mA s$^{-1}$ <br> • CC mode, Current steps: 7, Time per step: 40 min | [7] |

In some works, a rheostat with fixed electrical resistance values is connected to the RED unit for a defined time condition, and the corresponding values of output voltage and current are recorded using an external voltmeter and amperemeter or by using a data logger for data capture [21–24]. Although this method is the simplest and cheapest way for RED characterization, there is no way to control or limit the voltage or the current that the load consumes. The resolution of the resulting polarization and power curves depends on the number of resistances used as well as if their values are suitable to the power supply capacity of the RED unit, which depends on its internal resistance. In other words, if the RED input parameters changes, the selected external resistance values may not be suitable for the new experimental conditions, so adapting to changes in test requirements with fixed resistors is a time-consuming task that requires many resistors. Avci et al. [24] used several fixed resistances for RED characterization, employing a linear regression method to determine $OCV_{Stack}$ as also $R_i$. However, since the selected resistance values, only a small portion of the power curve was experimentally determined, and a quadratic correlation was used to estimate the entire power curve and the corresponding $P_{d\,max}$ value. Although this method represents a valid approximation, $P_{d\,max}$ obtained does not represent a precise experimental value because of the limited experimental points used to obtain the quadratic correlation, which was a consequence of the selected resistance values for characterization.

Other authors have reported the use of potentiostat/galvanostat to perform a current scan from OCV condition to SCC condition at a defined scan rate, measuring the corresponding voltage values [25,26]. Moreover, potentiostat/galvanostat has been employed using chronopotentiometric methods defining a current demand range, divided in several steps at a specific time per step, and by this evaluate the output voltage at every current demand step [27,28]. The use of potentiostat/galvanostat is a more precise method to obtain a high resolution polarization and power curves, since the whole current–voltage relationship from OCV until SCC conditions can be obtained, at the same data capture conditions on every test. Still, the cost of these devices is considerably higher than fixed resistances method since there are normally used to perform several electrochemical tests besides power supply devices characterization.

Lastly, in other studies, a similar technique has been carried out by using an DC electronic load, where a current demand condition can also be defined. A DC electronic load is a programable test instrument designed to characterize DC power supplies by emulating multiple load profiles, offering high flexibility to adapting to changes on experimental conditions. Normally, these devices have several modes of operation, were the most common are constant current condition (CC mode), constant voltage condition (CV mode), constant resistance condition (CR mode), and constant power condition (CP mode) [30]. In case of RED characterization, the most relevant modes reported in the literature are CC mode, where a current demand is established and keep it constant until a new value is requested, and also changing current condition (I–V mode), where the demanded current is gradually increased at a defined scan rate [7,8,10,19,29].

While potentiostat/galvanostat and DC electronic load devices can be programmed to set data capture conditions on every experiment, the fixed resistances method requires an appropriate data capture software, which is not always reported in the literature and normally this task is performed manually by the operator using a digital multimeter, increasing the uncertainty related to human random errors in the results.

Focusing on characterization using a DC electronic load, while I–V mode is normally used for lab scale characterization of the RED process, where several process parameters are usually compared without considering an equilibrium condition, CC mode results represents a behave of a steady state regime in a long-term operation, which has its relevance for bench scale and pilot plant scale tests [7]. Nevertheless, as can be seen from Table 1, until now there is no direct or undirect consensus on employed method used for RED process characterization, which can lead to a degree of uncertainty in the comparison of results between different groups, despite the use of same input parameters.

In this study, the influence of data capture conditions on a lab scale RED process were analyzed by means of a DC electronic load module (DCELM), which works as an actuator/data capture system, allowing to the reducing of human random error on data capture compared to conventional fixed electrical resistance methods, as also representing a low-cost alternative to commercial potentionstat/galvanostat and electronic load devices. The system was developed on an Arduino platform, controlled by an interface in MATLAB® (Ver. R2022b) and operated in the CC mode. The total number of steps for data capture ($NS$) and the number of measurements per step ($\rho$) are the data capture setup (DCS) parameters that were modified to compare their influence on accuracy and uncertainty of the determined electrical parameters in RED. The described method in this study represents a quick and reliable tool for RED process analysis, which can used on lab scale conditions, as well as be optimized and adapted to steady state experiments.

## 2. Materials and Methods

### 2.1. Reverse Electrodialysis Unit Design (RU1)

A RED unit was designed and built, based on schematics and diagrams available in the literature [20,31], using Nylamid M as base material to conform the end plates of the prototype. A picture of the developed RED prototype (RU1) is presented in Figure S2. Figure S2a correspond to the manufactured endplates while Figure S2b correspond to an image of the prototype before assembly. The stack was composed of five cells and the RED unit parameters used for the experiments are summarized in Table 2.

**Table 2.** Considered RED unit parameters.

| RED Unit Parameters | RU1 |
| --- | --- |
| Effective area per cell | $0.0049\ m^2$ |
| Effective area of the stack | $0.0245\ m^2$ |
| Cell number ($N$) | 5 |
| Intermembrane distance ($\delta$) | 255 μm |
| CEM (outer membrane) | Nafion® 324 (Dupont, Wilmington, DE, USA) |
| CEM (stack) | Fuji Type 10 CEM (Fujifilm, Tilburg, The Netherlands) |
| AEM (stack) | Fuji Type 10 AEM (Fujifilm) |
| Electrode material | Pt/Ti mesh electrode |
| Torque applied | 2.5 N·m |

The effective area per cell was $0.0049\ m^2$. The intermembrane distance was defined by a silicon gasket with a PES mesh-type spacer of the same thickness ($\delta = 255$ μm). As for IEM, Fuji Type 10 membrane (Fujifilm Manufacturing Europe BV, the Netherlands) was used as CEM and AEM in the stack, while Nafion®324 (Dupont, Wilmington, DE, USA) was used as an outer CEM to reduce the permeation of the electrode rinse solution (ERS) from the electrode compartment to the stack, according to Scialdone et al. [16]. The properties of the employed IEM in the stack are reported in Table S1.

The ERS was composed of 0.05 M $K_4[Fe(CN)_6]$, 0.05 M $K_3[Fe(CN)_6]$ (Aldrich, purity > 99%) and 0.25 M NaCl (Fermont, Mexico City, Mexico. Composition > 99.5%) as supporting electrolyte in deionized water (Fermont, Mexico. Specific Conductance: $1.8 \times 10^{-6}\ \Omega^{-1}\ cm^{-1}$). A pH of 7.22 was determined for the ERS using a pHmeter PC45 (Conductronic, Puebla, Mexico). Titanium mesh with platinum coating (Pt/Ti mesh) was employed as working electrode material. The electrodes were prepared by sticking Pt/Ti mesh (Fuellcellstore, College Station, TX, USA) and an $\frac{1}{4}$ inch Ti bar (used as a connector) with a conductive epoxy resin SG-3100 S (MG Chemicals, Burlington, ON, Canada). Once the resin was applied, the electrodes were heated until 65 °C for 3 h in an oven (Memert, Schwabach, Germany), and they were cooled at room temperature. Next, they were covered by a nonconductive epoxy resin to capsulate the conductive epoxy resin area. This has the purpose of isolating and protecting the joint from corrosion. Figure S3 shows a picture of in-house built electrodes before their use. The electrodes were assembled to the end plates and sealed with an acrylonitrile–butadiene–styrene (ABS) solution (ABS saturated in ketone) and left to dry for 24 h.

### 2.2. DC Electronic Load Module for RED Applications

A DC electronic load module (DCELM) was designed and integrated for RED applications (a picture of the DCELM is provided in Figure S4). The DCELM is a current demand emulator device and data acquisition system designed to characterize reverse electrodialysis prototypes and power generation systems up to 0.5 Amperes. It is composed of three parts: a communication module, a data acquisition and signal conditioning module and a power module. Together, these components allow the generation of a voltage signal that activates the power circuit, which imposes a load to an external power supply to determine the voltage and current generated. The communication module allows communication between the DCELM and the user through an interface developed in MATLAB® for its control and data capture, as represented in Figure 3.

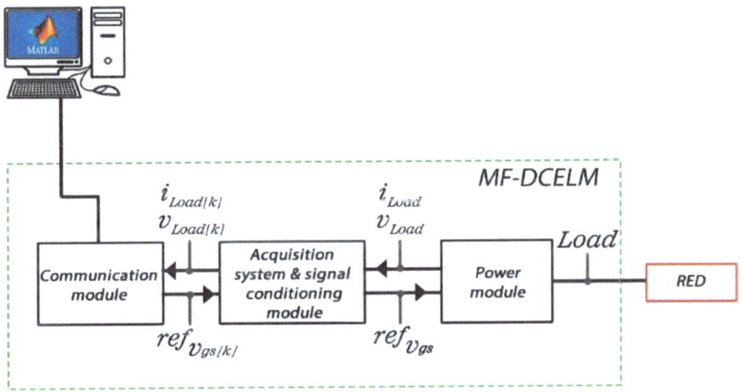


**Figure 3.** Schematic diagram of DCELM.

The power module is based on a directly polarized N-channel metal-oxide-semiconductor field-effect transistor (MOSFET), configurated to operate on its Ohmic region and by this obtain $i_{load}$ and $V_{load}$ (which represent the output current and voltage values of the power generation device). The MOSFET has three terminals: source ($s$), drain ($d$), and gate ($g$). The $g$ terminal controls the transistor activation, the $d$ terminal is connected to the positive terminal of the power supply device, and the $s$ terminal is connected in serial to a current sensor conformed by a shut—resistance of 0.01 $\Omega$, connected with the negative terminal of the power supply device. This is represented in Figure 4a.

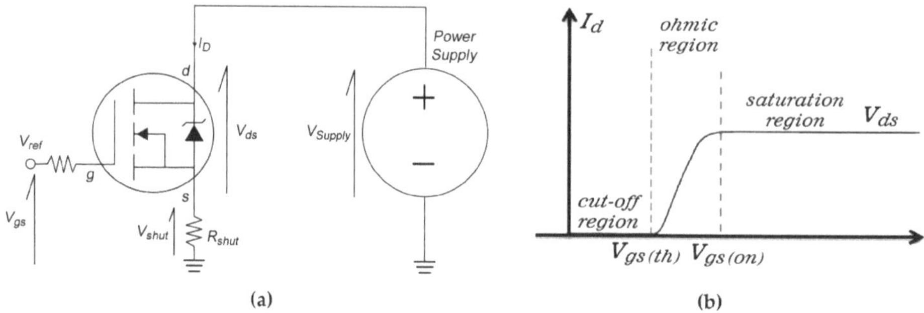



**Figure 4.** (**a**) Schematic diagram of the directly polarized circuit in the MOSFET and (**b**) operational regions and the current demand response of the MOSFET according to the voltage supply Vgs.

In general terms, the MOSFET has three operational modes, as it is described in Figure 4b.

Cut region: In this mode there is no current continuity, and the voltage difference between $d$ and $s$ terminals ($V_{ds}$) has a maximum value (OCV condition) and the voltage difference between $g$ and $s$ terminals ($V_{gs}$) is minor or equal to zero.

Ohmic region: In this region the MOSFET works as a nonlinear variable resistance whose value is a function of the $V_{gs}$ applied, allowing electrons to flow through the transistor and, by this, obtaining a theoretical drain current ($I_D$). This region is physically delimited by a threshold voltage in the $g$ terminal ($V_{gs}(th)$), which is the minimum voltage necessary for the electron flow to occur, and the saturation voltage in the $g$ terminal ($V_{gs}(on)$). As the $I_D$ increases, the resistance of the transistor decreases as well as the voltage $V_{ds}$. On the contrary, if $I_D$ decreases, the internal resistance of the transistor rises, increasing the voltage $V_{ds}$ at the transistor terminals.

Saturation region: At this condition $V_{gs} > V_{gs}(on)$, so the MOSFET overpasses the Ohmic region of operation and while $I_D$ remains constant, its value does not depend on the applied $V_{gs}$ anymore. The system conducts the maximum current value allowed by the transistor, where $V_{ds}$ is theoretically close to 0 V (SCC condition).

The mentioned parameters ($V_{gs}(th)$ and $V_{gs}(on)$) are proportionated by the transistor manufacturer and are characteristic of the MOSFET model employed (IRFZ44N in our experiments) and are reported in Table S2. The transconductance coefficient of the MOSFET was calculated and then was used to determine the relation between $V_{gs}$, which represents the supplied voltage to the MOSFET, and the $I_D$ demanded to the power generation device. By this, an operating region of the DCELM can be estimated by stablishing a $V_{gs}$ range between $V_{gs}(th)$ and $V_{gs}(on)$.

$I_D$ demanded can be divided into a total number of steps for data capture ($NS$), according to the desired number of experimental points for the characterization. To assure the stabilization of the output current and voltage signals among the transition between the current steps, a control step cycle was incorporated to set a number of repetitions at this condition, which is denominated number of measurements per step ($\rho$), in order to obtain an average value at the corresponding step condition in data capture.

On the other side, the data acquisition and signal conditioning module is integrated by a coupling circuit between the $g$ terminal and a 12- bits digital—analogical converter (DAC), which can generate voltage steps from $1.22 \times 10^{-3}$ V up to 1.5 V proportional to 500 mA, which is the maximum current demand of the device and by this generating the signals. It is composed by an ATmega2560 to generate a voltage reference ($ref_{V_{gs}}[\rho]$) through I2C communication with the analogical—digital converter (ADC), in order to measure the current and voltage response of the power generation device ($I_{Load}[\rho]$ and $U_{Load}[\rho]$). By this, the DCELM can obtain the output voltage of the RED unit at a defined current demand condition, dividing output current range into several steps and measurements per step and operate as a CC mode system.

### 2.3. The RED Process Characterization

Figure 5 shows a schematic representation of the lab scale RED process. Three working solutions were used: HC and LC solutions as feed waters and ERS. The feed waters flow in a continuous-flow mode through the system, while the ERS flow in a recycle loop between electrode compartments at the endplate and a dark container to isolate it from light.

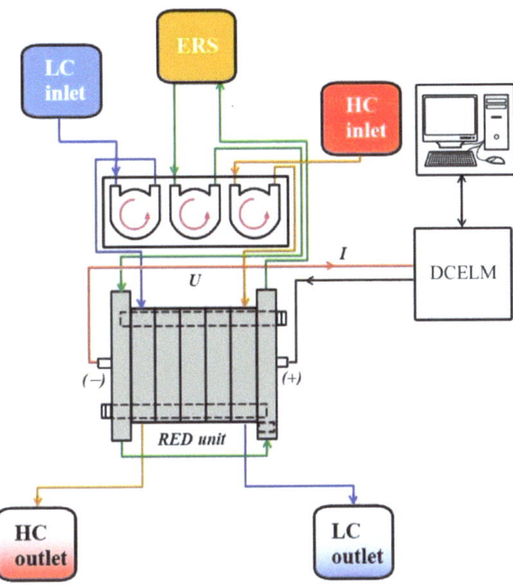


**Figure 5.** Schematic diagram of lab scale RED process.

All working solutions are fed through one peristaltic pump Masterflex L/S (Cole Parmer, USA) by assembling two pump heads to the same rotor. One of the pump heads has a double channel for HC and LC solutions, so the volumetric flow of each stream is the same, while the other has a single channel for ERS. Figure S5 shows a picture of the lab scale experimental setup of the RED test bench, on which the three main components are indicated: the working solutions system, the RED unit, and the DCELM. The feed parameters are shown in Table 3.

**Table 3.** Considered feed parameters.

| Feed Parameters | RU1 |
| --- | --- |
| High concentration solution (HC) | 0.5133 M |
| Low concentration solution (LC) | 0.0171 M |
| Electrode rinse solution (ERS) | $[Fe(CN)_6]^{-4}/[Fe(CN)_6]^{-3}$ 0.05 M/0.05 M and NaCl 0.25 M as supporting electrolyte |
| Temperature ($T$) | 298 K |
| Linear flow velocity ($v$) | 1.0 cm s$^{-1}$ (46 mL min$^{-1}$) |

Pure NaCl (Fermont, Mexico. Composition > 99.5%) solutions were used for all the experiments. The concentrations selected represent synthetic river water (0.0171 M) and synthetic seawater (0.5133 M) [32]. All solutions were prepared using deionized water (Fermont, Mexico. Specific Conductance: $1.8 \times 10^{-6}$ ohm$^{-1}$ cm$^{-1}$). For temperature control, a hot plate magnetic stirrer C-MAG HS7 (IKA, Königswinter, Germany) and glass thermometers were employed.

The DCELM was connected to the RED unit by a DB9 adaptor with alligator clips. Data capture was performed at a frequency of 0.5 Hz, equivalent to 2 s per measurement. In order to evaluate the differences on data capture setup (DCS) conditions, RED experiments were performed in the Ohmic region of the MOSFET. The resulting current output is divided in a total number of steps for data capture (*NS*) of 10, 50, and 100 with a number of measurements per step ($\rho$) of 10 and 20 for each DCS. For all experiments, a temperature of 298 K was used, the pump was calibrated to operate at a linear flow velocity of 1.0 cm·s$^{-1}$ and the system was operated for ten minutes at OCV conditions until the voltage signal was completely stable. After that, data capture begins and the output current and voltage raw data are recorded for its processing and plotting. The experiments were repeated three times for every experimental condition described.

### 2.4. Data Processing of RED Results

Based on the total number of steps for data capture (*NS*) and the number of measurements per step ($\rho$) selected on the DCS, the data raw captured were analyzed as follows. An average value of the output current and voltage raw data was calculated for each step condition, according to the $\rho$ selected. For the first step condition, the average values of current and voltage were obtained as is described in Equation (1):

$$M_s = \frac{\sum_{i=1}^{\rho} M_i}{\rho} \tag{1}$$

where $M_i$ and $M_s$ are the raw measurements and the resulting average measurement per step, respectively, (which can be output current or voltage) at the first step condition. Because of the structure of the MATLAB® script prepared for this work, for the following steps after first, one measurement less than the $\rho$ selected must be considered. Then, according to the $M_s$ of each step, considering the $\rho$ selected and the adjustment after the first step, the percentage of standard deviation per step (*%DS*) was calculated for each step as it is showed in Equation (2), and by this the repeatability of the measurements that composed a step was evaluated:

$$\%DS_i = \left( \left( \sqrt{\frac{\sum_{i=1}^{\rho}(M_i - M_s)^2}{\rho - 1}} \right) \div M_s \right) \times 100 \tag{2}$$

To estimate the general uncertainty related to the repeatability of measurements per step inside the operational region of the RED unit, a general percentage of standard deviation per step (*%GDS*) was considered as an average of individual *%DS* values between a selected output current range.

$$\%GDS = \frac{\sum_{i=1}^{S}(\%DS)_i}{S} \tag{3}$$

where $S$ is the number of steps considered for analysis and its value depends on the DCS selected and the resulting polarization and power curves. The $S$ values are described in Section 3 for every set of experiments. After this, the average general percentage of standard deviation per step (*%AGDS*) was obtained according to Equation (4), where $E$ is the number of experiments performed at a defined condition, in order to evaluate the average general uncertainty of the obtained output steps for all experiments performed under a certain condition.

$$\%AGDS = \frac{\sum_{i=1}^{E}(\%GDS)_i}{E} \tag{4}$$

The resulting average measurement per step, obtained by means of Equation (1), represents the obtained output current and voltage values per step and were used to obtain the corresponding $P_d$ values, using Equations (A10) and (A11) of Appendix A. After this, an average output step (*AOS*) value was calculated as its own standard deviation for average output step (*DAOS*) by means of Equations (5) and (6), respectively:

$$AOS = \frac{\sum_{n=1}^{E} M_n}{E} \tag{5}$$

$$DAOS = \sqrt{\frac{\sum_{n=1}^{E} (M_n - AOS)^2}{E - 1}} \tag{6}$$

where $M_n$ are the resulting output values per step and it can be referred to $I$, $U$, or $P_d$; $E$ is the number of experiments performed. The *AOS* values were used to plot the corresponding polarization and power curves at each DCS tested and the *DAOS* was used to evaluate the individual uncertainty related to the repeatability of the *AOS*, which values are expressed in terms of error bars. *DAOS* value was also obtained in terms of percentage (*%DAOS*), and a general percentage of standard deviation for average output steps (*%GDAOS*) was obtained.

$$\%GDAOS = \frac{\sum_{i=1}^{S} (\%DAOS)_i}{S}. \tag{7}$$

where $S$ is the number of steps considered for analysis selected in Equation (3), based on the obtained output current range. This parameter was used to evaluate the general uncertainty related to the repeatability of results in performance curves inside the operational region of the RED unit.

### 3. Results and Discussion
*Evaluation of Data Capture Setup*

The output current and voltage raw data obtained using RU1 at the described feed parameters in Table 3, were recorded by the MATLAB® interface, and then exported for processing and plotting to evaluate the influence of DCS on accuracy of electrical parameters. Table 4 describes the DCS studied and compared.

**Table 4.** Data capture setup employed for RED experiments.

| Data Capture Setup (DCS) | | Steps Considered for Analysis (S) |
| --- | --- | --- |
| N° Steps (NS) | Number of Measurements per Step ($\rho$) | |
| 10 | 10 | 4 |
| 10 | 20 | |
| 50 | 10 | 15 |
| 50 | 20 | |
| 100 | 10 | 30 |
| 100 | 20 | |

The DCELM was operated only in the Ohmic region of the MOSFET and Figure 6 shows the measurements that conform the steps considered for analysis (*S*) of one representative experiment at each DCS, between an output current range of approximately 0.007–0.06 A. The raw experimental data are grouped in different colors according to the obtained current and voltage step. For every DCS, two plots are shown: the left side describes current raw data vs. number of measurements while the right side describes voltage raw data vs. number of measurements. As can be seen, the resulting steps are delimited by the $\rho$ selected and one transition measurement between one step to the next it is present.

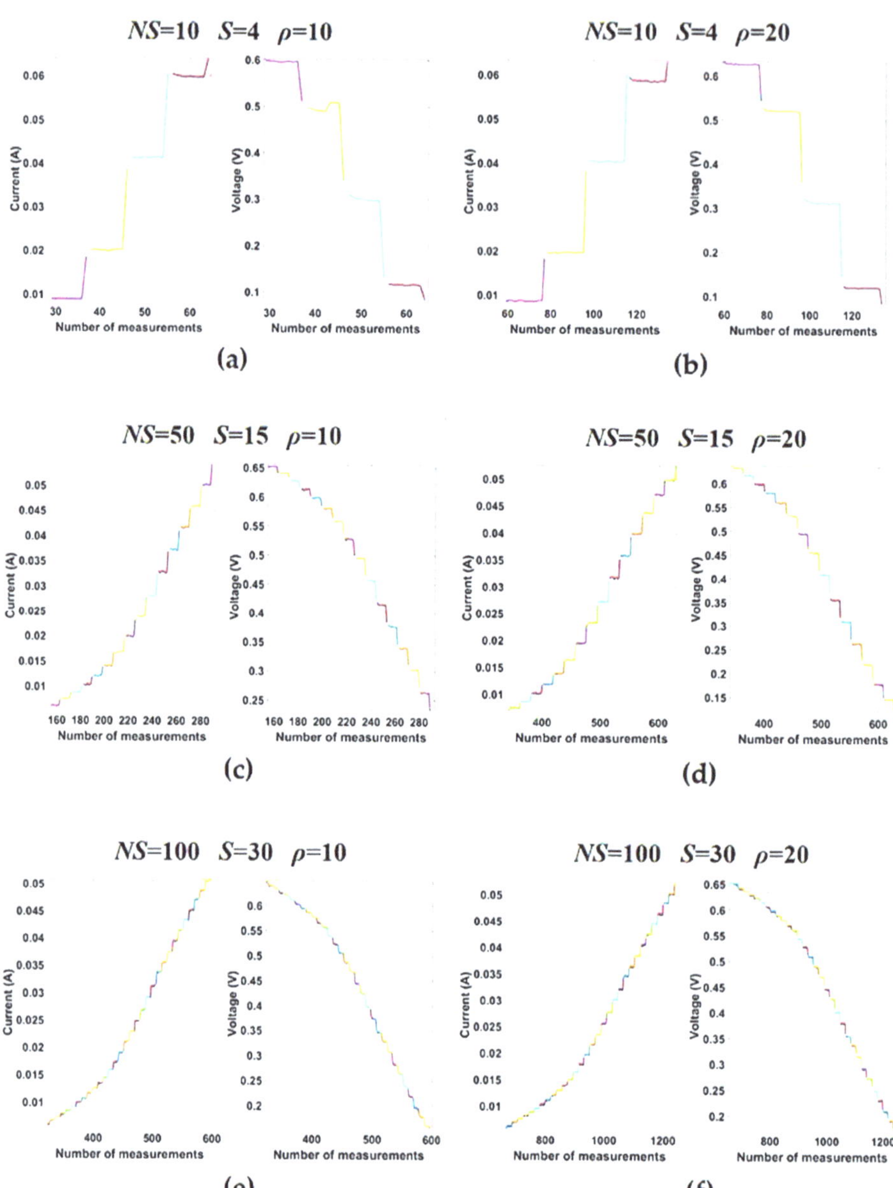

**Figure 6.** Obtained raw data for current (left side plots) and voltage (right side plots) vs. number of measurements, in a current range from 0.007–0.06 A, at different DCS: (**a**) $NS = 10$, $\rho = 10$; (**b**) $NS = 10$, $\rho = 20$; (**c**) $NS = 50$, $\rho = 10$; (**d**) $NS = 50$, $\rho = 20$; (**e**) $NS = 100$, $\rho = 10$ and (**f**) $NS = 100$, $\rho = 20$. ($NS$: Total number of steps for data capture; $S$: steps considered for analysis and $\rho$: Number of measurements per step). Data are grouped in different colors according to the corresponding current and voltage step.

In order to evaluate the influence of DCS on the repeatability of output current and voltage raw measurements per step, the %*DS* was calculated for every current and voltage step condition in a range of approximately 0.007–0.06 A for each experiment and then the

%*GDS* was obtained at this current range for every set of results. Later, the %*AGDS* of three experiments (*E* = 3) was obtained from the %*GDS* values and these results are presented in Table 5. According to them, the %*AGDS* decreases when a $\rho$ of 20 measurements per step is selected compared to a $\rho$ of 10 measurements per step, as well as if a higher number of steps is employed (10, 50, or 100 steps). This trend was attributed to the time it takes to the RED unit to achieve stability on its electromotive force, so as the $\rho$ increases, the uncertainty of obtained output steps will do it too, increasing the precision of results. In the same way, as a higher number of steps are used, the transition from one step to the next one will be shorter, allowing to the RED unit achieve a stability condition in a shorter time, enhancing repeatability and by this the precision of the resulting steps.

**Table 5.** Average general percentage of standard deviation per step (%*AGDS*) for current and voltage steps, obtained at each DCS.

| %*AGDS* | *NS* = 10<br>$\rho$ = 10 | *NS* = 10<br>$\rho$ = 20 | *NS* = 50<br>$\rho$ = 10 | *NS* = 50<br>$\rho$ = 20 | *NS* = 100<br>$\rho$ = 10 | *NS* = 100<br>$\rho$ = 20 |
|---|---|---|---|---|---|---|
| **Current** | 13.38 | 9.97 | 4.83 | 3.07 | 2.20 | 1.50 |
| **Voltage** | 9.08 | 5.90 | 1.88 | 1.70 | 1.30 | 1.01 |

For visualization and comparison of the electrical parameters at each DCS, the resulting output values per step, obtained by means of Equation (1), were used to determine the *AOS* with the results of three experiments (*E* = 3) using Equation (5). The calculated *AOS* were used to construct polarization and power curves at each DCS, as is shown in Figure 7. With the aim of assessing the individual uncertainty related to the repeatability of the *AOS*, the corresponding *DAOS* was calculated for every step plotted of each data set (the values of which are represented by error bars in Figure 7), according to Equation (6).

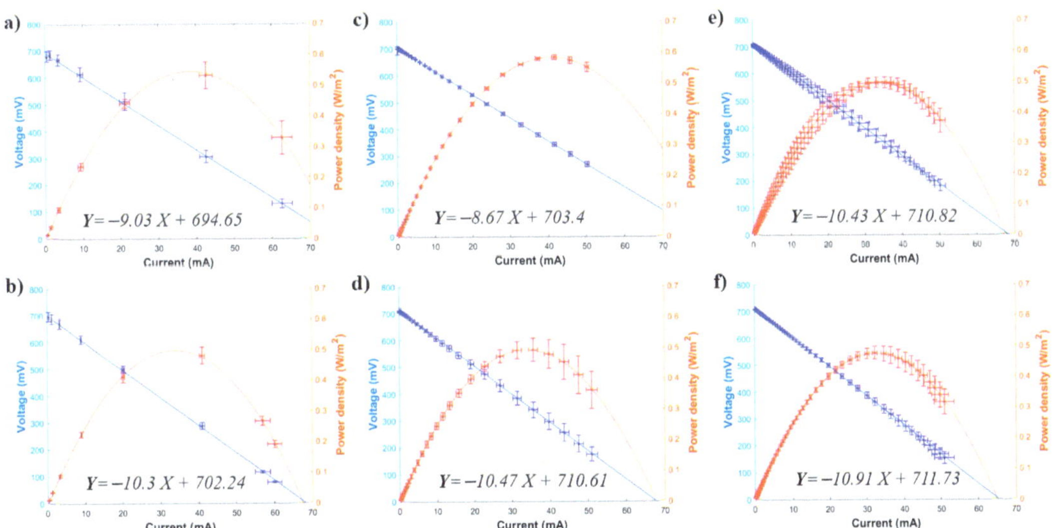


**Figure 7.** Voltage (mV) vs. current (mA) (blue color, **left**) and power density (W m$^{-2}$) vs. current (mA) (red color, **right**) obtained from the average output steps, at different DCS: (**a**) *NS* = 10, $\rho$ = 10; (**b**) *NS* = 10, $\rho$ = 20; (**c**) *NS* = 50, $\rho$ = 10; (**d**) *NS* = 50, $\rho$ = 20; (**e**) *NS* = 100, $\rho$ = 10 and (**f**) *NS* = 100, $\rho$ = 20.

Linear regression was used to calculate the slope of all polarization curves and by this obtain the internal resistance of RU1 under the described RED process parameters, according to Equation (A7) of the Appendix A. The corresponding linear regression and resulting values are also indicated. According to these, for all measurements conditions an $OCV_{stack}$ between 694–711 mV was obtained, as well as an $R_i$ between 8.6–10.9 $\Omega$. As for $P_{d\,max}$, the obtained results were between 0.47–0.58 W m$^{-2}$. The output current range for analysis (approximately 0.007–0.06 A) was selected based on the $P_{d\,max}$ obtained in the resulting power curves (Figure 7), who was around 40 mA, so our analysis focused on most of the steps before the top of the power curve (which is the maximum power generation zone) and only a few after, since from a technical perspective, it results more convenient to operate the RED unit at lower currents than at higher to generate the same power, because the system is more stable and the deviation values are lower. The $\%GDAOS$ for output current, voltage, and power density steps were obtained, at the mentioned output current range, using Equation (7). These values, as the rest of the considered electrical parameters, are indicated in Table 6.

**Table 6.** Electrical parameters and general percentage of standard deviation for average output steps ($\%GDAOS$) obtained at each DCS.

| Electrical Output Parameters | $NS = 10$ $\rho = 10$ | $NS = 10$ $\rho = 20$ | $NS = 50$ $\rho = 10$ | $NS = 50$ $\rho = 20$ | $NS = 100$ $\rho = 10$ | $NS = 100$ $\rho = 20$ |
|---|---|---|---|---|---|---|
| OCV$_{Stack}$ (mV) | 694.65 | 702.24 | 703.4 | 710.61 | 710.82 | 711.73 |
| OCV$_{Theo}$ (mV) | | | 838.4 | | | |
| P$_{d\,max}$ (W m$^{-2}$) | 0.5323 | 0.4806 | 0.5813 | 0.4928 | 0.4956 | 0.4766 |
| R$_i$ ($\Omega$) | 9.03 | 10.3 | 8.67 | 10.47 | 10.43 | 10.91 |
| $\bar{\alpha}$ (%) | 82.85 | 83.75 | 83.89 | 84.75 | 84.78 | 84.89 |
| %GDAOS—$I$ | 5.78 | 2.17 | 1.62 | 4.63 | 10.46 | 2.08 |
| %GDAOS—$U$ | 7.55 | 3.49 | 1.27 | 5.44 | 5.18 | 3.94 |
| %GDAOS—$P_{d\,max}$ | 8.38 | 5.02 | 1.26 | 7.10 | 8.60 | 4.63 |

Theoretical open circuit voltage ($OCV_{Theo}$) was calculated using the Nernst equation and based on its value the apparent permselectivity ($\bar{\alpha}$) was obtained. Both equations can be consulted in Appendix A.

For all cases, the $OCV_{Stack}$ (and consequently $\bar{\alpha}$) increases as a greater $NS$ and $\rho$ are selected, moving towards the calculated $OCV_{Theo}$ and specific permselectivity values provided by the manufacturer (presented in Table S1), firstly because the RED device has twice the time and measurements per step to stabilize the output current and voltage signals, and secondly because it allows the system to have lower changes on the transition between one step to the next, improving the stabilization among these and decreasing the uncertainty of obtained output steps. Moreover, as a greater number of steps the resolution of the resulting polarization and power curves its enhanced, allowing a more refined identification of the operational region and the $P_{d\,max}$ value. This would be translated in a higher trueness linear regression since it is composed by a greater number of output steps, so the resulting Y interception and slope are more representative of the $OCV_{Stack}$ and $R_i$ values, respectively.

In accordance with the results presented on Figure 7 and Table 6, for polarization and power curves obtained at $NS = 100$, the operational region has the best resolution, allowing to identify more precisely the maximum power generation zone and the $P_{d\,max}$ value. When selected $\rho$ is increased at this $NS$, the uncertainty of resulting electrical parameters decreases notoriously in both $\%AGDS$ and $\%GDAOS$ values. In this sense, the lowest $\%AGDS$ of all DCS studied was obtained using $NS = 100$ and $\rho = 20$, and in case of $AOS$, at this DCS the resulting $\%GDAOS$ for $I$, $U$ and $P_{d\,max}$ were lower than 5%. Hence, by considering the precision of the resulting output current and voltage steps and the uncertainty of $AOS$, a DCS of $NS = 100$ and $\rho = 20$ represents the most accurate and reliable approximation of electrical parameters of RU1 using CC mode at the described RED process parameters, among the DCS studied for lab scale RED characterization.

On the other hand, when a DCS of $NS = 10$ and $\rho = 20$ was used, the obtained $P_{d\,max}$ value was 0.4806 W·m$^{-2}$, while at $NS = 50$ and $\rho = 20$ the obtained result was 0.4928 W·m$^{-2}$, even though the $R_i$ obtained by linear regression was lower at the first DCS mentioned (10.3 $\Omega$ and 10.47 $\Omega$, respectively). When the resulting polarization and power curves are compared, in the first case ($NS = 10$, $\rho = 20$) the low $NS$ employed on data capture does not allow to define with more precision the maximum power generation zone, so even when the reported $P_{d\,max}$ represents the highest value among the $AOS$ obtained at this DCS, when the continuous line that corresponds to basic quadratic fitting is observed, this experimental point is located at the right side of the top of power curve, so its value might result in a underestimation of the real $P_{d\,max}$ under the described RED process parameters. In the second case ($NS = 50$ and $\rho = 20$), maximum power generation zone is more well defined and the resulting highest value in power curve is a more representative approximation of $P_{d\,max}$ in RED at those conditions. Nevertheless, as can be seen in error bars of resulting plots at this DCS, the uncertainty of $AOS$ was larger, and values obtained by linear regression would be expected to be less accurate.

Furthermore, the selected number of measurements on every step must be considered given that, as greater $\rho$ selected, the $\%DS$ decreases along the operational region of the RED unit, which is traduced in a lower $\%AGDS$ for resulting output current and voltage steps and represents higher accuracy on the results used to construct polarization and power curves, the same as the resulting $OCV_{Stack}$ and $R_i$, consequently. This could be observed when a DCS of $NS = 50$ and $\rho = 10$ was selected, given that although the lowest $\%GDAOS$ values for $I$, $U$, and $P_{d\,max}$ were obtained, in this scenario the $\%AGDS$ for current and voltage steps was higher than by using a larger $NS$ or $\rho$ values. This suggests a poorer stabilization of the output current and voltage raw data, which would cause a higher uncertainty of the obtained steps, producing an overestimation of $P_{d\,max}$. In this sense, when the $NS$ and $\rho$ selected increases, estimated $R_i$ trends to rise and $P_{d\,max}$ to fall, and a lower $\%AGDS$ is obtained as is showed in Tables 5 and 6.

The obtained results suggest that data capture conditions have a direct influence on the obtained electrical parameters and its accuracy. Thus, even when the most accurate results, among the studied DCS, were obtained at $NS = 100$ and $\rho = 20$ according to the employed methodology, it is clear that by exploring more DCS the accuracy of results can be improved. In case of apparent permselectivity, despite the fact that this parameter describes in a general form the ability of the membranes to allow the pass to counter-ions and inhibits it for co-ions, assuming that both CEM and AEM have the same value, the obtained $\bar{\alpha}$ at $NS = 100$ and $\rho = 20$ ($\bar{\alpha} = 84.89\%$) was lower than the specific permselectivities values proportioned by the manufacturer (reported in Table S1). In this sense, according to the trend of $\bar{\alpha}$ results in Table 6, if the $NS$ and $\rho$ are increased this could lead to an improvement on stabilization and repeatability of resulting steps, which would be translated into a lower uncertainty and higher accuracy of linear regression values such as $OCV_{Stack}$. In another aspect, it must also be considered that the employed electrodes in RU1 were not a commercial type, but in-house built. This is relevant since, although commercial Pt/Ti mesh was selected as electrode material, the type of joint employed between the Pt/Ti mesh and the Ti connector, based on a conductive epoxy resin and a conventional epoxy capsulate, might considerably increase the resistance associated with the electrode system, and by this the $R_i$ of the RED unit. This would increase the overpotential of redox reactions, reducing the resulting output voltage on the load and by this reducing the obtained $OCV_{Stack}$ and $\bar{\alpha}$. One way to confirm this last would be to include reference electrodes between the stack and the Pt/Ti mesh electrodes, to determine the OCV condition generated only by the membrane stack without considering the losses in working electrodes, as has been proved in the literature [20,26].

On the other hand, the DC electronic load module (DCELM) developed for this study can be adapted for bench scale or pilot plant scale RED devices, such as others power supply systems, such as Li-ion batteries or fuel cells, taking into account the following requirements: The power module must be scaled according to the current demand range required by the new power supply system, considering that it would need a power dissipation system according to the new demand, which implies adding a forced air or liquid cooling subsystem. Another aspect to be considered is that, in the case of signal acquisition and conditioning module, the direct measurement on the shut resistance (Figure 4a) should be replaced by an indirect measurement, such as a Hall effect sensor, to isolate the system and protect the circuit against high voltages that may be obtain from the power supply system.

## 4. Conclusions

The influence of data capture conditions on electrical parameters in reverse electrodialysis (RED) were analyzed employing a DC electronic load module (DCELM), designed and build it for this purpose, and operated in a constant current condition (CC mode). The developed DCELM represents a low-cost alternative to commercial potentiostat/galvanostat and electronic load devices, whose parameters can be controlled to optimize the accuracy of RED characterization. Several data capture setups (DCS) were tested for characterization of lab scale RED process, using an in-house built RED unit prototype (RU1), under the same process parameters. The described methodology in this study evaluates the precision and uncertainty of experimental results, being a quick and reliable option for obtaining a more accurate estimation of the electrical parameters in the RED process on lab scale conditions. In this sense, more DCS might be explored to perform steady state regime tests in benchmark scale or pilot plant scale RED devices. According to results, the accuracy of the resulting current and voltage steps can be enhanced by increasing the total number of steps for data capture ($NS$), since the repeatability of the measurements within each step increases because the system undergoes fewer changes during the transition from one step condition to the next. Furthermore, as a larger $NS$ is selected, the resolution of the resulting polarization and power curves improves. This enables a clearer identification of the maximum power generation zone in the power curve, as well as more trueness and representative results obtained through linear regression, because the resulting polarization curves are constructed from a greater number of average output steps ($AOS$). As for the number of measurements per step ($\rho$), larger values provide the system with additional time for stabilization and a greater number of measurements for step composition, leading to a reduction in the uncertainty of each step, and, by this, increasing its precision and the precision of the $AOS$ consequently. Among the employed DCS in this study, a condition of $NS = 100$ and $\rho = 20$ represented the most accurate setup for data capture at the described experimental conditions using RU1, since the average general percentage of standard deviation per step ($\%AGDS$), of current and voltage steps, were the lowest from all DCS, and the general percentage of standard deviation for average output step ($\%GDAOS$) for output current ($I$), output voltage ($U$), and power density ($P_{d\,max}$) were below a 5%, obtaining a high precision and a low uncertainty of the determined electrical parameters inside the operational region of the RED unit. Obtained results in this study suggest that accuracy of electrical parameters can be improved by optimizing the DCS parameters, according to the required specifications and the scale of RED prototypes. Regarding this last, since data capture conditions have a direct influence on results, it may be necessary to define and agree a standardized and feasible methodology of characterization of RED process, which make possible a more representative comparison of results among different research groups.

**Supplementary Materials:** The following supporting information can be downloaded at: https://www.mdpi.com/article/10.3390/en16217282/s1, Figure S1: Equivalent circuit of RED process; Figure S2: RED prototype (RU1) (a) Endplate and stack design, (b) Prototype before assembly; Figure S3: In-house built Pt/Ti mesh electrodes used for RED experiments; Figure S4: DC electronic load module (DCELM); Figure S5: Lab scale experimental setup of the RED test bench. Components of test bench: 1. Low concentration (LC) solution container, 2. High concentration (HC) solution container, 3. Electrode rinse solution (ERS) container, 4. Double head peristaltic pump, 5. RED prototype (RU1), 6. DC electronic load module (DCELM), 7. MATLAB® interface, 8. Hot plate for LC solution and 9. Hot plate for HC solution; Table S1: Properties of IEM used for RED experiments; Table S2: Parameters of the MOSFET IRFZ44N.

**Author Contributions:** Conceptualization, J.N.H.-P. and M.A.H.-N.; methodology, J.N.H.-P., M.A.H.-N. and J.G.-S.; software, M.A.H.-N.; validation, R.d.G.G.-H. and J.L.R.-R.; formal analysis, J.N.H.-P., M.A.H.-N. and R.d.G.G.-H.; resources, R.d.G.G.-H.; data curation, J.N.H.-P. and M.A.H.-N.;

writing—original draft preparation, J.N.H.-P. and M.A.H.-N.; writing—review and editing, R.d.G.G.-H., J.L.R.-R. and A.O.; visualization, J.N.H.-P. and M.A.H.-N.; supervision, R.d.G.G.-H. and A.O.; project administration, R.d.G.G.-H.; funding acquisition, R.d.G.G.-H. All authors have read and agreed to the published version of the manuscript.

**Funding:** This research was funded by CONAHCYT-SENER/Sustentabilidad Energética through the Centro Mexicano de Inovación en Energías del Océano (CEMIE-Océano), Grant No. 249795 and by "Proyectos de desarrollo tecnológico o innovación para alumnos del IPN 2021". Jesus Nahum Hernandez-Perez would like to thank IPN innovation project for students 2021 and the CONACYT for the support provided for project CEMIE Océano 249795. In addition, Rosa de Guadalupe Gonzalez-Huerta would like to thank the support granted by the L'oreal Foundation, UNESCO and the Mexican Academy of Sciences.

**Data Availability Statement:** Not applicable.

**Acknowledgments:** The authors want to express their gratitude to "Instituto Politecnico Nacional" (IPN) in Mexico, for the use of their facilities and resources to complete this research work. On the same way, to the Department of Chemical and Biomolecular Engineering of the University of Cantabria for the supervision, supporting and guidance on the preparation of this work. Jesus Nahum Hernandez-Perez would like to thank IPN innovation project for students 2021 and the CONACYT for the support provided for project CEMIE Océano 249795. In addition, Rosa de Guadalupe Gonzalez-Huerta would like to thank the support granted by the L'oreal Foundation, UNESCO and the Mexican Academy of Sciences.

**Conflicts of Interest:** The authors declare no conflict of interest. The funders had no role in the design of the study; in the collection, analyses, or interpretation of data; in the writing of the manuscript; or in the decision to publish the results.

## Nomenclature

| Nomenclature | Description | Unit |
| --- | --- | --- |
| AOS | Average output step | A, V or $W{\cdot}m^{-2}$ |
| DAOS | Deviation of average output step | A, V or $W{\cdot}m^{-2}$ |
| E | Number of experiments performed | - |
| $E_{Cell}$ | Theoretical electromotive force per cell | V |
| I | Output current | A |
| $I_D$ | Theoretical drain current | A |
| $M_i$ | Raw measurements | A or V |
| $M_n$ | Resulting output values per step | A, V or $W{\cdot}m^{-2}$ |
| $M_s$ | Resulting average value at a determined step condition | A or V |
| N | Number of cells | - |
| NS | Total number of steps for data capture | |
| OCV | Open circuit voltage | V |
| $OCV_{Stack}$ | Experimental open circuit voltage of the stack | V |
| $OCV_{Theo}$ | Theoretical open circuit voltage of the stack | V |
| $P_d$ | Power density | $W{\cdot}m^{-2}$ |
| $P_{d\,max}$ | Maximum power density | $W{\cdot}m^{-2}$ |
| $R_{Elec}$ | Electrode system resistance | $\Omega$ |
| $R_i$ | Internal resistance of the RED unit | $\Omega$ |
| $R_L$ | Load resistance | $\Omega$ |
| $R_{Stack}$ | Membrane stack resistance | $\Omega$ |
| S | Number of steps considered for analysis | - |
| SCC | Short circuit current | A |
| T | Temperature | K |
| U | Output voltage | V |
| $v$ | Linear flow velocity | $cm{\cdot}^{-1}$ |
| $V_{ds}$ | Voltage difference between $d$ and $s$ terminals | V |
| $V_{gs}$ | Voltage difference between $g$ and $s$ terminals | V |
| $V_{gs}(on)$ | Saturation voltage in the $g$ terminal | V |
| $V_{gs}(th)$ | Threshold voltage in the $g$ terminal | V |
| $\Delta G_{mix}$ | Change in Gibbs Free energy upon solutions mixing | kJ |

| $\bar{\alpha}$ | Apparent permselectivity | % |
| $\delta$ | Intermembrane distance | μm |
| $\rho$ | Number of measurements per step | - |
| %AGDS | Average general percentage of standard deviation per step | % |
| %DAOS | Percentage of deviation for average output step | % |
| %DS | Percentage of standard deviation per step | % |
| %GDAOS | General percentage of standard deviation for average output step | % |
| %GDS | General percentage of standard deviation per step | % |

## Abbreviations

| Abbreviations | Description |
| --- | --- |
| ABS | Acrylonitrile—Butadiene—Styrene |
| ADC | Analogical—digital converter |
| AEM | Anion Exchange Membrane |
| CC—mode | Constant current condition |
| CP—mode | Constant power condition |
| CR—mode | Constant resistance condition |
| CV—mode | Constant voltage condition |
| CEM | Cation Exchange Membrane |
| $d$ | Drain terminal |
| DAC | Digital—analogical converter |
| DCELM | DC-Electronic Load Module |
| DCS | Data Capture Setup |
| EMF | Electromotive Force |
| ERS | Electrode Rinse Solution |
| $g$ | Gate terminal |
| HC | High Concentration |
| IEM | Ion Exchange Membrane |
| I–V mode | Changing current condition |
| LC | Low Concentration |
| MOSFET | Metal-Oxide-Semiconductor Field-Effect Transistor |
| PES | Polyether Sulphone |
| Pt/Ti mesh | Titanium mesh with platinum coating |
| RED | Reverse Electrodialysis |
| RU1 | RED Unit 1 |
| $s$ | Source terminal |
| SGE | Salinity Gradient Energy |

## Appendix A

The Nernst Equation (A1) describes the electromotive force (EMF) available by each IEM according to the salinity gradient value [10,11,33].

$$E_{CEM} = \alpha_{CEM} \cdot \frac{R \cdot T}{z \cdot F} \cdot Ln \frac{a_{HC_{Na^+}}}{a_{LC_{Na^+}}} = \alpha_{CEM} \cdot \frac{R \cdot T}{z_{Na^+} \cdot F} \cdot Ln \frac{\gamma_{HC_{Na^+}} C_{HC_{Na^+}}}{\gamma_{LC_{Na^+}} C_{LC_{Na^+}}} \tag{A1}$$

$E_{CEM}$ is the EMF generated across the CEM, $\alpha_{CEM}$ is the membrane permselectivity, $z$ the valence of the ionic specie considered ($z = 1$ for $Na^+$ and $Cl^-$), $R$ the universal thermodynamic constant ($8.314\ J \cdot mol^{-1}\ K^{-1}$), $T$ the temperature in K, $F$ the Faraday constant ($96\,485\ C \cdot mol^{-1}$), and $a_{HC_{Na^+}}$ and $a_{LC_{Na^+}}$ are the activities of $Na^+$ on the HC and LC solutions, respectively. This last values can be obtained as the product of activity coefficients ($\gamma_{HC_{Na^+}}$, $\gamma_{LC_{Na^+}}$), which value is equal to 1 for ideal solutions, and the concentration of sodium ion on HC and LC solutions ($C_{HC_{Na^+}}$, $C_{LC_{Na^+}}$) in $mol \cdot m^{-3}$ [10,11,33]. The Nernst equation can also be expressed in terms of $Cl^-$ for $E_{AEM}$, the addition of both $E_{CEM}$ and $E_{AEM}$ give as result the cell potential ($E_{CEM} + E_{AEM} = E_{Cell}$) [5,10,11]. The theoretical electrical potential of the RED unit at the open circuit voltage condition ($OCV_{Theo}$) can be obtained as by $E_{Cell}$ times the number of cells ($N$):

$$OCV_{Theo} = N \cdot E_{Cell} \qquad (A2)$$

The amount of energy that can be obtained on the RED unit its limited by the available EMF and its internal resistance ($R_i$). In principle, $R_i$ its conformed by the Ohmic ($R_{Ohm}$) and non-Ohmic ($R_{non\text{-}Ohm}$) components of the RED stack resistance ($R_{Stack}$) and the electrode system resistance ($R_{Elec}$). All these parameters are expressed in Ohms:

$$R_i = R_{Stack} + R_{Elec} = N \cdot R_{Ohm} + R_{non-Ohm} + R_{Elec} \qquad (A3)$$

The non-Ohmic component of $R_i$ is associated with diffusional boundary layer (DBL) effects and concentration changes across the intermembrane space. On the other hand, the Ohmic components (Equation (A4)) are referred to intrinsically resistance of CEM and AEM ($R_{CEM}$ and $R_{AEM}$) and the resistance of the flow channels ($R_{HCC}$ and $R_{LCC}$) in $\Omega \cdot cm^2$, the term $A_{mem}$ refers to effective area of membrane expressed in $cm^2$ [5,11,18].

$$R_{Ohm} = \frac{1}{A_{mem}} \cdot [R_{AEM} + R_{CEM} + R_{HCC} + R_{LCC}] \qquad (A4)$$

The resistance of the flow channels can be obtained using Equations (A5) and (A6), where $\delta$ is the intermembrane space (in cm), $C_{HC}$ and $C_{LC}$ are the molar concentrations of the high concentration and low concentration solutions ($mol \cdot cm^{-3}$), $\sigma$ is the molar conductivity of the species in solution ($S \cdot cm^{-1} \, mol^{-1}$), and $\varepsilon$ is the obstruction factor, a coefficient that increases the resistance in consideration of the negative effects of the spacer to ion transfer [33,34]. From the literature, $\sigma$ value is $0.08798 \, S \cdot cm^{-1} \, mol^{-1}$ [34].

$$R_{HCC} = \varepsilon \frac{\delta}{\sigma \cdot C_{HC}} \qquad (A5)$$

$$R_{LCC} = \varepsilon \frac{\delta}{\sigma \cdot C_{LC}} \qquad (A6)$$

When an external load ($R_L$) is connected to the RED unit, the voltage output ($U$) in Volts can be calculated as the open circuit voltage condition ($OCV_{Theo}$), less voltage drop across the internal resistance of the RED unit:

$$U = OCV_{Theo} - I \cdot R_i \qquad (A7)$$

where $I$ is the electrical current (in Ampers) generated by the RED unit. By the other side, $U$ can also be calculated from the voltage drop on the external load ($R_L$):

$$U = I \cdot R_L \qquad (A8)$$

Since Equations (A7) and (A8) are equivalent, we can resolve the $I$ term and by this obtain the theoretical current value at different values of external load. By this the theoretical value of $I$ can be obtained [10,11,33]:

$$I = \frac{OCV_{Theo}}{R_i + R_L} \qquad (A9)$$

The gross power ($P_g$) obtained over $R_L$ can be calculated from the product of current and voltage output, according to Equation (A10) [33]:

$$P_g = I \cdot U \qquad (A10)$$

$P_g$ can be normalized by dividing its value by the total active area of a cell in the stack, obtaining the power density ($P_d$). Here, $A_{mem}$ is the membrane active area per cell and $N$ is the number of cells [5,11,18]:

$$P_d = \frac{P_g}{N \cdot A_{mem}} \tag{A11}$$

Linear flow velocity ($v$) is defined as the fluid speed inside the flow compartments formed in the intermembrane space, as is expressed in $m \cdot s^{-1}$ according to Equation (A12) [11]:

$$v_i = \frac{\phi_i}{\varepsilon \cdot N \cdot W \cdot \delta} \tag{A12}$$

where $\phi_i$ is volumetric flow of HC or LC feed stream in $m^3$, $\varepsilon$ is the spacer porosity, $N$ is the number of cells, $W$ is the wide of active area on membrane in m, and $\delta$ is the intermembrane space in m. Based on stack dimensions and number of cells, it is possible to define an operative $v$ and estimate the volumetric flow required for HC and LC solutions, so Equation (A13) can be reordered as:

$$\phi_{HC} = \phi_{LC} = v \cdot \varepsilon \cdot N \cdot W \cdot \delta \tag{A13}$$

The apparent permselectivity ($\bar{\alpha}$) is a parameter that describes in a general form the ability of the membranes to allow the pass to counter-ions and inhibits it for co-ions (i.e., for CEM $Na^+$ is the counter-ion while $Cl^-$ is a co-ion) [5,20]. The $\bar{\alpha}$ value can be calculated as in Equation (A14), where $OCV_{Stack}$ represents the experimental OCV value.

$$\bar{\alpha} = \frac{OCV_{stack}}{OCV_{Theo}} \times 100 \tag{A14}$$

## References


1. Osorio, A.F.; Arias-Gaviria, J.; Devis-Morales, A.; Acevedo, D.; Velasquez, H.I.; Arango-Aramburo, S. Beyond electricity: The potential of ocean thermal energy and ocean technology ecoparks in small tropical islands. *Energy Policy* **2016**, *98*, 713–724. [CrossRef]
2. Sims, R.E.H. Renewable energy: A response to climate change. *Sol. Energy* **2004**, *76*, 9–17. [CrossRef]
3. Pawlowski, S.; Crespo, J.; Velizarov, S. Sustainable power generation from salinity gradient energy by reverse electrodialysis. In *Electro-Kinetics across Disciplines and Continents: New Strategies for Sustainable Development*; Springer International Publishing: New York, NY, USA, 2015; pp. 57–80. [CrossRef]
4. Post, J.W.; Veerman, J.; Hamelers, H.V.M.; Euverink, G.J.W.; Metz, S.J.; Nymeijer, K.; Buisman, C.J.N. Salinity-gradient power: Evaluation of pressure-retarded osmosis and reverse electrodialysis. *J. Membr. Sci.* **2007**, *288*, 218–230. [CrossRef]
5. Tufa, R.A.; Pawlowski, S.; Veerman, J.; Bouzek, K.; Fontananova, E.; di Profio, G.; Velizarov, S.; Crespo, J.G.; Nijmeijer, K.; Curcio, E. Progress and prospects in reverse electrodialysis for salinity gradient energy conversion and storage. *Appl. Energy* **2018**, *225*, 290–331. [CrossRef]
6. Reyes-Mendoza, O.; Alvarez-Silva, O.; Chiappa-Carrara, X.; Enriquez, C. Variability of the thermohaline structure of a coastal hyper-saline lagoon and the implications for salinity gradient energy harvesting. *Sustain. Energy Technol. Assess.* **2020**, *38*, 100645. [CrossRef]
7. Yasukawa, M.; Mehdizadeh, S.; Sakurada, T.; Abo, T.; Kuno, M.; Higa, M. Power generation performance of a bench-scale reverse electrodialysis stack using wastewater discharged from sewage treatment and seawater reverse osmosis. *Desalination* **2020**, *491*, 114449. [CrossRef]
8. Gómez-Coma, L.; Abarca, J.A.; Fallanza, M.; Ortiz, A.; Ibáñez, R.; Ortiz, I. Optimum recovery of saline gradient power using reversal electrodialysis: Influence of the stack components. *J. Water Process Eng.* **2022**, *48*, 102816. [CrossRef]
9. Micale, G.; Cipollina, A.; Tamburini, A. Salinity gradient energy. In *Sustainable Energy from Salinity Gradients*; Elsevier: Amsterdam, The Netherlands, 2016; pp. 1–17. [CrossRef]
10. Ortiz-Imedio, R.; Gomez-Coma, L.; Fallanza, M.; Ortiz, A.; Ibañez, R.; Ortiz, I. Comparative performance of Salinity Gradient Pow-er-Reverse Electrodialysis under different operating conditions. *Desalination* **2019**, *457*, 8–21. [CrossRef]
11. Veerman, J.; Vermaas, D.A. Reverse electrodialysis. In *Sustainable Energy from Salinity Gradients*; Elsevier: Amsterdam, The Netherlands, 2016; pp. 77–133. [CrossRef]
12. Cahe, S.; Kim, H.; Hong, J.G.; Jang, J.; Higa, M.; Pishnamazi, M.; Choi, J.Y.; Walgama, R.C.; Bae, C.; Kim, I.S.; et al. Clean power generation from salinity gradient using reverse electrodialysis technologies: Recent advances, bottlenecks, and future direction. *Chem. Eng. J.* **2023**, *452*, 139482. [CrossRef]
13. Veerman, J.; Saakes, M.; Metz, S.J.; Harmsen, G.J. Reverse electrodialysis: Evaluation of suitable electrode systems. *J. Appl. Electrochem.* **2010**, *40*, 1461–1474. [CrossRef]

14. Jang, J.; Kang, Y.; Han, J.H.; Jang, K.; Kim, C.M.; Kim, I.S. Developments and future prospects of reverse electrodi-alysis for sa-linity gradient power generation: Influence of ion exchange membranes and electrodes. *Desalination* **2020**, *491*, 114540. [CrossRef]
15. Scialdone, O.; Guarisco, C.; Grispo, S.; Angelo, A.D.; Galia, A. Investigation of electrode material—Redox couple systems for reverse electrodialysis processes. Part I: Iron redox couples. *J. Electroanal. Chem.* **2012**, *681*, 66–75. [CrossRef]
16. Scialdone, O.; Albanese, A.; D'Angelo, A.; Galia, A.; Guarisco, C. Investigation of electrode material—Redox couple systems for reverse electrodialysis processes. Part II: Experiments in a stack with 10–50 cell pairs. *J. Electroanal. Chem.* **2013**, *704*, 1–9. [CrossRef]
17. Lee, S.Y.; Jeong, Y.J.; Chae, S.R.; Yeon, K.H.; Lee, Y.; Kim, C.S.; Jeong, N.J.; Park, J.S. Porous carbon-coated graphite electrodes for energy production from salinity gradient using reverse electrodialysis. *J. Phys. Chem. Solids* **2016**, *91*, 34–40. [CrossRef]
18. Mei, Y.; Tang, C.Y. Recent developments and future perspectives of reverse electrodialysis technology: A review. *Desalination* **2018**, *425*, 156–174. [CrossRef]
19. Ortiz-Martínez, V.M.; Gómez-Coma, L.; Tristán, C.; Pérez, G.; Fallanza, M.; Ortiz, A.; Ibañez, R. A comprehensive study on the effects of operation variables on reverse electrodialysis performance. *Desalination* **2020**, *482*, 114389. [CrossRef]
20. Długołęcki, P.; Gambier, A.; Nijmeijer, K.; Wessling, M. Practical potential of reverse electrodialysis as process for sustainable energy generation. *Environ. Sci. Technol.* **2009**, *43*, 6888–6894. [CrossRef]
21. Roldan-Carvajal, M.; Vallejo-Castaño, S.; Álvarez-Silva, O.; Bernal-García, S.; Arango-Aramburo, S.; Sánchez-Sáenz, C.I.; Osorio, A.F. Salinity gradient power by reverse electrodialysis: A multidisciplinary assessment in the Colombian context. *Desalination* **2021**, *503*, 114933. [CrossRef]
22. Sandoval-Sánchez, E.; de la Cruz-Barragán, Z.; Miranda-Hernández, M.; Mendoza, E. Effect of Gaskets Geometry on the Performance of a Reverse Electrodialysis Cell. *Energies* **2022**, *15*, 3361. [CrossRef]
23. Tedesco, M.; Scalici, C.; Vaccari, D.; Cipollina, A.; Tamburini, A.; Micale, G. Performance of the first reverse electrodialysis pilot plant for power production from saline waters and concentrated brines. *J. Membr. Sci.* **2016**, *500*, 33–45. [CrossRef]
24. Avci, A.H.; Rijnaarts, T.; Fontanova, E.; Di Profio, G.; Vankelecom, I.F.V.; De Vos, W.M.; Cursio, E. Sulfonated polyethersulfone based cation exchange membranes for reverse electrodialysis under high salinity gradients. *J. Membr. Sci.* **2020**, *595*, 117585. [CrossRef]
25. Mehdizadeh, S.; Yasukawa, M.; Kuno, M.; Kawabata, Y.; Higa, M. Evaluation of energy harvesting from discharged solutions in a salt production plant by reverse electrodialysis (RED). *Desalination* **2019**, *467*, 95–102. [CrossRef]
26. Veerman, J.; Saakes, M.; Metz, S.J.; Harmsen, G.J. Reverse electrodialysis: Performance of a stack with 50 cells on the mixing of sea and river water. *J. Membr. Sci.* **2009**, *327*, 136–144. [CrossRef]
27. Benneker, A.M.; Rijnaarts, T.; Lammertink, R.G.H.; Wood, J.A. Effect of temperature gradients in (reverse) electrodialysis in the Ohmic regime. *J. Membr. Sci.* **2018**, *548*, 421–428. [CrossRef]
28. Güler, E.; Elizen, R.; Vermaas, D.A.; Saakes, M.; Nijmeijer, K. Performance-determining membrane properties in reverse electrodialysis. *J. Membr. Sci.* **2013**, *446*, 266–276. [CrossRef]
29. Gómez-Coma, L.; Ortiz-Martínez, V.M.; Carmona, J.; Palacio, L.; Prádanos, P.; Fallanza, M.; Ortiz, A.; Ibañez, R.; Ortiz, I. Modeling the influence of divalent ions on membrane resistance and electric power in reverse electrodialysis. *J. Membr. Sci.* **2019**, *592*, 117385. [CrossRef]
30. Electronic Load Fundamentals. Available online: https://www.keysight.com/us/en/assets/7018-06481/white-papers/5992-3625.pdf (accessed on 18 October 2023).
31. Hong, J.G.; Chen, Y. Nanocomposite reverse electrodialysis (RED) ion exchange membranes for salinity gradient power generation. *J. Membr. Sci.* **2014**, *460*, 139–147. [CrossRef]
32. Marin-Coria, E.; Silva, R.; Enriquez, C.; Martínez, M.L.; Mendoza, E. Environmental assessment of the impacts and benefits of a salinity gradient energy pilot plant. *Energies* **2021**, *14*, 3252. [CrossRef]
33. Veerman, J.; Saakes, M.; Metz, S.J.; Harmsen, G.J. Reverse electrodialysis: A validated process model for design and optimization. *Chem. Eng. J.* **2011**, *166*, 256–268. [CrossRef]
34. Yip, N.Y.; Vermaas, D.A.; Nijmeijer, K.; Elimelech, M. Thermodynamic, energy efficiency, and power density analysis of reverse electrodialysis power generation with natural salinity gradients. *Environ. Sci. Technol.* **2014**, *48*, 4925–4936. [CrossRef]

MDPI AG
Grosspeteranlage 5
4052 Basel
Switzerland
Tel.: +41 61 683 77 34

*Energies* Editorial Office
E-mail: energies@mdpi.com
www.mdpi.com/journal/energies